The Anthropocene

'The Anthropocene' is one of the most important scientific concepts of our time. It is also one of the most contested. This book offers the most advanced yet accessible introduction to the complex and often heated debates about the meaning, nature, and implications of the Anthropocene.

Drawing on the work of physical and human geographers, alongside geographically-informed contributions from natural and physical scientists, sociologists, historians, anthropologists, philosophers, and others, this book maps and critically evaluates the implications of the Anthropocene for the discipline of Geography and for geographical thinking across other fields. It asks: Does the Anthropocene mark the start of the greatest environmental crisis in history? Which humans are most responsible? What impacts is it having on human well-being and the natural world? What needs to be done to mitigate its threats, and who will bear the costs? How does 'the Anthropocene' challenge the conduct of research, policy, and activism around issues like environmental justice, resilience, nature conservation, or decolonisation? Is 'the Anthropocene' even the best name for this new age?

Essential reading for university students and researchers in Geography and beyond, this book also appeals to general readers interested in understanding how human actions are reshaping the Earth and what this means for the future of people and planet.

Martin Mahony is an Associate Professor in Human Geography in the School of Environmental Sciences, University of East Anglia, UK. His background is in human geography and science & technology studies (STS), and his research concerns the co-production of space, knowledge, and power in the history and politics of environmental change.

Noel Castree is a Professor of Geography at Manchester University, UK; he is also an Honorary Professorial Fellow at UTS in Sydney. His main research interests are the political economy of environmental change and, most recently, the role of expertise in shaping human responses to Earth-wide changes to the biophysical world. He is the author of *What Future For The Earth?* (2026).

Key Ideas in Geography
Series editors

Noel Castree
University of Manchester

Audrey Kobayashi
Queen's University

The *Key Ideas in Geography* series will provide strong, original, and accessible texts on important spatial concepts for academics and students working in the fields of geography, sociology, and anthropology, as well as the interdisciplinary fields of urban and rural studies, development, and cultural studies. Each text will locate a key idea within its traditions of thought, provide grounds for understanding its various usages and meanings, and offer a critical discussion of the contribution of relevant authors and thinkers.

Non-representational Theory
Paul Simpson

Climate Change
Mike Hulme

Space
Peter Merriman

Home, 2nd Edition
Alison Blunt and Robyn Dowling

Nationalism
David Kaplan and Kathryn Hannum

More-than-Human
Jamie Lorimer and Timothy Hodgetts

The Anthropocene
Martin Mahony and Noel Castree

For more information about this series, please visit: www.routledge.com/series/KIG

The Anthropocene

Martin Mahony and Noel Castree

Routledge
Taylor & Francis Group
LONDON AND NEW YORK

Designed cover image: © Getty Images

First published 2027
by Routledge
4 Park Square, Milton Park, Abingdon, Oxon OX14 4RN

and by Routledge
605 Third Avenue, New York, NY 10158

Routledge is an imprint of the Taylor & Francis Group, an informa business

British Library Cataloguing-in-Publication Data
A catalogue record for this book is available from the British Library

ISBN: 978-0-367-75137-1 (hbk)
ISBN: 978-0-367-75418-1 (pbk)
ISBN: 978-1-003-16246-9 (ebk)

DOI: 10.4324/9781003162469

Typeset in Times New Roman
by KnowledgeWorks Global Ltd.

For Orla and Leo, and the worlds you make.

- M.M.

Contents

Figures

Tables

Boxes

Acknowledgements

This book is the result of an accidental collaboration. It emerged from a conversation between Noel, co-editor of the Routledge Key Ideas in Geography book series, and Martin. Martin then created a book prospectus back in 2020, which Noel commented on and suggested some alterations to. Martin was subsequently offered a contract to write *The Anthropocene*. However, later on, an opportunity arose for Noel to join him and write some parts of the book. Despite living at a distance – Martin was in England and Noel in Australia for most of the long period of writing – the (initially unplanned) collaboration has been most enjoyable and even, dare we say, 'easy' (in the best sense of the word). It has allowed us to learn a great deal, to read widely and to think big, though always mindful of the limits of our understanding and experience.

Martin wishes to thank Noel for being so receptive to the initial plans for the book, and for his valuable early editorial input which then, delightfully, grew into authorial input. Helen Pallett has been a constant source of wisdom, encouragement, and inspiration, ever since that walk by the River Wensum during the early days of the COVID-19 pandemic when a sketch of the book was discussed for the first time. Martin also wishes to thank all the students who have taken his 'Human Geography in the Anthropocene' module at UEA over the last few years – the discussions, ideas, feedback, puzzled looks, and eureka faces have all been central to how this book was conceived. As have all the discussions with those who have contributed to the module's delivery over the years – Dave McLaughlin, Casper Laing Ebbensgaard, Irene Lorenzoni, Elliot Honeybun-Arnolda, Sean Irving, Erik Hartin, everyone at the Sainsbury Centre, and, of course, Helen. The 3S Research Group at UEA has been, as ever, a wellspring of ideas and solidarity. Lastly, Martin wishes to thank his family for all the love and support, especially Orla and Leo, whose arrival helped bring it all into focus.

Noel is most grateful to Martin for the chance to work together. He would also like to acknowledge Nick Fowler and Greg Martin for humorously interrupting extended periods of solitary writing (of this book and another, entitled *What Future For The Earth?* [2026]). He dedicates his contributions to *The*

Anthropocene to his son (Tom), his daughter (Felicity), his mum (Christine), and to the one and only RJO.

We are both incredibly grateful to the editorial team and Routledge for stewarding this project along, and to the three manuscript reviewers who kindly lent their time and their insight to help us strengthen the text. We are deeply thankful. Any errors or omissions remain, of course, our own.

Prologue: The Anthropocene and Geography

Welcome to the Anthropocene

This is a phrase you've probably heard or read before – perhaps on the front of a magazine, in a scientific article, or at the start of a lecture. It's a phrase bidding you acknowledge this new geological age ('-cene') defined by the global impacts that humanity ('anthropos') is having on things like the climate system, on biodiversity and ecosystems, on various biogeochemical cycles, and other elements of what is often now called the 'Earth System'. Humanity's role in changing the functioning of this system - in quite literally changing the face and workings of the planet - is now so powerful that its traces will be found in the rocks for millennia to come. Until recently, big shifts in Earth System functioning were driven by natural forces like changes in the planet's orbit or by plants evolving to release new gases into the atmosphere. Now, it's us. *We* are the planet-makers. Welcome to the Anthropocene, the *age of humans*.

So, what does it mean to be 'welcomed' into this new age? How do you think those writing or speaking the words 'welcome to the Anthropocene' intend to make you feel? For some, it's a phrase that might invoke fear – of humanity entering a period of dangerous and uncertain environmental change, in which it's difficult to imagine that the future will be anything other than a desperate struggle for 'civilization'. For others, being welcomed to the Anthropocene may be an optimistic or even thrilling experience – a moment to celebrate humanity completing its takeover of planet Earth, where human ingenuity and technological know-how can manage any negative side-effects and risks, and where people can get on with the wider task of perfecting the planet as a home fit for this new master-species. But sometimes the phrase 'welcome to the Anthropocene' is spoken with a bitter irony – why would one welcome anyone into something so *dire*? At other times, the phrase is intended as an invitation to think anew about the human relationship with nature, and to replace fear and risk aversion with a confidence in the human ability to become successful planetary stewards.

Across that spectrum of fear and celebration lies a whole host of other possible responses to the idea that we, the human species, now inhabit the

Anthropocene. Our own personal and collective responses to the concept reveal much about our individual and shared values about the ideal relationship between people and nature, about our predisposition to risk, about our understanding and interpretation of history, and about our hopes and fears for the future.

Since the term first entered scientific and public discourse around the turn of the present century, it's attracted widespread commentary, criticism, and debate. Within the discipline of Geography there have been special issues of journals such as the *Annals of the American Association of Geographers*, the *Geographical Review,* and *Geographical Research.* New interdisciplinary journals have appeared such as *Anthropocene* and *The Anthropocene Review*. And countless books have been written on the subject from a range of disciplinary perspectives, from Earth System Science to anthropology, and from geology and geomorphology to art, architecture, and literary criticism to legal studies. There are also several popular books and TV documentaries about the subject. Part of the novelty of the concept of the Anthropocene is how it's led to new conversations across disciplines about people-nature relationships and about the environmental crisis writ large. This is because it's so encompassing, referring to how humans impact on – and are impacted by – the atmosphere, biosphere, cryosphere, hydrosphere, and pedosphere. It offers a shared language and an overarching framing of human-environment relationships which has led to people like Earth System scientists, geographers, historians, artists and philosophers sharing platforms, talking to each other, and swapping ideas in a way that hasn't happened before. There's lots of disagreement between them of course, as this book will show. But the very fact that these cross-disciplinary conversations are taking place, and that areas of agreement and convergence are being found, is an exciting and noteworthy development.

So, amidst this cacophony of Anthropocene talk, writing, and publishing, what's the point of this book? Well, it tries to do two things that haven't really been done before. Firstly, it offers to geographers – whether students or researchers – a way of navigating this abundance of Anthropocene debate. Think of it as something of a field guide to what some, like Jamie Lorimer, Andrew Curley, and Sara Smith call the 'Anthropo-scene' – this new field of debate, argument, and collaborative research about what the Anthropocene means, and about what it means to be human in this new age (Curley and Smith 2024; Lorimer 2017). We place particular emphasis on the implications of this 'scene' for geographical thought and research – and this links to our second aim, which is to critically examine the engagement of geographical researchers and writers themselves with the concept, albeit in a wider multidisciplinary context.

We don't attempt to survey everything that's been written on the Anthropocene – that would make for a very long and probably quite boring book! Rather, we seek to draw out the *geographical significance* of the Anthropocene. The Anthropocene idea refers to epic alterations to the surface of the Earth – to its physical and human geographies. Recent years have seen Anthropocene scholars and researchers from across the sciences, social sciences and humanities offering new arguments and findings on some core geographical topics – space, place, scale, globalisation, landscapes and networks – at the same time as geographers themselves have been debating the implications of the Anthropocene for their own discipline – a discipline which, for some, is defined by a core focus on human-environment relationships within and between different regions and countries. This book therefore maps and evaluates the implications of the Anthropocene for the discipline of Geography and for geographical thinking in other disciplines, drawing on the work of physical and human geographers as well as the geographically-inflected work of natural and physical scientists, sociologists, historians, anthropologists, philosophers, and others. It thereby covers both the *discipline* and wider *discourse* of G/geography, noting that the former is a shaper – but not determinant – of the latter, the latter being wider than the former.[1]

It is important for us to point out here the geographical skew of our own knowledge. We draw primarily on geographical scholarship written in English, representing our own linguistic biases, as well as the geography of the discipline of Geography, which has historically been, and remains, an academic discipline with a far greater presence in English-speaking parts of the world than in others. We have endeavoured to address this skew where we can, but we invite readers to view this text not as a comprehensive overview of Anthropocene geographies but as one emanating from a particular time and place and from particular networks of geographical knowledge production – networks which we hope will continue to diversify and embrace ever greater levels of epistemic pluralism.

That said, we hope that the book will achieve at least two things:-

1 for students and researchers in Geography, we hope that it will provide a useful guide to this new, fast-changing and lively terrain of thought and scholarship, and a prompt for re-thinking the role that Geography and geographers play in consequential debates about human-environment relationships; and
2 for those engaging with the Anthropocene from a different disciplinary starting point, we hope this text will offer a useful guide to thinking geographically about the Anthropocene, as well as provide some inspiration for new cross-disciplinary thinking about humanity's environmental predicament. For the Anthropocene is geographical all the way up, down, and out: we miss important things if we ignore this fact.

The overarching intention of this book is to offer a *critical geography of the Anthropocene*. It does so by engaging with the Anthropocene at two levels – as a set of *material conditions*, and as an inherently geographical *idea* about those conditions. On the first level, the book examines the processes that have led us into this new geological age. It explores the geographies of colonialism, industrialisation, and consumer capitalism, which are commonly taken as the key drivers of the planetary-scale transformations underway today. And it engages with the material realities of living in the Anthropocene in the present: with the unequal distribution of risks and harms; with struggles to govern human-environment relations in a moment of immense and uncertain change; and with the efforts of artists, activists, and communities to find new ways of living in and working with Anthropocene conditions. We call this a *critical* geography because we're interested in work which addresses the root causes – the 'how?' and 'why?' – of environmental transformations (such as colonialism or capitalism, for instance), alongside the (more voluminous) work which simply describes the 'what?' and 'when?' of environmental change. Being critical here means attending closely to how human-caused environmental change is, and always has been, bound up with the use and abuse of social power, with the marginalisation of certain communities and the hegemony of others, and with the hugely unequal nature of human exposure and vulnerability to things like pollution, toxicity, ecosystem collapse, and changing climates.

So, a critical geography of the Anthropocene in this sense means examining how environmental changes are driven by some of us, though not all of us, and how impacts and consequences of that environmental change are felt more acutely by some rather than others. And this brings us to the second level at which we seek to develop a critical account of the Anthropocene and its geographies: the Anthropocene as an *idea* or a particular representation of reality. It is a concept that refers to humanity as a whole – *the Anthropos*. The story of the Anthropocene, as commonly told (see Chapter 1), is of humanity acting as a species or 'unconscious' collective actor. The undifferentiated 'we' and 'us' that populates much Anthropocene discourse (including the first few paragraphs of this chapter) has been roundly challenged by critical scholars who rightly insist that it is meaningless to attribute world-shaping agency to humanity as a whole. Again, it's only *some* of us, not all of us who are driving planetary change. So, is the Anthropocene concept really fit for purpose when it comes to naming and describing 'our' current environmental predicament, with all its inequalities and geographical variability? We will spend a good portion of our space here engaging critically with the idea of the Anthropocene – interrogating where the notion came from, how it fits within a longer tradition of thinking about human-environment relations in Geography and beyond, and exploring and questioning how the idea is being mobilised to rethink and reorientate the science and politics of environmental change.

These two levels of critical engagement are, of course, linked. Throughout the book we'll emphasise what geographer Neil Smith (2011) called the 'tight dialectic' between geographical (and indeed any environmental) knowledge making and material processes of geographical and environmental change (see also Millar and Mitchell 2017). That is to say, the knowledge and ideas that people concoct about nature and the environment are deeply consequential for the way environmental change happens, and for the ways in which people seek to govern or ameliorate it. How people see 'nature' – for example as something fragile and precious to be protected, or as a standing stock of potentially profitable resources – shapes what people do to that perceived nature (Castree 2005). Likewise we might, for example, question if the heralding of the Anthropocene implies, as the American environmentalist-entrepreneur Stewart Brand once famously put it, that 'We are as gods and have to get good at it' – i.e. inhabitants of a human-dominated planet, which, like it or not, requires clever and conscious management. Or perhaps the heralding of the Anthropocene implies a recognition that we're not apart from 'nature' (and certainly not in control of it), but rather completely dependent on other species and nonhuman systems for our collective survival: the need here is for humility, rather than hubris; to dial down the fantasies of control and to get better at living alongside other beings. These are two extremes of the debate about what the idea of the Anthropocene really means and are no doubt caricatures of different commentators' positions – as we'll go on to explore, there's much more nuance and complexity to the Anthropocene debate than we can do justice to in an introductory chapter.

Those two positions nonetheless emphasise a key point long made by feminist theorists and philosophers of science, and summarised by Harvard University sociologist Sheila Jasanoff: 'the ways in which we know and represent the world (both nature and society) are inseparable from the ways in which we choose to live in it' (Jasanoff 2004, 2). In other words, our knowledge is imbued with our values and shapes our practices, rather than being a 'neutral' lens through which to view to the world.[2] Like Neil Smith, Jasanoff has emphasised how the production of environmental knowledge – including the master categories used to organise that knowledge, like 'the Anthropocene' – is influenced by its socio-political context; and how that knowledge, in turn, affects the way that people and societies use, shape, transform, and govern environments. So, in offering a critical geography of the Anthropocene, we develop an account of the tight dialectic (Smith) between, or 'co-production' (Jasanoff) of, the material conditions of the Anthropocene and the knowledge claims that are made about it. At each step, we aim to contextualise the science of environmental change in different times and places, and to show how the historical geography of environmental change is also an historical geography of ideas and science about the nonhuman world. In thinking about the contributions that geographers are making and might make to understanding and living in the Anthropocene now and into the future, we attend closely to how different geographical knowledges 'perform' – i.e. what effects do they have in the world? What kind of socio-environmental futures might they help bring into being? Who do they empower, and who might they marginalise? While we don't wish to overemphasise the power of geographical knowledge to change hearts and minds, we nonetheless emphasise that no knowledge is neutral, especially when it's created by certified 'experts' with university positions; all knowledge has power, whether potential or actual, and that to be an agent of geographical knowledge in the Anthropocene – whether as a student, a researcher or an experienced professor – is a great privilege, with great responsibilities (Derickson 2025). That's why, towards the end of the book, we explore geographical scholarship that's experimented with radical new ways of living and being in the world, whether that's at the level of community organising or in the governance of global sustainability.

Our overall aim, then, is to offer a critical geography of the Anthropocene which can help readers better understand the root causes of the environmental crisis, but which also offers ideas and resources for building better, more just social and environmental futures. Our aim, to paraphrase cultural critic Raymond Williams (1980), is not just to make environmental despair more convincing, but to help make hope possible.

Overview of the book

In the next chapter (Chapter 1), 'The Anthropocene Event', we introduce the concept more fully and explore its scientific and conceptual origins. As geologists move towards thinking about the Anthropocene as an 'event' in the geological timescale, we discuss the 'Anthropocene event' as the period in the early twenty-first century when scientists, other academics, and environmentalists began debating whether the planet had entered a new age. Rather than a prosaic process of geological record keeping, this has been a moment of profound rethinking of human-nature relationships, of the relationships between past and present, between societies and cultures, between colonising and colonised peoples, and between science and politics.

While the Anthropocene proposition has rightly been subject to much criticism and contestation, this chapter makes the case that the Anthropocene event has seen a much richer set of interdisciplinary debates than have heretofore been achieved in the contexts of, for example, climate change or biodiversity loss. By agglomerating such ideas and phenomena into a new 'master narrative', the Anthropocene idea tends towards abstraction and is riddled with blind spots; nonetheless, critical responses to it have done important work in emphasising how environmental change needs to be considered as thoroughly bound-up with histories and geographies of colonialism, racism, capital accumulation, and class formation, of science and technology, changing aesthetic sensibilities and shifting ontologies and epistemologies of 'nature'. As such, radically different versions or narratives of the Anthropocene have proliferated, each with very different moral lessons for life on a volatile planet. The Anthropocene therefore invites environmental researchers and thinkers of all stripes, but perhaps most particularly geographers, to re-appraise the structures and boundaries of their knowledge-making practices.

Chapter 2 takes up this challenge which the Anthropocene poses to Geography and to thinking geographically. We start by briefly positioning the

Anthropocene concept within longer lineages of environmental thought, and within the 'geographical tradition', defined as an 'experiment' in keeping the study of nature and society under one umbrella (Livingstone 1993). In its enfolding of local and global processes of socioenvironmental change, its blurring of ontological boundaries and its insistence on new modes of knowledge-making which transcend Geography's own human/physical divide, we show how 'the Anthropocene proposition is geographical to the core' (Castree 2015, 72), and how the Anthropocene event thus represents an exciting, if challenging, new phase of the geographical experiment. The chapter then identifies the major 'geographic dimensions' of the Anthropocene, noting – however – that these aren't amenable to a single 'truthful' mode of analysis but can, instead, be understood in multiple ways. This adds extra complexity to already complex real-world geographical issues. There are several potential 'tight dialectics' and modes of 'coproduction' available. The chapter places geographical knowledge in a force-field of debates about how best to live on Earth, depending on what it is on Earth we value. We distinguish between 'anaemic' and 'substantive' knowledge of the world's geography; knowledge 'thin' and 'thick'. Along the way, the chapter also asks whether the Anthropocene concept is really that new.

Part II of the book, 'Geo-histories', features our critical examination of the historical geographies of the Anthropocene. In Chapter 3, we start from the proposition of physical geographers Mark Maslin and Simon Lewis (2015) that the Anthropocene truly began in the early 17th century, when the effects of massive depopulation and reforestation in the Americas, following European colonisation, led to a discernible change in the global climate as well as a marked globalisation of ecologies. This proposition has proven popular with human geographers who similarly see processes of European colonial expansion as critical to understanding the planetary transformations of the Anthropocene. For example, for historical geographer Jason W. Moore, these processes are central to how a new system of capitalist production and exchange emerged, with far-reaching consequences for a new global 'world-ecology' (Moore 2015). For British geographer Kathryn Yusoff, the onset of chattel slavery lay at the heart of a new 'White Geology' which parsed the world into the 'human' and the 'inhuman', the latter a category of matter rich with properties ready for exploitation – whether that be gold, coal, or human flesh (Yusoff 2024). By engaging with the multiple ways in which geographers, historians, and Earth System scientists invoke the period of European colonial expansion in arguments about the Anthropocene, in this chapter, we explore how imperialism and colonialism are being rethought as processes of not just human domination, but also of the transformation of nonhuman natures.

Chapter 4 traces this story into the period of the Industrial Revolution. Whether pointing to the invention of the steam engine or to the reorganisation of labour relations and economic production, this period in history serves for many Anthropocene thinkers as a key moment of epochal transformation. Yet many historical geographers and historians have emphasised the continued significance of European empires in processes of industrialisation, pointing to how the economic surpluses generated by the slavery economy underpinned capital investments in western Europe, as well as the 'ecological debt' – in the form of resources and agricultural land – accrued by western European nations as populations expanded and moved to new industrial cities. The debate over the Anthropocene has prompted new reflection on how the 'industrial revolution' was not just a product of clever inventions but was enmeshed in, and contingent upon, global geographies of circulating finance, resources, commodities, and energy. As such, re-appraising this period of planetary transformation has urgent relevance for contemporary debates about racial inequalities and the political, cultural, and economic legacies of slavery and empire. New concepts like 'fossil capital' have been advanced to help us grasp the world-changing significance of a global capitalist system fuelled by things like coal and oil.

As the final chapter in this section, Chapter 5, 'Geographies of the Great Acceleration', examines the geographies of the period since the end of World War II. Graphs of everything from carbon dioxide emissions and crude oil production to the number of McDonald's restaurants are used to emphasise that the second half of the twentieth century really was quite different from everything that came before it. The processes which are pointed to are profoundly geographical – various graphical representations speak of economic and cultural globalisation, the shift to a new, oil-based energy regime, new cultures of travel and mobility, changing societal relationships to food, rapid urbanisation, and new spaces of consumption. Yet while World War II is often taken as a marker of an epochal break, geographers and historians have argued that the roots of such transformations lie much deeper. In particular, the new relationships between nation, state and citizen which emerged in certain countries during World War I shaped subsequent developments in consequential ways, while the emergence of a new form of 'consumer capitalism' was not just an inevitable stage of capitalist development, but was something deliberately engineered from the ruins of two world wars in order to steer global economies towards a new mode of accumulation. Nonetheless, the Great Acceleration was also an era of accelerated resistance to the increasingly visible environmental consequences of economic growth. However, this was no simple enlightenment; new geographies of knowledge emerged, which often pitted scientists against each other in a struggle over the relative valuation of economic expansion and environmental integrity.

Part III of the book, 'Geo-politics', examines the implications of the Anthropocene for understanding the geographies of political responsibility and power across a range of spatial scales. Chapter 6, 'Scaling Justice in the Anthropocene', explores work in Geography and related disciplines focussed on 'environmental justice' and examines how this concept might guide thinking about the politics of the Anthropocene. We'll introduce conceptualisations of environmental justice grounded in theories of distribution, recognition, procedural and epistemic justice, using international case studies of toxic pollution to explore how it is that some communities always seem to be at the sharp end of the 'slow violence' of environmental degradation, while others are spared. We then ask what it means to seek 'planetary justice', which urges us to think about justice across borders, across time, and beyond the human. These conceptualisations provide promising avenues for thinking about and realising justice in the Anthropocene. Nonetheless, work by feminist and decolonial geographers has emphasised that as well thinking across scales of space and time, we need to retain a focus on local contexts and circumstances of harm, and on how intersections of multiple forms of inequality create particular patterns of environmental injustice.

Chapter 7 builds on ideas introduced in Chapter 6 to explore how human-nonhuman relationships are being rethought and remade in the context of wildlife management and conservation. The chapter traces how attitudes to wildlife have co-evolved with the historical transformations mapped in Part II. From colonial game reserves and so-called 'fortress conservation' through to recent experiments in more relational, co-productionist modes of wildlife governance, the chapter traces how changing ideas about nature and people have found expression in different geographies of conservation, preservation, restoration, protection, and co-habiting. It examines how the Anthropocene idea has prompted new thinking about spacing wildness, from E.O. Wilson's 'Half Earth' proposal to Jamie Lorimer's calls for more experimental approaches to conservation, which abandon illusions of control and embrace the uncertainties and surprises which characterise any dealings with the 'multispecies agencies' of the wild. Consistent with our insistence that knowledge is constructed and has consequences, in this chapter we show that different ideas about wildlife and its future geographies both co-exist and compete so as to actively shape the material realities they purport to describe with honesty and accuracy.

Chapter 8 then zooms out to consider the geographies of efforts to govern human-environment relationships more broadly in the Anthropocene. The proposed onset of the Anthropocene has stimulated new thinking about the political geographies of environmental governance, and about how conventional assumptions about geopolitics may be upturned by life on a rapidly changing planet. As ice caps melt, seas rise, forests contract, and biodiversity declines, the *geo* of geopolitics can no longer be thought of as a stable, passive backdrop to human affairs. Instead, the *geo* is itself now a lively, three-dimensional, and perhaps unpredictable actor in planetary politics. We examine two emerging 'governing logics' for this lively planet, which we conceive of as assemblages of actors, institutions, entities, ideas and technologies that are transforming not only the conduct but also the scope of political action in the Anthropocene. 'Geo-management' and 'resilience' have both emerged in tandem with the notion of the Anthropocene, and offer different but comparable prescriptions for how politics on a lively, interconnected planet should be conducted. While 'resilience' offers a potentially valuable experimental approach to grappling with the inherent uncertainties of life in the Anthropocene, the logic of 'geo-management' proposes that humanity can and perhaps should embrace its role as planetary environment-makers by adopting geoengineering technologies to subdue the planetary liveliness being unleashed by human actions.

The **final part** of the book, 'Geo-futures', examines how geographers are engaged with envisioning and making futures. In Chapter 9, we explore how the Anthropocene idea has gained great traction in the arts, with writers and visual artists responding in creatively and intellectually generative ways to new propositions about the state of the planet. The chapter explores work by artists such as Edward Burtynsky and Fabrice Monteiro, as well as theorisations of an emerging 'Anthropocene aesthetics' by geographers such as David Matless, Caitlin DeSilvey, and Mia Bennett. We look at ways in which cultural geographers have engaged directly with artists working in an environmental grain, not just through a stance of critique, but through new experimental collaborations which cut across conventional art/science distinctions, and which upend implicit hierarchies which still commonly position the arts as a mode of 'science communication', rather than as a field of theory and practice which can speak back to, inform and enrich environmental research. From post-industrial ruination to regenerative ecologies, we argue that a new aesthetics of the Anthropocene is emerging, which challenges conventional depictions of space and time in environmental art and literature.

The final chapter, 'Placing the Future', asks: who gets to imagine the future and speak for it? Why do some future visions become authoritative and 'travel' further than others? And how do future visions reproduce or challenge existing imaginative geographies and hegemonic modes of spatial order? Situating existing constructions of Anthropocene futures geographically and historically prompts us to consider how more diverse modes of future-work can be fostered, at multiple scales. But following Australian geographer Lesley Head's (2016) argument that hope is something to be practiced as much as felt, the chapter explores spaces where futures are not just being imagined or predicted, but actively prefigured and made. The prefigurative politics of community activism around food, energy, sustainable cities, and alternative economies have become new objects of geographical inquiry and sites of collaboration, with new research practices emerging 'to excavate, encounter and extend reparative possibilities for alternative futures', by working with and alongside concerned others (Gibson et al. 2015, ii). This fostering of new modes of economic and social life is occurring in the shadow of heretofore insufficient action by central governments and intergovernmental institutions; the chapter therefore examines the inevitable question of whether and how such initiatives need to be 'scaled up' to encompass broader publics and ecologies, or rather whether our political imagination needs to be re-scaled, to abandon conceits of centralised control and instead embrace the rhizomatic possibilities of community-driven transformations. Within those networked transformations are many possible roles for geographers to play

We also add a short **epilogue**. There we share our own considered, personal perspective on Anthropocene futures and how best to narrate them, extending in brief the discussion in the previous chapter.

How to use this book

We've written this book so that it has an overall narrative about the material geography of the Anthropocene, and about the far-reaching implications of the concept for the discipline (and wider discourse) of Geography, which aims to understand and shape that material geography. In that sense, the book bears reading in a linear fashion. Nonetheless, each chapter is designed to stand on its own, and we hope that this will prove useful for those readers who prefer to focus on particular topics, or on the implications of the Anthropocene for different geographical sub-disciplines and topics. For readers wishing to read selectively, we recommend Chapters 1 and 2 and then whichever additional chapter/s you find interesting. From Chapter 3 onwards, study questions for student readers are offered to help guide discussion and analysis of some key debates.

A major theme of the book – which has proven a challenge for us authors to manage without contradiction – is that the Anthropocene is a question of *both* reality and how we choose to represent it. Getting to grips with the Anthropocene is about fully acknowledging the fact of Earth System change. But it also requires us grappling with the fact that we have choices about how we register, and respond to, this change. There's thus a chronic tension between a One World argument that is grounded in hard scientific evidence about planetary change, and a Many Worlds argument that accepts there are competing narratives about what the realities of the Anthropocene are all about, and where our future lies.

However you read this book, we hope that once you've spent some time with it, you'll have a new appreciation of what it means to live, work, and study in the Anthropocene, and of what it means to be a geographical thinker in this challenging new world. However we view it, there's no doubt that the Earth's geography is altering rapidly. The challenge is to find intelligent and just ways to shape that geography for the good of people and the non-human realm.

Notes

1 By 'Geography' we refer to the academic discipline, with all its histories, norms, traditions, concepts, ways of thinking and working, etc. By 'geography' we refer to the wider world out there – physical and human environments, the things that folks working in Geography and, of course, lots of other fields like to study.

2 We might frame this too as what Dorreen Massey (2004) called the 'entanglement' of the analytical and the political.

References

Castree, N. 2005. *Nature*. London: Routledge.

Castree, N. 2015. "The Anthropocene: A Primer for Geographers." *Geography* 100: 66–75.

Curley, A., and S. Smith. 2024. "The Cene Scene: Who Gets to Theorize Global Time and How Do We Center Indigenous and Black Futurities?" *Environment and Planning E: Nature and Space* 7 (1): 166–88.

Derickson, K. 2025. "Building Response-Able Abstractions With Gullah/Geechee Nation." *The Professional Geographer* 77 (4): 481–86.

Gibson, K., D. B. Rose, and R. Fincher. 2015. *Manifesto for Living in the Anthropocene*. Brooklyn: Punctum.

Head, L. 2016. *Hope and Grief in the Anthropocene : Re-Conceptualising Human-Nature Relations*. London: Routledge.

Jasanoff, S. 2004. "The Idiom of Co-Production." In *States of Knowledge: The Co-Production of Science and Social Order*, edited by Sheila Jasanoff. London: Routledge.

Lewis, S. L., and M. A. Maslin. 2015. "Defining the Anthropocene." *Nature* 519 (7542): 171–80.

Livingstone, D. N. 1993. *The Geographical Tradition: Episodes in the History of a Contested Enterprise*. Oxford: Wiley-Blackwell.

Lorimer, J. 2017. "The Anthropo-Scene: A Guide for the Perplexed." *Social Studies of Science* 47 (1): 117–42.

Massey, D. 2004. "Geographies of Responsibility." *Geografiska Annaler: Series B, Human Geography* 86 (1): 5–18.

Millar, S. W. S., and D. Mitchell. 2017. "The Tight Dialectic: The Anthropocene and the Capitalist Production of Nature." *Antipode* 49 (S1): 75–93.

Moore, J. W. 2015. *Capitalism in the Web of Life: Ecology and the Accumulation of Capital*. London: Verso.

Smith, N. 2011. "Uneven Development Redux." *New Political Economy* 16 (2): 261–65.

Williams, R. 1980. "The Politics of Nuclear Disarmament." *New Left Review* 124: 25–42.

Yusoff, K. 2024. *Geologic Life: Inhuman Intimacies and the Geophysics of Race*. Durham: Duke University Press.

Part I

Welcome to the Anthropocene

1 The Anthropocene Event

Enter 'Anthropocene'

It was in the Mexican city of Cuernavaca that the late Paul Crutzen, the famous atmospheric chemist, Nobel Laureate, and discoverer of the ozone 'hole', reputedly had his 'Eureka!' moment. At a conference of the International Geosphere-Biosphere Programme (IGBP) in the year 2000, Crutzen got increasingly agitated by the use of the term 'Holocene' to describe the current environmental state. The Holocene, the period since the last ice age defined by relative environmental stability, just didn't seem to make sense anymore given the rapidly growing evidence of human-driven changes to the 'critical zone' (the thin, global layer that supports life). Exasperated, Crutzen blurted out that 'we no longer live in the Holocene … we live in … the Anthropocene!' After the conference, Crutzen discovered that another scientist, the American freshwater ecologist Eugene F. Stoermer, had already been using the term, albeit informally, since the 1980s. Crutzen got in touch, and the pair did what mild-mannered academics usually do when they discover they've been independently working on the same ideas – agree to write a paper together (Schwägerl 2015). The paper (Crutzen and Stoermer 2000) was a slim affair. Just over a page in length and simply titled 'The Anthropocene', it appeared in the internal newsletter of the IGBP and was mainly designed to be a provocation to the rather esoteric science of stratigraphy – the science of defining and delineating geological ages. Grounded 'in a dry litany of numbers', the authors 'could hardly have imagined the immense pull it would have on scientific and scholarly imaginations' (Jasanoff 2020, 341). The idea began to ripple out beyond the rarefied networks of Earth System Science (ESS) and the IGBP, and Crutzen quickly clocked that they were onto something (Trischler 2016).

In 2002, Crutzen penned a similarly short but rather bolder statement of his hypothesis for one of the most widely read scientific journals, *Nature*. His piece, 'Geology of mankind', argued that humans have become a gigantic force and are set to remain so for thousands of years. Accordingly, a new geological epoch should be designated (Crutzen 2002). Crutzen punchily

DOI: 10.4324/9781003162469-2

recounted a 'rapid expansion' of the human population, as well as 'per capita exploitation of Earth's resources'. The methane-producing cattle population had reached some 1.4 billion, perhaps half the Earth's land surface was being exploited by humans, dams and river diversions were commonplace, artificial nitrogen fixation for agriculture exceeded natural processes, atmospheric greenhouse gas concentrations were at their highest level for 400 millennia. In one paragraph, Crutzen offered a sobering litany of human transformation and destruction – but he was also quick to point out that 'these effects have largely been caused by only 25% of the world population' (Crutzen 2002, 23). Nonetheless, navigating this new age would require unprecedented human cooperation:

> A daunting task lies ahead for scientists and engineers to guide society towards environmentally sustainable management during the era of the Anthropocene. This will require appropriate human behaviour at all scales, and may well involve internationally accepted, large-scale geo-engineering projects, for instance to 'optimize' climate. At this stage, however, we are still largely treading on *terra incognita*.
>
> (Crutzen 2002, 23)

This quote from Crutzen's article illuminates a number of key features of what it means to both define and live in the Anthropocene. His mention of 'appropriate human behaviour at all scales' speaks to what the literary scholar Timothy Clark calls 'scalar derangement' – suddenly, as the realities of climate change and other global environmental transformations have thrust themselves upon us, we – as both citizens and scholars – have had to come to grips with how mundane actions taken at a very 'local' scale have geophysical, political, and ethical implications at national, international, and planetary scales. Put the kettle on, make a cup of tea, contribute to the intensification of hurricanes in the Caribbean, or of habitat destruction and nutrient depletion in the East African highlands. How does that tea taste now? Everything is connected it seems, horizontally across space, and 'vertically', from the local to the global and back again. 'The overall force', Clark suggests, 'is of an implosion of scales, implicating seemingly trivial or small actions with enormous stakes' (Clark 2012b, 152).

These actions and behaviours, even 'small' ones, must be made more 'appropriate', Crutzen suggests. How so? Crutzen speaks of scientists and engineers 'guiding' society towards a more sustainable future. This sounds quite reasonable – who wouldn't want people with technical expertise on hand to help societies navigate complex situations like climate change or, for example, a coronavirus pandemic? But some have wondered whether there's a risk, in emphasising the power of technical experts to direct societal decision-making, of potentially shutting out democratic debate and forms of accountability (Lövbrand et al., 2015). Might devolving more power to experts mean that decisions are taken which don't enjoy democratic consent? And might those experts be somehow 'biased' towards certain kinds of solutions that fit their own particular worldview? Note the prominence of 'geoengineering' in Crutzen's statement – doing things like injecting reflective aerosol particles into the upper atmosphere in order to cool the planet and 'optimise' it for human habitation. Geoengineering often looms large in the corners of Anthropocene debate (Baskin 2019). For many thinkers, it is the logical outcome of humanity's new technological power over the natural world and a rational solution to the slow-motion efforts of democratic politics to deal with challenges like decarbonising energy systems.

We deal more directly with these politics of governing the Anthropocene in Chapter 8, but for now we'll just observe one more key feature of Anthropocene discourse that Crutzen prefigured back in 2002: 'we are still largely treading on *terra incognita*'. Despite the apparent promise of surefooted technoscientific guidance, and of ways of engineering the planet to suit human needs, Crutzen also makes plain that the Anthropocene is in many ways defined by uncertainty. The scientific community may be certain that large-scale, rapid environmental changes are happening, but what they might mean for the planet and for humanity in the future, what the world might look like in 100 or 200 years' time, no one can say with any certainty. Indeed, a big part of the scientific justification for the Anthropocene proposition is that the Earth System and global climate have a demonstrated tendency to 'flip' between states on relatively quick timescales (at least in geological terms). The concern for scientists like Steffen et al. (2018) is that humans might be pushing the Earth towards another such irreversible 'tipping point'; where such a point might lie, and what it might mean to cross it, is uncertain. As such, precautionary action should be taken to avoid it, the argument goes, and options like geoengineering should be held in reserve in case things really go awry. This argument comes through strongly in the geoscience concept of 'planetary boundaries' which, in their Holocene state, define a 'safe operating space for humanity' according to Johan Rockström, Will Steffen, and others (Rockström et al. 2009; Steffen et al. 2015).

So, the Anthropocene is an age of widespread human transformation of the planet. It's also an age of scalar derangement, of deep uncertainty, and of a powerful and related tension between democratic and 'technocratic' forms of politics, as societies collectively try to come to terms with a new set of planetary conditions. These (and other) key features of the Anthropocene have created and striated debate across academic disciplines for over two decades now, with no sign of slowing down. As we move deeper into the Anthropocene, these debates about what it means to define and live in such a new age are intensifying. As we write, we're living through what we might call an 'Anthropocene event' in culture, scholarship, politics, and more – a moment in time in which human relationships with the nonhuman world are being radically reconfigured, both in the material world and in the world of ideas. To begin understanding the stakes of this reconfiguration, we first need to dig a little deeper into where the idea of the Anthropocene came from.

The origins of Anthropocene thinking

Understanding life on a volatile planet: Earth System Science

When Paul Crutzen heralded the Anthropocene in Mexico in 2000, he was speaking as a member of an emerging scientific 'super-discipline' – ESS (Clifford and Richards 2005). ESS seeks to observe, model, and predict the interactions of things like the biosphere, the atmosphere, the oceans, and the cryosphere, as well as the biophysical impacts of human activities on those systems. While 'traditional' scientific disciplines had largely concerned themselves with discrete parts of the 'Earth system' (such as rocks or the atmosphere), from the 1950s onwards, a community of scientists developed which sought to 'join the dots' between these things (Castree 2019). In part, this was a product of Cold War military patronage and the desire for ever-more detailed geophysical knowledge of the oceans through which nuclear submarines would sail, or of the regions of the atmosphere through which atomic radiation might circulate (Doel 2003; Turchetti and Roberts 2014). At the same time, ecologists had been developing a 'complexity science' concerned with understanding sudden, nonlinear shifts in ecological systems (Grove 2018; Holling 1973; see also Chapter 8). Thinking on a planetary scale while employed at NASA during the 1960s, British scientist James Lovelock developed his influential 'Gaia hypothesis' that the Earth could be thought of as a self-regulating organism, within which biological lifeforms regulated the composition of the atmosphere, on multi-millennial timescales, to maintain conditions conducive to life (Lovelock and Margulis 1974). The arguments of Lovelock and the American microbiologist Lynn Margulis attracted some controversy and criticism, but were important in dislodging a purely geophysics-based conception of the 'global environment' and emphasising the central role of biota within the complex web of interconnections and feedback loops that constituted the Earth System (Steffen, Richardson, and Rockström 2020).

The term 'Earth System Science' was coined a few years later within NASA as it sought the best way to put to work its newfound capabilities in Earth observation. The so-called 'Bretherton diagram' was one product of this work, and it became an influential visual depiction of the newly conceived 'Earth System' (Figure 1.1).

This diagram 'epitomised the rapidly growing field of ESS through its visualisation of the interacting physical, chemical and biological processes that connect components of the Earth System', and even though the complex world of 'human activities' was bracketed off in one box, connected to the rest

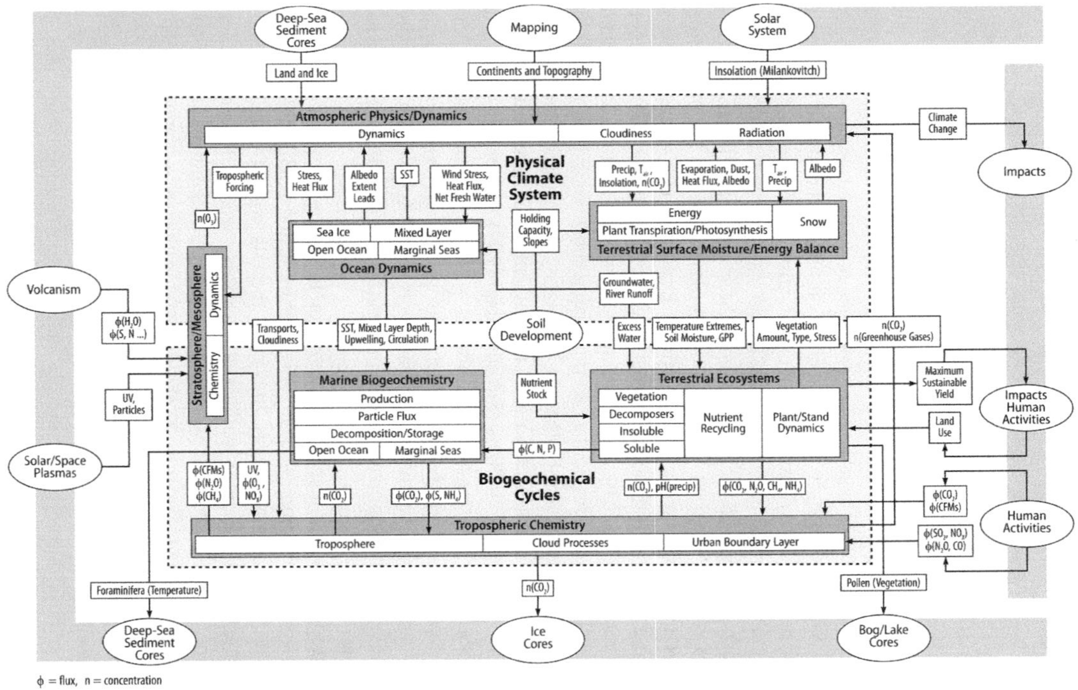

Figure 1.1 The 'Bretherton diagram' of the Earth System. (Adapted from NRC (1986).)

of the system through three forcings of land-use change, CO_2 emissions and other pollutants, it represented an important recognition that 'human activities were a significant driving force for change in the system' (Steffen, Richardson, and Rockström 2020, 59). As the report carrying the diagram put it, 'the people of the Earth are no longer simple spectators of the drama of Earth evolution but have become active participants' (NRC 1986, 5)

Following a few more conferences and statements of scientific intent, in 1987 the IGBP was set up to help coordinate and advance research on the interlinked components of the Earth System. Initial priorities included the international coordination of work on the global carbon cycle, the transformation of terrestrial ecosystems, and the hydrological cycle, with attention focused on improving observations, building sophisticated models, and developing a synthesis of scientific knowledge that could ultimately support policy-making and environmental governance (Seitzinger, Gaffney, and Brasseur 2015). The IGBP didn't pay much attention to the dynamics of the human drivers of environmental change – it didn't really open the box marked 'human activities' on the Bretherton diagram – and so other programmes like the International Human Dimensions Programme on Global Environmental Change were set up to give greater prominence to social science (including human geography) research into society's many interactions with the various components of the Earth System. But ESS as an organising rubric took hold most forcefully within the IGBP, and by the late 1990s, key figures like German physicist Hans Joachim Schellnhuber were making emphatic statements about the field's key synthesised findings (Schellnhuber 1999). Put briefly, these were (1) that the Earth is a 'far-from-equilibrium planet' that sometimes achieves stability, as in the Holocene, but which also 'periodically veers into cascading, self-amplifying change'; and (2) more unsettlingly, that 'human action is a likely trigger for imminent planetary reorganization' (Clark and Szerszynski 2021, 79).

It was the revolutionary implications of these findings that made Paul Crutzen so uneasy about what he was hearing at the IGBP meeting in the year 2000 (Simon and Thomas 2022). All the talk was of accelerating human impacts on the natural world, and yet people were still using the term 'Holocene'. A few decades' worth of scientific development in ESS had started to make the Holocene – a relatively stable climatic and environmental period – seem rather passé as a descriptor of the current environmental state. But during the first five to six years of the Anthropocene's career as a concept, kick-started by Crutzen's initial publications, conversations were largely restricted to the ESS community and key authors like Crutzen, Will Steffen, Jill Jaeger, and Veerabhadran Ramanathan. For these scientists, the Anthropocene was a useful new way of conceptualising and identifying the human role in Earth System evolution, particularly in relation to feedback loops between subsystems and the potential for 'tipping points' to be crossed, and their implications for sustainability and human security (e.g. Eckhart and Krafft 2005).

But the ESS community was not fully engaged with the question of how and why 'the Anthropocene' might be formally deemed a new geological epoch (Brauch 2021). Over the ensuing years though, a range of geologists entered the debate about what Crutzen had tantalisingly called (for geologists at least) the 'geology of mankind'. For them, the key question was not just about the magnitude of human impacts on the Earth System, but whether that impact would be so visible in the rocks as to warrant the formal designation of a new slice of geological time.

The Anthropocene in the rocks: searching for the stratigraphic signal

To be formally recognised as a new geological epoch, the Anthropocene must be detectable in the layers of rock that collectively document Earth's history. The sciences of chronostratigraphy (relating layers of rock, or strata, to time) and of geochronology (the techniques of dating rock formations) are relatively recent, and it was only from the 1960s onwards that an internationally standardised Geological Time Scale (GTS) was developed. The GTS functions as a kind of cognitive 'infrastructure' for Geology and other sciences of the past (such as paleoclimatology and archaeology), providing a shared set of reference points and a common language for dividing up and comprehending Earth history, on timescales ranging from centuries to millions of years (Swanson 2016).

In 2008, the Stratigraphy Commission of the Geological Society of London responded to the growing interest in and use of the term Anthropocene in the scientific literature with a vote on whether the term should be considered for eventual formalisation as a unit of the GTS. As a national body, the Society had no direct influence over the GTS, but it could express an opinion, which it did strongly, led by the University of Leicester geologist Jan Zalasiewicz (Zalasiewicz, Williams, and Smith 2008). Following this, the International Commission on Stratigraphy (ICS), geology's official timekeepers, suggested the formation of an Anthropocene Working Group (AWG) to formally examine the case for this new unit of geological time.

While such Working Groups are usually made up solely of geological specialists for the time period in question, the ICS recognised that the Anthropocene was different. The period in question overlapped, by definition, with human history, with all its inherent complexities. Therefore, alongside geologists and stratigraphers were to sit not only Earth System scientists (such as Paul Crutzen), but also geographers, archaeologists, historians, and even a specialist in international law (Zalasiewicz et al. 2018). The task of the AWG was to evaluate whether the Anthropocene met the conditions for addition to the GTS. This investigation of the 'geological Anthropocene' was quite different to the work of the ESS community: rather than being defined by 'holistic analysis of all human impacts on Earth', the geological Anthropocene is defined by the question of 'whether those impacts have produced suitable signals in the stratal record'; i.e. whether one can observe traces of those impacts in the rocks. To qualify, such a signal must be unique, it must be global in extent and temporally synchronous, and it must have the potential to be preserved well into the future, such that imagined future geologists would be able to spot it as clearly as their present-day colleagues can (Zalasiewicz, Waters, and Head 2017, 289).

Although the AWG was uniquely interdisciplinary compared to the rest of the ICS, it still had to observe the precise, institutionalised rules of stratigraphy (Lundershausen 2018; Nichols and Gogineni 2018). On that front it seemingly failed. In March 2024, the International Union of Geological Sciences confirmed that ICS had voted to reject the Anthropocene proposal (IUGS 2024). Some geologists had expressed unease about identifying a new layer in rocks in 'real time', while that layer is still being formed, rather than centuries or millennia hence when its stratigraphic composition can be more definitively evaluated (Autin and Holbrook 2012; Edgeworth et al. 2019). Others worried that the need to identify a globally synchronous signal as the start of the Anthropocene meant that the geographical and temporal variability of anthropogenic impacts is overlooked and de-emphasised, hindering a full understanding of the new nature of human–environment relations (Gibbard et al. 2021). Some very public arguments and recriminations were aired over whether the ICS had followed the right voting rules, and conversely about perceived shortcomings in the AWG's work (IUGS 2024; McCarthy et al. 2025). The debates and disagreements reflect a wider uneasiness detectable in the stratigraphic community about the 'politics' of declaring a new Anthropocene epoch. A discipline which was used to quietly adjudicating on the nature of geological changes in the deep past was suddenly charged with adjudicating a decision with profound political and philosophical implications (Damianos 2025; Waters et al. 2025).

Nonetheless, the ICS and IUGS don't deny the reality of the planetary transformations being identified by the ESS community; their concern was whether the proposal conformed to the established rules of stratigraphy. The IUGS statement lends its support to an emerging notion that the Anthropocene might be better considered a geological 'event' (Edgeworth et al. 2023), 'similar to the great transformative events in Earth history such as the Great Oxygenation, the Cambrian Explosion, or the Great Ordovician Biodiversification events'. Such 'events' do not require formal ratification as a stratigraphic unit, but nonetheless are an accepted part of geological theory and knowledge. Perhaps that is what 'The Anthropocene' will become – 'an informal non-stratigraphical term', which will 'continue to be widely used not only by Earth and environmental scientists, but also by social scientists, politicians and economists, as well as by the public at large. As such, it will remain an invaluable descriptor in human-environment interactions' (IUGS 2024, 2).

Navigating the Anthropocene event

While geologists were arguing about whether the Anthropocene is an epoch or an event, the concept travelled widely across the natural sciences as a way of 'throwing a rope' around a great variety of environmental challenges and pressures, and linking them together under one conceptual umbrella (Castree 2021). At the same time, scholars in the social sciences and humanities were grappling with the concept and its implications for thinking about human-environment relationships through time and across space (Brauch 2021), and the concept began to get more public visibility too through its use by activists (Wright et al. 2018), journalists (Sklair 2020), and cultural institutions (Robin et al. 2014) as a new way to describe the scale, urgency, and interconnectedness of humanity's environmental problems.

Here then is a different sense of an 'Anthropocene event' – not only a period of rapid change in the Earth System but also a cultural, political, intellectual 'event' which has seen people grappling with the enormity of global environmental challenges and coming to terms with humanity's apparent new role in shaping the Earth System. Rather than just a prosaic process of geological record keeping, this has been a moment of profound rethinking of human–nature relationships, of the relationships between past and present, between present and future, between societies and cultures, between colonising and colonised peoples, and between science and politics.

As the Anthropocene concept has travelled, it's been subject to much criticism and contestation; but it's also sparked new conversations and collaborations within and between disciplines (including Geography – see, e.g. Butler (2021) – in a way which other environmental ideas and concepts simply haven't. Many of those criticisms have addressed how 'the Anthropocene' idea tends towards abstraction and universalisation, and how it is riddled with historical and geographical blind spots (Davies 2016, Chapter 2). As we touch upon in the next chapter, this is characteristic of other modern ideas and imaginaries of the global environment which claim universality but which need to be understood as thoroughly situated in the politics and culture of Western modernities (Rickards 2015). The Anthropocene is a concept which folds humanity – with all its inequalities and disparate histories and experiences – together into a singular actor, at least nominally; critical responses to it have therefore done important work in emphasising how environmental change needs to be considered as thoroughly bound up with histories and geographies of colonialism, racism, capital accumulation, and class formation, changing aesthetic sensibilities and shifting ontologies and epistemologies of 'nature'. Critical responses have sought to either appropriate and re-interpret the Anthropocene concept to better reflect such realities, or they have rejected the concept altogether as a way of understanding human–environment relationships.

As such, there's been an explosion of Anthropo*cenes* – i.e. various versions of the concept, or stories told with the concept, which are distinct from and often in direct conflict with each other in the claims they make about the fundamental nature of human–environment relationships. In essence, the Anthropocene has quickly become what we might call an 'essentially contested concept' (Gallie 1955) – different understandings of the Anthropocene contain within them sets of implicit normative and evaluative claims, with different ethical and political implications. As cultural theorist Claire Colebrook puts it, '[t]he Anthropocene is not only a geological declaration, but a series of diagnoses that generate imperatives' (Colebrook 2016, 82). One job of the rest of this book is to parse these different versions of the Anthropocene, with their different diagnoses and different imperatives, their different political and ethical implications, and to examine what they mean for geographical scholarship. One way to start that job, and to get a handle on this explosion of thinking that characterises the 'Anthropocene event', is to map out the different *narratives* of the Anthropocene that are currently in circulation in academic research, in popular culture, and in environmental activism.

Narrating the Anthropocene

Anthropocene scholars and critics have recognised that the stories 'we tell ourselves about environmental crises', including the emerging story 'of humanity's place on earth and its presence within geological time', crucially determine 'how we understand how we got here, where we might be headed, and what we need to do' (Davis and Todd 2017, 763–64). This argument recalls an influential article from 1992 in which environmental historian and geographer Bill Cronon argued that how we choose to tell stories of environmental change is deeply significant. The stories we tell shape collective understanding of events and processes, and even shape the politics of what we might do about them. He made four key observations about environmental stories:

- They attribute a certain value to the state of things at the beginning and end of the story;
- they offer a particular focus and 'framing' that highlights some things, while backgrounding others – key people, events, and processes;
- they carve time into sequences, pinpoint certain moments and periods, turning points and key forces, while downplaying others;
- they indicate, implicitly or explicitly, what the main causal factors are, what the key moral lessons might be, and which actors need to do what to produce a suitable conclusion to the story (Bonneuil 2015).

In short, stories are sense-making devices and are plural. They speak to the diverse ways that humans interpret and experience the world. To illustrate his argument, Cronon looked at how historians have narrated the transformation of the US Great Plains in the nineteenth and twentieth centuries, which culminated in the 'Dust Bowl' event of the 1930s. Cronon identified two distinctive narratives of these events: a *progressive* narrative and a *declinist* narrative. The *progressive* narrative tended to bemoan the prairie management practices of the native Americans and presented the settlers' wheatfields and railways as a great improvement and welcome 'progress'. They praised the settlers' efforts to tame a resistant and unproductive nature and suggested that overcoming challenges like the Dust Bowl would be dependent on further embracing the power of modern science and technology to tame natural forces. On the other hand, the *declinist* narrative instead insisted that the Great Plains simply couldn't support the demands placed on the environment by the greedy settlers, with their high-tech agriculture and capitalist markets. Instead of calling for *more* modernisation, these narratives tended to call for new forms of state-led management based on new, ecological principles.

Cronon's fundamental point was that the same events could be subjected to very different interpretations. The telling of a story – both in popular culture and in academic research – includes making choices about what happened, why, who was to blame, and what should be done about it (Cronon 1992). The choices, in part, reflect value judgements about how the world ought to be, and these may be implicit and uneven unconscious judgements.

In a similar vein, the historian Christophe Bonneuil (2015) identified four dominant narratives of the Anthropocene, something altogether larger than the Dust Bowl. But like the Dust Bowl narratives identified by Cronon, those identified by Bonneuil each tell very different stories about what's happened, why, who's responsible, and what might happen next. We've adapted these to reflect how narratives have evolved since Bonneuil did his analysis – including the addition of a further one. Below we describe the five narratives briefly, and highlight how academic geographers have begun mobilising these different versions of the Anthropocene in their own work – points which we develop further in the following chapters.

1. The standard narrative

In many ways, the dominant narrative of the Anthropocene, offered in the main by scientists like Paul Crutzen, Will Steffen, and colleagues, is summarised by Bonneuil (2015, 18), like this:

> Since about 1800, 'we', the human species, have inadvertently altered the Earth system at a geological scale. *Anthropos* did so through three stages that can best be documented through quantitative global environmental data. The key causal forces are population growth, economic growth and expansion of international exchange. But a revolution (to be compared only to the Copernican or Darwinian revolutions) occurred recently: Earth system scientists have made *anthropos* aware, at last, of the danger. And, if only policy-makers would act on the basis of sound science, these scientists have the knowledge to lead humanity towards a sustainable future.

Let's think about some key features of that narrative, in relation to Cronon's observations noted above. The implication is that there was a reasonably steady state of human–environment relations before 1800, at which point things started changing rapidly. Key turning points are identified as the onset of the Industrial Revolution, the intensification of economic and population growth since 1945, and a more recent moment of environmental enlightenment provided by the revolutionary findings of ESS. The story is 'framed' – it gives selected emphasis – in relation to the activities of humanity as a whole – the *anthropos*; we learn about the trajectory of a singular 'human enterprise', from its origins in 'our long past as hunter-gatherers to the Great Acceleration' (Steffen, Persson, and Deutsch 2011, 746). And this is a morality tale – there are heroes (scientists uncovering the new reality of our collective situation) and villains (the policymakers who won't act on that knowledge); and the moral lesson is that (Earth System) scientists need to be afforded new power to guide and steward humanity towards a more sustainable future.

Critics have argued that this narrative betrays an awful lot of silences and omissions. Absent is recognition of the long history of human knowledge and concern about anthropogenic environmental change (which we'll explore in Chapters 3–5); there's little acknowledgement that environmental change has been driven by particular people, institutions, economic systems, and political decisions, rather than the agency of the human species as a whole; and likewise, this isn't a narrative which tells us very much about how the effects of environmental change have been and will likely continue to be felt disproportionately by the poor and the marginalised.

These 'gaps' in the standard narrative have spurred the development of other ways of understanding and narrating the Anthropocene, as we explore below (and throughout this book). Nonetheless, this is a narrative with significant power and influence. Academic analysts like John McNeill (2001) and Dipesh Chakrabarty (2019, 2009) have used it to argue that historians urgently need to rethink the terms of their discipline, in particular to consider the human population as a whole as a new kind of historical actor. Human geographers too have embraced this new mode of species-thinking and the revolutionary implications of ESS for thinking about the nature of human life on a lively planet (e.g. Clark 2012a; Clark and Szerszynski 2021), although the majority of human and indeed physical geographers engaged in Anthropocene debate have tended to align themselves more with alternative narratives and formulations, as we explore below (see also Chapter 2).

2. *The post-natural Anthropocene*

If the 'standard narrative' is about humanity transgressing nature's limits and pushing the Earth System into dangerous new states, then our second narrative, which we call the 'post-natural' narrative, suggests instead that humanity has been re-shaping and re-making the natural world for centuries and that we just need to get better at it.

We could have called this narrative either 'post-natural' or 'post-Nature', because it concerns both 'nature' as a set of nonhuman material things and 'Nature' as a philosophical idea. Let us explain. The small-n threads of this narrative emphasise how humanity has, since it took up agriculture, been re-organising the nonhuman world for its own benefit. Biogeographer Erle Ellis, for example, has shown how human societies thousands of years ago transformed land and ecosystems on vast scales (Ellis et al. 2013), with implications not just for the fate of natural ecosystems, but for the climate system too (Ruddiman 2013). Current anthropogenic environmental changes are a continuation of humanity expanding or 'scaling-up' its 'ecological niche', and re-fashioning the face of the Earth for its own needs (Dalby 2020). Scholars like Ellis and Ruddiman – the latter argues that greenhouse gas emissions from early agriculture caused climatic changes long *before* fossil fuel burning got underway – therefore suggest that the Anthropocene is almost co-extensive with the Holocene, and that a pristine, untouched, nonhuman 'nature', on a planetary scale, has long since ceased to exist. This differs from many 'green' thinkers and activists who point to more recent losses, incursions, and alterations of nature, ones they lament.

This is where we get to the large-N threads of the narrative. If the Anthropocene represents the end of nature as a material thing, then it perhaps also marks the end of Nature as an idea (Purdy 2015; Wapner 2014). Biogeographer Erle Ellis belongs to a strand of thinking variously described as eco-modernism or eco-pragmatism which accuses 'earlier environmentalism of romantically idealising a pristine nature that exists apart from people, and of irrationally rejecting technology as a fix to save the planet' (Bonneuil 2015, 25). Eco-modernists argue that from a starting point of clumsy land-use transformation in support of early agriculture, the progressive, technology-enabled intensification of food and energy production points towards a future where each person's ecological footprint can be radically reduced, and catastrophic environmental change avoided. In so doing, humanity may even learn to build a 'Good Anthropocene' (Bennett, Solan, and Biggs 2016). The key to such an outcome will not be the protection of a mythical, pristine Nature, but rather a more enlightened engineering of nonhuman nature through the machinations of liberal capitalism and advanced technology (Dalby 2016; Wright et al. 2018). As such, for two of its most influential proponents, eco-modernism:

> offers a positive vision of our environmental future, rejects Romantic ideas about nature as unscientific and reactionary, and embraces advanced technologies, including taboo ones, like nuclear power and genetically modified organisms, as necessary to reducing humankind's ecological footprint.
> (Shellenberger and Nordhaus 2014, np.)

Figure 1.2 The literature of ecomodernism commonly features images like this, of a futuristic green cityscape where 'nature' is not something remote and separate from humanity, but is instead thoroughly enmeshed with human-made technologies and architecture.

One might reasonably wonder how this narrative really differs from what we've called the 'standard narrative', which likewise emphasises the need for high-tech solutions and an enlightened stewardship of the planet (Figure 1.2). But if, following Cronon (1992), we consider how the two narratives sequence and attribute value to different states of human-environment relations, we can see more clearly how they differ. The eco-modernist insistence on humanity essentially taking control of the planet thousands of years ago departs radically from the competing idea of 'humans as recent destroyers of a pristine nature', with a clear moral lesson about the enhancement of 'humanity's role as sustained and permanent stewards of the biosphere' (Ellis et al. 2013, quoted in Bonneuil 2015, 25). If the standard narrative speaks of the failure of the modern project to subdue and control nature, then the 'post-natural' narrative offers a contrary radicalisation of modernity. Recall the quotation from influential environmentalist, futurist and de-extinction advocate Steward Brand in the Prologue: 'We are as gods and might as well get good at it'[1]; likewise, British environmentalist Mark Lynas argues in his book *The God Species* that 'Nature no longer runs the Earth. We do. It is our choice what happens from here' (Lynas 2011, 8).

This line of thought sits in a sometimes-uneasy alliance with a variant of the 'post-natural' narrative which has its origins in constructivist philosophy and critical social science, and which has been influential in Geography. For example, in his 1993 book *We Have Never Been Modern*, French philosopher-anthropologist Bruno Latour argued that modernity is founded on a powerful myth. The West's 'modern constitution', Latour argued, rests on a fundamental ontological and metaphysical distinction between the natural and the cultural, which persists despite the evident profusion of hybrid entities which span and muddle the boundary. Think, for instance, of the atmosphere, a battery-farmed hen, a genetically modified strain of rice, or a mosquito that has evolved and adapted to urban environments like the London Underground system. These entities cannot be neatly bracketed as either 'natural' or 'cultural', and the modernist delusion that one could feasibly do so has fundamentally weakened (Western) societies' abilities to deal with the inherent risks and consequences of these proliferating hybrids (Latour 1993).

Instead of seeing 'nature' as something external to something called 'culture', and of seeing modernity as the steady emancipation of humanity from the restrictions and dangers of the natural world, Latour and cognate thinkers urge us to understand the recent history of humanity as a period of ever-deepening connections and hybridisations between the 'natural' and the 'cultural'. This line of argument about the human-nonhuman entanglements of the Anthropocene – essentially *'twas ever thus'* – resonates strongly with that of the eco-modernists, and Latour himself contributed to the journal of the most prominent eco-modernist institution, the environmental think-tank The Breakthrough Institute. A range of what we might call geo-philosophers have offered similar accounts of the entanglement of the human and the non-human, which have been influential in Geography. Informed by writers like Donna Haraway (2008) and Elizabeth Grosz (2008), geographical scholarship in this vein has regarded the Anthropocene as something which 'forces upon those of us who had forgotten it the recognition that we are thoroughly *earthly* creatures....simultaneously emergent from, dependent upon and at times vulnerable to a plethora of living and non-living phenomena' (Castree 2014, 456). For geographers like Nigel Clark and Kathryn Yusoff, this means attending not only to how human life and subjectivity are entangled with the fate of nonhuman beings (like plants, animals, and insects), but also with the thoroughly *inhuman* – those 'biophysical forces capable of destroying us and utterly indifferent to our existence' (Castree 2014, 456; see Clark 2011; Yusoff 2020). In this version of the narrative, the 'end of Nature' means radically re-thinking the fundamental question of what it means to be human on an increasingly volatile planet.

For a geographer like Jamie Lorimer, arguments about the ever-deepening entanglement of the human and the nonhuman are a prompt for re-thinking how we might do better at living alongside the species with which we share the planet – from the charismatic megafauna of conservation initiatives (Lorimer 2012) to the microbes which inhabit our own bodies (Lorimer 2017). For example, taking a cue from conservation biologists that the idea of completely 'natural' biomes – or even wilderness (see Cronon 1996) – is a myth, and that the *natures* we see, even in nature reserves, are products of both human and nonhuman agency, Lorimer examines how conservation could and should happen in more diverse spaces (like urban 'brownfield' sites or among industrial ruins; see also Gandy 2013). Freed from an unchanging 'nature' as a benchmark for successful conservation, and from a will to precisely control ecological processes, humans should engage in much more open-ended, experimental efforts at conserving the wild, Lorimer suggests (Lorimer 2015). For thinkers in this vein, it is this kind of disposition and practice that is the appropriate response to both the end of n/Nature, and of the uncertainties which characterise this phase of modernity – something we explore in more depth in Chapter 7. But while the eco-modernists may be confident in the ability of rational management and high technology to improve human relationships with nonhumans, human geographers like Lorimer, Yusoff, and Clark are united in calling for a rather more modest stance in the face of the inhuman forces and profound uncertainties of life in the Anthropocene. This stance echoes that of classical environmentalism, represented by American activist and public intellectual Bill McKibben. For McKibben, n/Nature is indeed a thing of the past, but this is no reason to double down on the approach which brought an end to n/Nature in the first place.

3. *The eco-catastrophist narrative: Anthropocene end times*

Our third narrative takes a rather different tack. While largely accepting the scientific arguments of the standard narrative, what might be called an 'eco-catastrophist' narrative rejects the progressivist overtones of both of our first two narratives[2] and narrates the Anthropocene as the end-game for industrial civilisation (see, e.g. Read and Alexander 2019; Scranton 2015). According to this narrative,

> the move into the Anthropocene is a long story of unsustainable practices, resource depletion, transgressed 'planetary boundaries',[3] and increased complexity creating new vulnerabilities paving the way to tipping points and a planetary state shift.
>
> (Bonneuil 2015, 26)

In many ways, this story of impending collapse is not new. In the 1970s Western scientists and some policymakers began to worry about 'limits to growth' of population, and whether the earth's stocks of resources could withstand perpetual economic growth (Meadows 1972). But this isn't so much about *stocks* as about *flows* – specifically the capacity of Earth's biogeophysical systems to cope with accelerated flows of things like carbon, water, and nitrogen without flipping into new states. For example, there are plentiful stocks of fossil fuels to warm the planet by more than 12°C by 2300 (IPCC 2014); the challenge is to limit the flow of gases like CO_2 into the atmosphere before certain thresholds, boundaries, or tipping points are transgressed and the climate system moves into a state which renders human civilisation as we know it untenable.

Proponents of the standard narrative have, more recently, been more vocal about the possibilities of catastrophic environmental change and societal collapse (Kemp, Xu, and Depledge 2022). But where the catastrophist version of the story really diverges from the standard narrative is in its scepticism towards green technologies as the potential saviour. While standard narrative people might aim to develop a quantitative risk-management orientation towards scenarios of catastrophe, and advocate for rapidly reducing emissions by deploying advanced technologies and even geoengineering to bring more benign future scenarios into being, adherents to the catastrophist narrative see environmental change as the inevitable brick wall facing industrial civilisation, and emphasise the profound unknowability of the future behaviour of a complex Earth System made up of multiple interconnecting sub-systems and feedback loops. As such, analysts of fragile socio-ecological systems, as well as activist groups like Extinction Rebellion, contend that radical change is inevitable and, indeed, desirable, in the form of new kinds of local socio-economic organisation and the 'deep adaptation' of social structures to cope with a volatile future (Bendell and Read 2021). The catastrophists argue for relinquishing the pretensions of technoscientific solutions to our environmental predicament, and instead advocate for developing radically new social and economic systems at local scales that can perhaps help stave off catastrophe, but also, and crucially, provide a new model of social organisation for the period after the fall. While in the first two narratives civil society is largely a mute bystander while experts call the shots, this narrative sees civil society and local communities as important sources of 'solutions', albeit solutions that are less about the adoption of novel technologies and more about experimentation in new ways of living which emphasise communitarianism (Arias-Maldonado 2022), relationships of mutuality and kinship with both human and nonhuman others (Haraway 2015), circular flows of money and material goods (Calisto Friant, Vermeulen, and Salomone 2020), and human flourishing detached from the drive for financial accumulation (Reichel and Perey 2018; Richter 2019).

While many geographers are sceptical about the more catastrophic visions of socio-ecological futures, and about the notion that it's simply 'too late' to avoid them (e.g. Hulme 2020), others have laboured towards the other end of this narrative, working with similar ideas about the importance of giving voice and prominence to communities who are prefiguratively developing alternative ways of living. These alternatives may be explicitly about delaying the apocalypse, providing a model for human life after it, or may simply represent communities grappling with the practical and political challenges of building alternative modes of production, exchange, and consumption under capitalism. For example, Pallett, Chilvers, and Hargreaves (2019) offer a new 'map' of the diversity of ways in which communities in the U.K. are *already* pioneering new means of making and using energy, in the face of an apparent ignorance of such innovation on the part of policymakers and large sections of the environmental research community. Going further into the realm of prefiguratively building alternative modes of socioeconomic life, Gibson, Rose, and Fincher (2015) offer a manifesto for living in the Anthropocene that explores the implications of feminist geo-philosopher Val Plumwood's suggestion that 'We will go onwards in a different mode of humanity, or not at all' (Plumwood 2007). Building on earlier research in alternative economic geographies, this strand of work investigates how communities can be built outside of patriarchal, racist, and consumerist social orders; communities which can enable both human and nonhuman flourishing. Together this research demonstrates how 'human geographers can be active participants in unlocking the potential for new socio-ecological alternatives beyond the mainstream' through 'active involvement with those already creating different modes of living' (Castree 2014, 458). We come back to the role of geographers in fashioning new worlds in our final chapter.

4. *The eco-Marxist narrative*

Our fourth narrative takes issue with how all the other three (albeit mostly the first two) tell the story of the Anthropocene in terms of the actions of a somewhat undifferentiated human population. Instead, this narrative proposes that the chief protagonist of the story is a particular 'super-human' actor: *capital* (Foster 2022). Capital is the continuous transformation of money into yet more money by way of various investments. Capital is all about *circulation* and *expansion*. The growth of capital, and of capitalist modes of economic production and exchange, has not only created huge levels of economic inequality, but is also uniquely responsible for pushing the planet into the Anthropocene (Royle 2016). While the human population has increased 10-fold since 1700 and may well therefore be considered the crucial variable for understanding the Anthropocene, the value of capital has increased 134-fold (Bonneuil and Fressoz 2016). As such, 'instead of undifferentiated population and economic growth, processes of dispossession and commodification associated with the logic of capitalist expansion, along with the mechanisms of imperial domination, are the essential causal forces' (Bonneuil 2015, 28). This narrative has been developed by historians and historical geographers who have reconstructed how imperial powers like Great Britain, France and the United States grew their wealth through the plundering of resources from other parts of the planet, providing the necessary wealth, food and energy to enable the take-off of industrialism and the burning of fossil fuels whose environmental legacies we are grappling with to this day (e.g. Malm 2016; Moore 2015). As we explore in Chapter 3, historical geographer and sociologist Jason W. Moore has done important work in tracing the unsustainability of what he calls the capitalist 'world-ecology': the systems through which capitalism organises the transformation of nature into wealth, with profits accumulating in certain places and environmental harms accumulating in others.

The moral-political message of this narrative is quite clear – rather than seeking technoscientific 'fixes' which can enable capitalism to continue despite things like climate change (Surprise 2018), or re-organising social and economic life on local grounds, we need to pursue thoroughgoing structural change which tackles the social inequalities and environmental unsustainability of capitalism *together*. 'System change not climate change' has become the rallying-cry of Leftist environmentalism, and activists have adopted elements of Marxist analysis and critique to inform strategies of environmental resistance (Klein 2014). Political geographers and students of geopolitics have been active in interpreting the politics of the Anthropocene through a Marxist lens of class conflict (e.g. Huber 2022) and in further exploring the links between imperialism, capitalism, and environmental change (Dalby 2020; Grove 2019). For thinkers in this vein, the very notion of the 'Anthropocene' is a sleight of hand, a typically capitalist trick of universalising problems evidently caused by capitalism itself – i.e. blaming the whole human species for something caused by a very small majority of the population: the capitalist elite. As such, the 'Capitalocene' has been proposed as a more appropriate way of precisely identifying the nature and causes of this putative new geological age (Moore 2016).

5. *The colonial Anthropocene*

For many Marxist theorists, the rise of capitalism cannot be understood in isolation from the practices of imperialism and colonialism. But while many such thinkers would still locate the origins of the Anthropocene story in the nineteenth-century industrial revolution and the emergence of 'fossil capital' as a new world-making (and perhaps world-breaking) force (Malm 2016; see Chapter 4), others have argued that without the preceding centuries of European imperialism, the industrial revolution might not have even happened. As we explore in more detail across Chapters 3 and 4, historians and historical geographers have done the maths and worked out that the industrial revolution was only made physically possible by the spoils of empire-building. Physical geographers have also presented evidence to suggest that the first truly 'global' human impacts on the environment likewise pre-dated the industrial revolution, and were closely associated with European colonial expansion. But this wasn't just about physical resources and transformations – it was also about intellectual revolutions which divided the world by race, gender, hierarchies of 'civilization', 'nature' and 'culture' – intellectual frameworks which were used to justify colonial exploitation of both peoples and environments (Yusoff 2024). Theorists of *racial capitalism* have argued that racism is historically constitutive of capitalist modes of production, and that we are still grappling with the geographical legacies of, for example, plantation slavery in how inequality is reproduced and in how environmental harms are disproportionately visited upon marginalised communities (McKittrick 2011; Pulido 2017).

For many thinkers convinced by this narrative then, the Anthropocene didn't just start with the industrial revolution; rather, it is the earlier onset of European colonialism, with all its material, economic, and ideological effects, which is taken as the key inflection point. The key actor here isn't necessarily *capital*, but *colonialism and coloniality*. We can understand colonialism here as being the varied practices by which control is exercised over others' territory, over peoples, and over resources. But while the era of European colonial expansion might be formally over, forms of *coloniality* persist – this term describes the ideologies of colonialism, as expressed in its forms of knowledge (e.g. in claims about the racial superiority of some groups), and in the socio-political orders and hierarchies that have outlived formal colonial arrangements but which continue to marginalise and disenfranchise certain communities and regions (Kaltmeier et al. 2026; Quijano 2000). These ideas have proven influential in Geography in recent years, as scholars have grappled both with the origins of our environmental crises in colonial relations, as well as with how those relations continue to structure how societies are unequally impacted by, and able to respond to, environmental challenges (Sultana 2022).

These sorts of ideas have informed a growing recognition of how Western environmental movements – including groups like Extinction Rebellion (Bell and Bevan 2021) – have often been guilty of overlooking not only the racial inequalities of environmental change but also the long history of non-white and colonised communities struggling against the conjoined forces of capitalism, imperialism and environmental destruction (Davis and Todd 2017; Montrie 2018; Whyte 2017). For a growing number of scholars and activists, the Anthropocene is fundamentally a colonial phenomenon, and a just and sustainable future must not only be a post-capitalist future but also a decolonial future.

This conjunction of anti-colonial scholarship and activism around environmental change is arguably one of the most significant developments in response to the diagnosis of the Anthropocene, particularly in Geography, and is something we dwell on at numerous points throughout the book. For now, let's take a step back and think about how these five narratives compare.

Narratives compared

In the Table 1.1 below, we've compiled some of the key features of each of the five narratives described above. These are necessary simplifications, and we don't suppose to be able to contain the bewildering diversity of Anthropocene thought within the pages of this chapter, let alone in a 5 x 6 table. Nonetheless,

Table 1.1 Anthropocene narratives compared

Narrative	*Offered by*	*Chief actor(s)*	*Knowledge claim*	*Start date*	*What to do?*
Standard narrative	Earth System scientists, some historians	Undifferentiated human species	The Earth System is inherently unstable and close to a tipping point	Industrial Revolution or Great Acceleration	Scientific steering; use advanced technology; large-scale geoengineering as last resort
Post-natural Anthropocene	Eco-modernists Geo-philosophers	Humans and nonhumans in consort	We need to unravel the myth of nature/culture divide	Agricultural revolution	Better stewardship Multi-species kinship
Eco-catastrophist	Socio-ecological systems science; activists	Western Civilisation	Earth System complexity creates unknowable risks	Industrial revolution	Deep adaptation; radical localism; ecological justice, social equality
Eco-Marxist	Political economists, historians, activists	Capital and private business	Capital operates through 'world-ecologies'	Industrial Revolution	Structural change via political revolution; achieve social and environmental justice
Colonial Anthropocene	Historians, critical geographers, anti-colonial environmental activists	Colonialism and coloniality	The origins of the Anthropocene lie in colonialism, and its effects are heavily structured by ongoing coloniality	Onset of European colonialism	Decolonisation of culture, and structural change via political revolution; build respect for the non-human realm

we think this provides a useful guide and comparison of some major lines of Anthropocene thought, which can give some orientation for what comes next in the book.

In the table we have:

- identified who the key communities are who are using and promoting each narrative;
- identified the 'chief actors' within the narrative who are given the role of main protagonist;
- suggested that each narrative makes a key 'knowledge claim' – a big *epistemological* statement about what we know (or what we can't know); or a big *ontological* statement about how we should understand the world;
- pointed out where in human history each narrative locates the start of the Anthropocene;
- briefly accounted for the 'solutions' which are advocated as part of the implicit or explicit moral lessons of each narrative.

To rephrase this, narratives about the Anthropocene offer, in more or less detailed ways, an account of who or what *caused* the trajectory away from the Holocene at a particular date, of its current and/or future *impacts* (on people and the non-human world), of what the impacts *mean* (how they are judged to be significant and from whose perspective) and of what sorts of *actions* are necessary, prudent or desirable (aka mitigation and adaptive responses, be they incremental or more radical). To narrate the Anthropocene – to seek to represent it faithfully – is, of necessity, to make a set of rather grand claims. These claims about the world are also claims upon all those who would encounter them: they are both attention-grabbing and, as it were, attention-seeking. As we'll explain in the next chapter, there are geographies to all those things mentioned in various Anthropocene narratives, though some narrators barely seem to recognise them or else explore them only superficially. The geographies matter: they are not mere 'empirical decoration' on the 'structure' of our new age. They are the very fabric of the new world we are creating.

The politics of the Anthropocene

Each of the narratives just summarised makes a set of political claims. More formally, each contains *political presuppositions*, lodges explicit *political statements* about the past and present, and announces *political goals* for the future. Throughout this book we will return to these political dimensions, at times amplifying their meaning by way of concepts such as 'biopolitics', 'ontological politics', and 'necropolitics'. It's thus worth us saying something here about what the complicated word 'politics' means, how it relates to a closely connected phenomena which is 'social power', and how it relates to 'morality', 'ethics', and 'values' (we define other key terms we use frequently in this book in the Glossary)

We are political animals, as the saying goes. We often think of politics as those activities that governments undertake at home or overseas. Technically, in this context, 'politics' is the process of deciding how to govern, to what ends, at what cost, and to whose advantage. But there is a much broader definition of politics that extends far beyond the realm of government institutions, or big P politics. Here politics describes those variegated processes whereby different groups of people in arenas ranging from the home to their workplaces to public spaces to social media platforms – vie to have their needs, wants, and desires acknowledged, articulated, and supported. In the broadest sense, then, politics is about which values will guide how different people live together. In democratic societies, this happens through debate and open disagreement – for instance, the animal rights movement sprang up in the United States and elsewhere in the early 1970s, while in the 1990s campaigns against genetically modified crops became prominent quite rapidly, in both cases shaping public sentiment and policy. In autocratic societies, meanwhile, diverse political agendas get silenced and must submit to those imposed from on high (a sort of 'anti-politics', if you will). For something to become 'political', therefore, is more than it becoming a focus of *government* attention. It becomes political as soon as pretty much *anyone* seeks to make it a public issue for discussion leading, in time, to a decision to alter modes of thinking and action or else to sustain existing modes in the face of criticism. In recent years, entrenched 'white privilege', transphobia, ongoing misogyny, AI, 'unchecked immigration', and many other things have been politicised, within and beyond the realms of government. Voices outside government can become noted 'political voices', such as the Swedish climate activist Gerta Thunberg. All politics involves attempts to assign and uphold values about how aspects of the world ought to be treated and organised. Values are attributions of worth and importance. Values vary, and this is the root of diversity and conflict in political thought and action.

How does politics link to social power? Social power is the capacity of some actors to legitimately or illegitimately impose certain things on many others or to offer them (or make available) certain things. The things can be values, preferences, opinions, actions, material constraints, material opportunities, punishments, and so on. Power can work via persuasion, via publicly accepted legal procedures, by reasoned consensus among the affected parties, by the threat of force, the use of force, and so on. There are various types and mechanisms of social power. Such power is often asymmetric, in that some actors possess it while many others do not, or only possess it to a limited degree. The influential geographer Doreen Massey (2005) analysed these asymmetries as 'power-geometries'. Within such power geometries, those with less power may 'resist' and 'protest' against those with more. Clearly, sovereign governments possess a great deal of power relative to their citizens and the various places and environments within their borders. Usually, but not always, this power is perceived as 'legitimate' by those over whom it is exercised. Very powerful governments, like the USA, can project power far and wide *beyond* their borders. But social power is also possessed by those outside the realms of government – think of the power of major religions over their most devoted followers, of large firms like Amazon over their workers and consumers, of wealthy shareholders over companies they own a major stake in, or corporate executives like Elon Musk who hold majority stakes in their companies. Often, the power exercised by these agents is contested. Overall, then, social power is Janus-faced. Some forms of social power are necessary to ensure the well-being of a great many people and the material environments – consider, for instance, how tax revenues (which most adults are obliged to pay by law) get channelled into social welfare and environmental protection programs. But other forms of social power can be exercised in harmful ways, as with those criminal gangs who traffic people, drugs, or exotic species in order to make money illegally.

Finally, if social power is one vector for the enactment of politics – within and beyond the governmental arena – how does it link to morality and ethics? These two words are often used as synonyms. Broadly, they refer to the norms and principles that together specify 'right treatment' of other people or elements of the material world. The words 'wrong' and 'right' are ways of passing moral and ethical judgment. There are many different understandings of right and wrong. Moral and ethical concepts include rights, fairness, equality, entitlements, duties, and so on. Morality and ethics tend to be public rather than purely private matters (Raghuram 2016). They're often placed under the umbrella term of 'justice' (which takes different forms – see Chapter 6) and have greatest practical force when part of the law or when sedimented in everyday practices of human behaviour (Popke 2006), such as when people are habituated to not resolving disagreements through physical force and by obeying the rule of law. But moral and ethical questions pertain to more than justice: they also pertain to what is 'good' for any given society, community, or (if you're an environmentalist) non-human phenomena (like the Great Barrier Reef). Together, such considerations bear upon the question of what would constitute a 'Good Anthropocene'. The 'right' and the 'good' are the two great domains of moral-ethical reasoning and action.

Clearly, a great deal of political claims-making and action involves morality and ethics (Figure 1.3). Indeed, political disagreements are often moral and ethical to the core, with arguments and evidence adduced to support conflicting visions. Powerful political actors can enforce certain forms of morality and ethics, as we see when governments pass laws about farm and laboratory animal welfare. But politics isn't reducible to moral and ethical arguments: other values are at stake. Not all political decisions rest on solid ethical or moral solid ground and occur, instead, because a certain circumstance or a

Figure 1.3 Political and moral claims-making: an animal rights protester in New York City, September 2025. (Source: Shutterstock.)

powerful actor led to a (possibly unpopular or harmful) political decision being taken. In addition, the words morality and ethics suggest something pure and elevated, a set of ideals if you will (of the sort furnished by accomplished political philosophers such as Martha Nussbaum). By contrast, politics is that often messy process whereby morally and ethically inflected choices get made that involve 'winners' and 'losers', both human and non-human, depending on the situation. In politics, there's often a trade-off of different values, some of them not especially 'moral' (such as billionaires pressuring governments into giving their companies tax breaks in the name of 'freedom').

Throughout this book, the words politics, power, morality, ethics, and associated terms (such as value and justice) will be used in a range of ways. We hope the broad discussion just offered allows some sense to be made of this variety by readers. The advent of the Anthropocene raises wide-ranging political questions about what matters in our lives and how we will, together, work to realise those values through our decisions and actions. The 'facts' of our changing world do not tell us what to think, even as they inspire new thoughts and plans – in part because we view the facts through filters that tell us how to think and what to see when we look beyond ourselves. In our 'all-affected' world of hyperconnected people and physical environments, it follows that politics, power, morality, and ethics can't proceed in isolation: decisions taken in one time and place get tangled up with other places and times.

Conclusions

From the earlier table, we can see how there are several Anthropocenes in circulation – plural ways of understanding, thinking about, and responding to the indisputable recent intensification of human impacts on the nonhuman world. The Anthropocene concept has its origins in Western science, particularly in the field of ESS and in the discipline of Geology. It's a concept which stresses the extraordinary magnitude, scope, and scale of human impacts on what is now called the Earth System, and geologists are currently busy working out whether those impacts are sufficient to warrant the official designation of a new geological epoch or event. But as the Anthropocene has travelled into other knowledge communities, including Physical and Human Geography, key claims have been disputed, and other stories told. In the ensuing chapters, we'll explore these different versions of the Anthropocene, examining both their divergence from the 'standard narrative' and their implications for geographical study and for the politics and ethics of living on a transformed planet.

In the next chapter, we examine the Anthropocene through a geographical lens. In what ways is the proclaimed Anthropocene age intrinsically *geographical*? Has this geography been fully recognised by key thought-shapers in the Anthropocene debate? How have geographers represented the Anthropocene in their publications on the topic? How, if at all, might the discipline of Geography engender insights into the Anthropocene that are distinctive and valuable relative to other fields of knowledge? Rephrased, we can ask about the implicit and explicit claims about a changing world geography that are embedded in the narratives we summarised above. These narratives appear, unequally, within the *discipline* of Geography, but also exist as part of a wider *discourse* that contains geographical content and claims. We also ask whether the Anthropocene concept is really anything new, tracing some precursors and identifying where the Anthropocene idea, and its underpinning science, departs from previous ways of understanding human–environment relations.

This next chapter, along with the present one, sets the 'Anthropo*scene*' for the rest of this book. Together, the chapters offer concepts and venture claims that operate throughout the rest of this book.

Notes

1 Brand's "might as well get good at it" remark in 1968 was originally reflective of the then-new capacities of humans to view the earth from space, with the God's-eye view of satellite vision. He has more recently updated his oft-quoted remark as "we *have* to get good at it", in response to the urgency of climate change (Brockman 2009). This re-phrasing is indicative of how eco-modernist styles of thought are not necessarily sanguine or indeed sceptical about the risks of climate change and other environmental changes, even if their 'solutions' may depart from those of other, more recognisable brands of environmental thought.

2 That is, concerning the inherent value of progressively more knowledge and more control over nonhuman systems, or of more economic growth and more high technology.
3 See Rockström et al. (2009) and Brown (2017).

References

Arias-Maldonado, M. 2022. 'Politics in the Anthropocene'. In *Altered Earth: Getting the Anthropocene Right*, edited by Julia Adeney Thomas, 160–81. Cambridge: Cambridge University Press.

Autin, W. J., and J. M. Holbrook. 2012. "Is the Anthropocene an Issue of Stratigraphy or Pop Culture?" *GSA Today* 22 (7): 60–61.

Baskin, J. 2019. *Geoengineering, the Anthropocene and the End of Nature*. Cham: Springer.

Bell, K., and G. Bevan. 2021. "Beyond Inclusion? Perceptions of the Extent to Which Extinction Rebellion Speaks to, and for, Black, Asian and Minority Ethnic (BAME) and Working-Class Communities." *Local Environment* 26 (10): 1205–20.

Bendell, J., and R. Read, eds. 2021. *Deep Adaptation: Navigating the Realities of Climate Chaos*. Oxford: Wiley.

Bennett, E. M., M. Solan, R. Biggs, et al. 2016. "Bright Spots: Seeds of a Good Anthropocene." *Frontiers in Ecology and the Environment* 14 (8): 441–48.

Bonneuil, C. 2015. "The Geological Turn: Narratives of the Anthropocene." In *The Anthropocene and the Global Environmental Crisis: Rethinking Modernity in a New Epoch*, edited by C. Hamilton, Christophe Bonneuil, and François Gemenne, 17–31. Abingdon: Routledge.

Bonneuil, C., and J. B. Fressoz. 2016. *The Shock of the Anthropocene*. London: Verso.

Brauch, H. G. 2021. "The Anthropocene Concept in the Natural and Social Sciences, the Humanities and Law – A Bibliometric Analysis and a Qualitative Interpretation (2000–2020)." In *Paul J. Crutzen and the Anthropocene: A New Epoch in Earth's History*, edited by Susanne Benner, Gregor Lax, Paul J. Crutzen, Ulrich Pöschl, Jos Lelieveld, and Hans Günter Brauch, 289–438. The Anthropocene: Politik—Economics—Society—Science. Cham: Springer International Publishing.

Brockman, J. 2009. 'WE ARE AS GODS AND HAVE TO GET GOOD AT IT: Stewart Brand Talks About His Ecopragmatist Manifesto'. Edge Magazine, https://www.edge.org/conversation/stewart_brand-we-are-as-gods-and-have-to-get-good-at-it Accessed 14 May 2026

Brown, K. 2017. "Global Environmental Change II: Planetary Boundaries - A Safe Operating Space for Human Geographers?" *Progress in Human Geography* 41 (1): 118–30.

Butler, D. R. 2021. *The Anthropocene*. Abingdon: Routledge.

Calisto Friant, M., W. J. V. Vermeulen, and R. Salomone. 2020. "A Typology of Circular Economy Discourses: Navigating the Diverse Visions of a Contested Paradigm." *Resources, Conservation and Recycling* 161 (October): 104917.

Castree, N. 2014. "Geography and the Anthropocene II: Current Contributions." *Geography Compass* 8 (7): 450–63.

Castree, N. 2019. 'Earth System Science'. In *International Encyclopedia of Geography*, 1–16. Oxford: John Wiley & Sons, Ltd.

Castree, N. 2021. "Framing, Deframing and Reframing the Anthropocene." *Ambio* 50 (10): 1788–92.

Chakrabarty, D. 2009. "The Climate of History: Four Theses." *Critical Inquiry* 35 (2): 197–222.

Chakrabarty, D. 2019. "The Planet: An Emergent Humanist Category." *Critical Inquiry* 46 (1): 1–31.

Clark, N. 2011. *Inhuman Nature: Sociable Life on a Dynamic Planet.* In *Inhuman Nature: Sociable Life on a Dynamic Planet.* London: SAGE.

Clark, N. 2012a. "Rock, Life, Fire: Speculative Geophysics and the Anthropocene." *Oxford Literary Review* 34 (2): 259–76.

Clark, N., and B. Szerszynski. 2021. 'Planetary Multiplicity, Earthly Multitudes: Interscalar Practices for a Volatile Planet'. In *Narratives of Scale in the Anthropocene.* Abingdon: Routledge.

Clark, T. 2012b. 'Scale'. In *Telemorphosis: Theory in the Era of Climate Change*, edited by Tom Cohen, 148–66. London: Open Humanites Press.

Clifford, N., and K. Richards. 2005. "Earth System Science: An Oxymoron?" *Earth Surface Processes and Landforms* 30 (3): 379–83.

Colebrook, C. 2016. "What Is the Anthropo-Political?" In *Twilight of the Anthropocene Idols*, edited by T Cohen, C Colebrook, and J Hillis Miller, 81–125. London: Open Humanites Press.

Cronon, W. 1992. "A Place for Stories: Nature, History, and Narrative." *The Journal of American History* 78 (4): 1347–76.

Cronon, W. 1996. 'The Trouble With Wilderness; or, Getting Back to the Wrong Nature'. In *Uncommon Ground: Rethinking the Human Place in Nature*, edited by William Cronon, 69–90. London: Norton.

Crutzen, P. J. 2002. "Geology of Mankind." *Nature* 415 (January): 23.

Crutzen, P. J., and E. F. Stoermer. 2000. "The Anthropocene." *Global Change Newsletter* 41: 17–18.

Dalby, S. 2016. "Framing the Anthropocene: The Good, the Bad and the Ugly." *The Anthropocene Review* 3 (1): 33–51.

Dalby, S. 2020. *Anthropocene Geopolitics: Globalization, Security, Sustainability.* Ottawa: University of Ottawa Press.

Damianos, A. 2025. "Anthropocene Angst: Authentic Geology and Stratigraphic Sincerity." *Social Studies of Science* 55 (3): 444–64.

Davies, J. 2016. *The Birth of the Anthropocene.* Oakland: University of California Press.

Davis, H., and Z. Todd. 2017. "On the Importance of a Date, or Decolonizing the Anthropocene." *Acme* 16 (4): 761–80.

Doel, R. E. 2003. "Constituting the Postwar Earth Sciences: The Military's Influence on the Environmental Sciences in the USA after 1945." *Social Studies of Science* 33 (5): 635–66.

Eckhart, E., and T. Krafft, eds. 2005. *Earth System Science in the Anthropocene.* Dordrect: Springer.

Edgeworth, M., E. C. Ellis, P. Gibbard, C. Neal, and M. Ellis. 2019. "The Chronostratigraphic Method Is Unsuitable for Determining the Start of the Anthropocene." *Progress in Physical Geography: Earth and Environment* 43 (3): 334–44.

Edgeworth, M., P. Gibbard, M. Walker, D. Merritts, S. Finney, and M. Maslin. 2023. "The Stratigraphic Basis of the Anthropocene Event." *Quaternary Science Advances* 13: 100088.

Ellis, E. C., J. O. Kaplan, D. Q. Fuller, S. Vavrus, K. Klein Goldewijk, and P. H. Verburg. 2013. "Used Planet: A Global History." *Proceedings of the National Academy of Sciences* 110 (20): 7978–85.

Foster, John Bellamy. 2022. *Capitalism in the Anthropocene: Ecological Ruin or Ecological Revolution*. New York: NYU Press.

Gallie, W B. 1955. "Essentially Contested Concepts." *Proceedings of the Aristotelian Society* 56: 167–98.

Gandy, M. 2013. "Marginalia: Aesthetics, Ecology, and Urban Wastelands." *Annals of the Association of American Geographers* 103 (6): 1301–16.

Gibbard, P. L., A. M. Bauer, M. Edgeworth, et al. 2021. "A Practical Solution: The Anthropocene Is a Geological Event, Not a Formal Epoch." *Episodes Journal of International Geoscience* 44 (3): 319–28.

Gibson, K., D. B. Rose, and R. Fincher. 2015. *Manifesto for Living in the Anthropocene*. Brooklyn: Punctum.

Grosz, E. 2008. *Chaos, Territory, Art: Deleuze and the Framing of the Earth*. New York: Columbia University Press.

Grove, J. V. 2019. *Savage Ecology: War and Geopolitics at the End of the World*. Durham NC: Duke University Press.

Grove, K. 2018. *Resilience*. London: Routledge.

Haraway, D. 2015. "Anthropocene, Capitalocene, Plantationocene, Chthulucene: Making Kin." *Environmental Humanities* 6: 159–65.

Haraway, D. J. 2008. *When Species Meet*. Minneapolis: University of Minnesota Press.

Holling, C. S. 1973. "Resilience and Stability of Ecological Systems." *Annual Review of Ecology and Systematics* 4: 1–23.

Huber, M. T. 2022. *Climate Change as Class War: Building Socialism on a Warming Planet*. London: Verso Books.

Hulme, M. 2020. "Is It Too Late (to Stop Dangerous Climate Change)? An Editorial." *WIREs Climate Change* 11 (1): e619.

IPCC. 2014. *Climate Change 2013: The Physical Science Basis*. Cambridge: Cambridge University Press.

IUGS. 2024. *The Anthropocene*. International Union of Geological Sciences. https://www.iugs.org/_files/ugd/f1fc07_40d1a7ed58de458c9f8f24de5e739663.pdf?index=true

Jasanoff, S. 2020. "Ours Is the Earth: Science and Human History in the Anthropocene." *Journal of the Philosophy of History* 14 (3): 337–58.

Kaltmeier, O., L. R. Ellermeier, E. M. Rummelhoff, O. Sierra Cháves, and A. Volmer, eds. 2026. *Contesting the Anthropocene: Latin American Perspectives Beyond Coloniality and Capitalism*. Abingdon: Routledge.

Kemp, L., C. Xu, J. Depledge, et al. 2022. "Climate Endgame: Exploring Catastrophic Climate Change Scenarios." *Proceedings of the National Academy of Sciences* 119 (34): e2108146119.

Klein, N. 2014. *This Changes Everything: Capitalism Vs. The Climate*. New York: Simon and Schuster.

Latour, B. 1993. *We Have Never Been Modern*. Cambridge: Harvard University Press.

Lorimer, J. 2012. "Multinatural Geographies for the Anthropocene." *Progress in Human Geography* 36 (5): 593–612.

Lorimer, J. 2015. *Wildlife in the Anthropocene: Conservation After Nature*. Minneapolis: University of Minnesota Press.

Lorimer, J. 2017. "Parasites, Ghosts and Mutualists: A Relational Geography of Microbes for Global Health." *Transactions of the Institute of British Geographers* 42 (4): 544–58.

Lövbrand, E., S. Beck, J. Chilvers, et al. 2015. "Who Speaks for the Future of Earth? How Critical Social Science Can Extend the Conversation on the Anthropocene." *Global Environmental Change* 32 211–18.

Lovelock, J. E., and L. Margulis. 1974. "Atmospheric Homeostasis by and for the Biosphere: The Gaia Hypothesis." *Tellus* 26 (1–2): 2–10.

Lundershausen, J. 2018. "Marking the Boundaries of Stratigraphy: Is Stratigraphy Able and Willing to Define, Describe and Explain the Anthropocene?" *Geo: Geography and Environment* 5 (1): e00055.

Lynas, M. 2011. *The God Species: Saving the Planet in the Age of Humans*. Washington: National Geographic Society.

Malm, A. 2016. *Fossil Capital: The Rise of Steam-Power and the Roots of Global Warming*. London: Verso.

Massey, D. 2005. *For Space*. Sage.

McCarthy, F. M. G., M. J. Head, C. N. Waters, and J. Zalasiewicz. 2025. "Would Adding the Anthropocene to the Geologic Time Scale Matter?" *AGU Advances* 6 (2): e2024AV001430.

McKittrick, K. 2011. "On Plantations, Prisons, and a Black Sense of Place." *Social & Cultural Geography* 12 (8): 947–63.

McNeill, J. R. 2001. *Something New Under the Sun: An Environmental History of the Twentieth-Century World*. London: Norton.

Meadows, D. H. 1972. *The Limits to Growth: A Report for the Club of Rome's Project on the Predicament of Mankind*. New York: Universe Books.

Montrie, C. 2018. *The Myth of Silent Spring*. Berkeley: University of California Press.

Moore, J. W. 2015. *Capitalism in the Web of Life: Ecology and the Accumulation of Capital*. London: Verso.

Moore, J. W. 2016. *Anthropocene or Capitalocene? Nature, History, and the Crisis of Capitalism*. Edited by Jason W. Moore. Oakland: Pm Press.

Nichols, K., and B. Gogineni. 2018. "The Anthropocene's Dating Problem: Insights from the Geosciences and the Humanities." *The Anthropocene Review* 5 (2): 107–19.

NRC. 1986. *Earth System Science: Overview: A Program for Global Change*. Washington: National Academies Press.

Pallett, H., J. Chilvers, and T. Hargreaves. 2019. "Mapping Participation: A Systematic Analysis of Diverse Public Participation in the UK Energy System." *Environment and Planning E: Nature and Space* 2 (3): 590–616.

Plumwood, V. 2007. "*A Review of Deborah Bird Rose's 'Reports from a Wild Country: Ethics for Decolonisation*." *AHR* 42 (August): 1–4.

Popke, J. 2006. "Geography and Ethics: Everyday Mediations Through Care and Consumption." *Progress in Human Geography* 30 (4): 504–12.

Pulido, L. 2017. "Geographies of Race and Ethnicity II: Environmental Racism, Racial Capitalism and State-Sanctioned Violence." *Progress in Human Geography* 41 (4): 524–33.

Purdy, J. 2015. *After Nature: A Politics for the Anthropocene*. Cambridge: Harvard University Press.

Quijano, A. 2000. "Coloniality of Power and Eurocentrism in Latin America." *International Sociology* 15 (2): 215–32.

Raghuram, P. 2016. "Locating Care Ethics Beyond the Global North." *ACME: An International Journal for Critical Geographies* 15 (3): 3.

Read, R., and S. Alexander. 2019. *This Civilisation Is Finished: Conversations on the End of Empire - and What Lies Beyond.* Melbourne: Simplicity Institute Publishing.

Reichel, A., and R. Perey. 2018. "Moving Beyond Growth in the Anthropocene." *The Anthropocene Review* 5 (3): 242–49.

Richter, K. 2019. 'Struggling for Another Life: The Ontology of Degrowth'. *Transtext(e)s Transcultures 跨文本跨文化. Journal of Global Cultural Studies* 14 (December).

Rickards, L. 2015. "Critiquing, Mining and Engaging Anthropocene Science." *Dialogues in Human Geography* 5 (3): 337–42.

Robin, L., D. Avango, L. Keogh, N. Möllers, B. Scherer, and H. Trischler. 2014. "Three Galleries of the Anthropocene." *The Anthropocene Review* 1 (3): 207–24.

Rockström, J., W. Steffen, K. Noone, et al. 2009. "Planetary Boundaries: Exploring the Safe Operating Space for Humanity." *Ecology and Society* 14 (2): 32.

Royle, C. 2016. 'Marxism and the Anthropocene'. *International Socialism* 151: 23. https://isj.org.uk/marxism-and-the-anthropocene/

Ruddiman, W. F. 2013. "The Anthropocene." *Annual Review of Earth and Planetary Sciences* 41 (1): 45–68.

Schellnhuber, H. 1999. "Earth System" Analysis and the Second Copernican Revolution." *Nature* 402 (December): 19–22.

Schwägerl, C. 2015. "A Concept With a Past." In *Welcome to the Anthropocene: The Earth in Our Hands*, edited by Nina Möllers, Christian Schwägerl, and Helmuth Trischler. Munich: Deutsches Museum.

Scranton, R. 2015. *Learning to Die in the Anthropocene: Reflections on the End of a Civilization*. San Francisco: City Lights Books.

Seitzinger, S. P., O. Gaffney, G. Brasseur, et al. 2015. "International Geosphere-Biosphere Programme and Earth System Science: Three Decades of Co-Evolution." *Anthropocene* 12: 3–16.

Shellenberger, M., and T. Nordhaus. 2014. *On Becoming an Ecomodernist*. Oakland: The Breakthrough Institute.

Simon, Z. B., and J. A. Thomas. 2022. "Earth System Science, Anthropocene Historiography, and Three Forms of Human Agency." *Isis* 113 (2): 396–406.

Sklair, L., ed. 2020. *The Anthropocene in Global Media: Neutralizing the Risk*. London: Routledge.

Steffen, W., et al. 2018. 'Trajectories of the Earth System in the Anthropocene'. *Proceedings of the National Academy of Sciences of the United States of America* 115 (33): 8252–59.

Steffen, W., Å. Persson, L. Deutsch, et al. 2011. "The Anthropocene: From Global Change to Planetary Stewardship." *AMBIO* 40 (7): 739–61.

Steffen, W., K. Richardson, J. Rockström, et al. 2015. "Planetary Boundaries: Guiding Human Development on a Changing Planet." *Science* 347 (6223): 1259855.

Steffen, W., K. Richardson, J. Rockström, et al. 2020. "The Emergence and Evolution of Earth System Science." *Nature Reviews Earth & Environment* 1 (1): 1.

Sultana, F. 2022. "The Unbearable Heaviness of Climate Coloniality." *Political Geography* 99: 102638.

Surprise, K. 2018. "Preempting the Second Contradiction: Solar Geoengineering as Spatiotemporal Fix." *Annals of the American Association of Geographers* 108 (5): 1228–44.

Swanson, H. A. 2016. "Anthropocene as Political Geology: Current Debates Over How to Tell Time." *Science as Culture* 25 (1): 157–63.

Trischler, H. 2016. "The Anthropocene: A Challenge for the History of Science, Technology, and the Environment." *NTM International Journal of History and Ethics of Natural Sciences, Technology and Medicine* 24 (3): 309–35.

Turchetti, S., and P. Roberts, eds. 2014. *The Surveillance Imperative: Geosciences During the Cold War and Beyond*. Dordrecht: Springer.

Wapner, P. 2014. "The Changing Nature of Nature: Environmental Politics in the Anthropocene." *Global Environmental Politics* 14 (4): 36–54.

Waters, C. N., J. Zalasiewicz, M. J. Head, G. N. Schäfer, F. M. G. McCarthy, and S. D. Turner. 2025. "Response to Damianos—Anthropocene Angst: Authentic Geology and Stratigraphic Sincerity." *Social Studies of Science* 55 (3): 465–78.

Whyte, K. 2017. "Indigenous Climate Change Studies: Indigenizing Futures, Decolonizing the Anthropocene." *English Language Notes* 55 (1–2): 153–62.

Wright, C., D. Nyberg, L. Rickards, and J. Freund. 2018. "Organizing in the Anthropocene." *Organization* 25 (4): 455–71.

Yusoff, K. 2020. "The Inhumanities." *Annals of the American Association of Geographers* 111 (3): 663–76.

Yusoff, K. 2024. *Geologic Life: Inhuman Intimacies and the Geophysics of Race*. Durham: Duke University Press.

Zalasiewicz, J., C. Waters, and M. J. Head. 2017. "Anthropocene: Its Stratigraphic Basis." *Nature* 541 (7637): 7637.

Zalasiewicz, J., C. Waters, C. Summerhayes, and M. Williams. 2018. "The Anthropocene." *Geology Today* 34 (5): 177–81.

Zalasiewicz, J., M. Williams, A. Smith, et al. 2008. "Are We Now Living in the Anthropocene." *GSA Today* 18 (2): 4.

2 Geographies of the Anthropocene

Geo-graphing the Anthropocene

The Anthropocene concept is very obviously geographical (as, by implication are some of the alternative concepts we'll examine in this book, such as the Capitalocene). It denotes major changes to the entire surface of the Earth, what some geoscientists these days call 'the critical zone', where biotic and abiotic phenomena interact. But what's involved in elucidating the geography of the Anthropocene? In what does this geography consist?

As we've noted, the Anthropocene concept originates in two areas of geoscience. This immediately suggests that the variegated and connected geography of our new age – or at least its physical geography – is strongly emphasised by the Earth System Scientists and stratigraphers who've elaborated the Anthropocene proposition over the last 20 years or so. However, this is not the case. As we'll explain below, this is symptomatic of much expert discussion of the Anthropocene to-date. An anaemic geography – a 'geography lite', if you will – often animates discussions of our emerging post-Holocene world.[1] Oddly, perhaps, members of the geographic profession sometimes contribute to this, while others outside the discipline of Geography can at times offer a richer sense of how and why the geographies of the Anthropocene matter. In what follows we distinguish between a 'thin' and a 'thick' understanding of the Anthropocene's geographies. While the former are quite easy for researchers to point to, the latter require careful elucidation. Any serious attempt to narrate the Anthropocene – it's start date, its causes, its impacts and suitable responses to its onset – needs, we argue, to convey a thick sense of the geography. This involves joining the dots among various specialist areas of research, though there's more than one way to do this.

In this chapter, we note that the 'discourse of geography', both thin and thick, operates not only within the eponymous discipline but in other disciplines too, at times, and also the wider societies that academic disciplines and inter-disciplines seek to serve. 'Discourse' here means any communication about the world's past, present, and future possible geographies. While a thick sense of the Anthropocene's geographies can certainly be conveyed

DOI: 10.4324/9781003162469-3

by analysts outside Geography, we point in this chapter to the advantages the discipline currently possesses in seeking to push beyond a 'geography lite' rendering of our emerging geological age. The discourse about the Anthropocene's geographies, that's to say, needs to move ever more in a 'thick' direction, with geographers key to this but not exclusively so. This is true whichever of the five Anthropocene narratives summarised in chapter one you adhere to. The cardinal geographical dimensions we identify in the following pages should be borne in mind by readers as they move through the chapters to come. Crucially, there's no one 'correct' or once-and-for-all way to characterise these dimensions – and we'll explain exactly why this complication applies towards the end of the chapter. But the key is the aim for 'thick' understandings, and the aim of academic disciplines, of course, is to push beyond superficial understandings of reality.

Before we explore geographies thin and thick, however, we begin by recounting the precursors of the Anthropocene idea. What terms preceded the one coined, separately, by Paul Crutzen and Eugene Stoermer prior to their jointly authored piece appearing in the IGBP newsletter a quarter of a century ago?

Is the Anthropocene idea new?

The Anthropocene was certainly a new word when coined at the turn of the new millennium. But prior to Crutzen and Stoermer, others had, from time to time, devised neologisms with a seemingly similar meaning. Let's briefly consider them, starting in the mid-to-late nineteenth century.

- The German zoologist Ernst Haeckel (1868) used the term 'Anthropozoic period' to define the 'era of man', characterised by 'the development and dispersal of the human organisms and their culture'. Soon after, the Italian geologist Antonio Stoppani (1873) also used the word 'Anthropozoic'. Presciently, he considered humankind to be a 'new telluric force which in power and universality may be compared to the greater forces of earth'.
- In the early twentieth century, the Russian biogeochemist Vladimir Vernadsky and the French philosopher, palaeontologist, and Jesuit priest Pierre Teilhard de Chardin both used the term 'noosphere' in various publications and lectures. This described a human realm of consciousness and intent with an ability to shape the biosphere and other elements of the Earth's surface.
- Later, in the early 1970s, the American systems analyst Jay Forrester (1971) promoted the idea of a 'world system' and his work informed the famous global-scale modelling undertaken in the best-selling book *The Limits To Growth* (Meadows et al. 1972). This book predicted future natural resource shortages as human demand outstripped a finite resource base. Around the same time, though not with special reference to the natural

environment, sociologist Immanuel Wallerstein began to develop 'world-systems theory'.

- Then in 1992 the American environmentalist and journalist Andrew Revkin coined the term 'Anthrocene' in a short book about climate change. He speculated that future generations might inherit a thoroughly remade natural environment at all points of the compass.

The various pre-cursors, spanning a century, each in their own way pointed to the growing ability of at least some societies to significantly reshape the natural environment at a planetary scale. As early as 1867, American George Perkins Marsh was expressing concern about the human impact on Earth. This concern escalated after World War II, the era of nuclear weapons proliferation, the Cold War, and an exploding world population. It led, via books authored by William Vogt, Fairfield Osborne, Rachel Carson, Paul Ehrlich, Barry Commoner, Murray Bookchin, and others, to the first Earth Day (1970) and the inaugural United Nations conference on the environment, held in Stockholm in 1972. But the precursor ideas listed above weren't supported by the evidence, computational models, or sophisticated systems theory that's underpinned the geoscience work conducted since around 2000. Indeed, the environmental philosopher Clive Hamilton and the historian Jacques Grinevald (2015) have maintained that the Anthropocene concept is unprecedented. For them, the reason why is twofold: earlier commentators, like Vernadsky, lacked the concept of the Earth System (which originates in 1970s research by geoscientist Jim Lovelock and others), and also lacked compelling data about significant historic and present-day Earth System change. This made their claims speculative and weakly demonstrated.

Given this, Hamilton and Grinevald have argued that the Anthropocene concept is genuinely novel (compare Steffen et al. 2011). Furthermore, they maintain that the Anthropocene began after 1945. Those who'd point to earlier periods, they argue, are confusing humanity's global environmental *signature* – which has been evident for centuries – with its ability to instigate *system change* at a planetary level. The latter, they argue, is a recently acquired ability in historical terms. The sudden onset of the Anthropocene, for them, is a sure sign of the unplanned nature and alarming size of humanity's ecological footprint. It's an event not merely of accelerating 'environmental change' but of momentous Earth System change. Earlier commentators may have foreseen the Anthropocene before the term was coined, but for Hamilton and Grinevald, they lacked the ability to demonstrate its onset. Relatedly, for these two authors, those geoscientists who pinpoint the Anthropocene's onset hundreds or even thousands of years ago – such as American palaeoclimatologist Bill Ruddiman (2003) – are mistaking anthropogenic signals in the long-term environmental record for Earth System change outside Holocene parameters.

Regardless, one thing is abundantly clear: never have so many geoscientists agreed that humanity's environmental footprint is so large and deep.

Thus, where some previous concepts may have contained similar ideas to the Anthropocene concept, they were marginal in their time. The Anthropocene concept, by contrast, has powerful geoscientific backing across several research fields.

Geography lite: the anaemic Anthropocene

The Anthropocene, as we've seen, is an epochal term. It denotes a planet-wide shift in the characteristics of the Earth System – a transition away from the Holocene conditions in which *homo sapiens* have vastly multiplied their numbers, invented steam engines, built cities like Tokyo and Sao Paulo, cleared vast tracts of forest, mined fossil fuels *en masse*, genetically altered crop species, detonated nuclear weapons, created plastics, and so on. So, the Anthropocene, at base, involves the replacement of a vast, complex, distinctive, inherited physical geography – of vegetation, ice cover, wild animal populations, river courses, near-surface minerals, etc. – by both design and accident. Humans, in a range of ways and especially since the mid-twentieth century, have created an enormous, complex, and distinctive geography of towns, cities, production facilities, farm land, transportation networks, energy infrastructures, and waste disposal streams and sites. This human geography has been, and will continue to be, forged from the natural 'goods and services' furnished by the Earth and will continue to reconfigure the planet's physical geography for decades to millennia. But as this physical geography changes over time, humans will need to adapt, such is the scale and power of some of the alterations that will unfold – global sea level rise is an obvious example and exceeds human control. In this sense, the Anthropocene designates an unprecedented coupling of physical and human geography: both will become more 'liquid' and less stable unless, somehow, humans club together so as to radically reduce their collective footprint on Earth.

In light of this, it might be thought that a fairly detailed characterisation of the human impact on land, air, and water, across the two hemispheres, would have been presented by the key geoscience teams who've promoted the Anthropocene concept since around 2000. However, for the most part, this is *not* the case. Instead, the geographic details have been left largely implicit. The teams have tended to abstract from the local and regional specifics – relating, say, to Western Europe and Central America – and to paint in broad brush strokes. This is evident in the Earth System diagram we reproduced in Chapter 1. It's also evident across any number of key publications by Earth System Scientists and by stratigraphers such as Jan Zalasiewicz. For instance, the early 'planetary boundaries' publications led by Johan Rockström and Will Steffen (Rockström et al. 2009; Steffen et al. 2015) defined a 'safe operating space for humanity' with reference to aggregate thresholds, such as greenhouse gas (GHG) concentrations in the world's atmosphere Then, in numerous publications examining potential stratigraphic markers for a Holocene-Anthropocene 'lower boundary', various members of the Anthropocene Working Group (the AWG, founded in 2009, with Zalasiewicz as chairman) have also tended to look for globally prevalent indicators – such as certain radioactive elements deposited worldwide in the 1950s after nuclear weapons tests.

The reason for this anaemic or thin physical geography is a good one: both Earth System Scientists and stratigraphers have been keen to detect *non-incremental global-scale changes that cross-cut all manner of different localities, environments, and continents* (see Box 2.1 for one of numerous examples). That is, rather than dwelling on sub-global details, the geoscientists have been on the lookout for planet-wide alterations outside the natural range of variability characteristic of the Holocene epoch (indeed, this is an absolute requirement for stratigraphers when designating a new phase in Earth's long history).

Box 2.1 'Thin' geographies of the Anthropocene

In late 2022, an international geoscience team published a widely read and widely reported peer review paper in the prestigious journal *Science* (David Armstrong McKay *et al.* 'Exceeding 1.5°C global warming could trigger multiple climate tipping points' *Science* 377: eabn7950). As the map below indicates, it focussed on 'tipping elements' that, in certain combinations, can 'flip' the whole Earth System into a new post-Holocene 'regime' if unduly perturbed. While each tipping element receives roughly equal treatment in the paper, there is little 'granularity' offered, and it is *aggregate* data for each one that anchors the analysis. The diagram below shows the atmospheric temperature thresholds for 'tipping' in each of the key regional-scale biophysical systems. Some systems will tip sooner than others, according to the paper's authors. Their's is very much a global, high-level analysis that is highly abstract (Figure 2.1).

A number of these geoscientists have been physical geographers, or at least geographers interested in human impacts on the non-human world (so-called society-environment geographers). For instance, over the years, British physical geographer John Dearing, British-American geographer Diana Liverman, American biogeographer Erle Ellis, and British physical geographer Frank Oldfield have each lent their names to key publications about the Anthropocene led by Earth System Scientists and/or by stratigraphers. Across these two geoscience communities, three key issues about the Anthropocene have preoccupied them: namely, what bodies of evidence are needed to detect

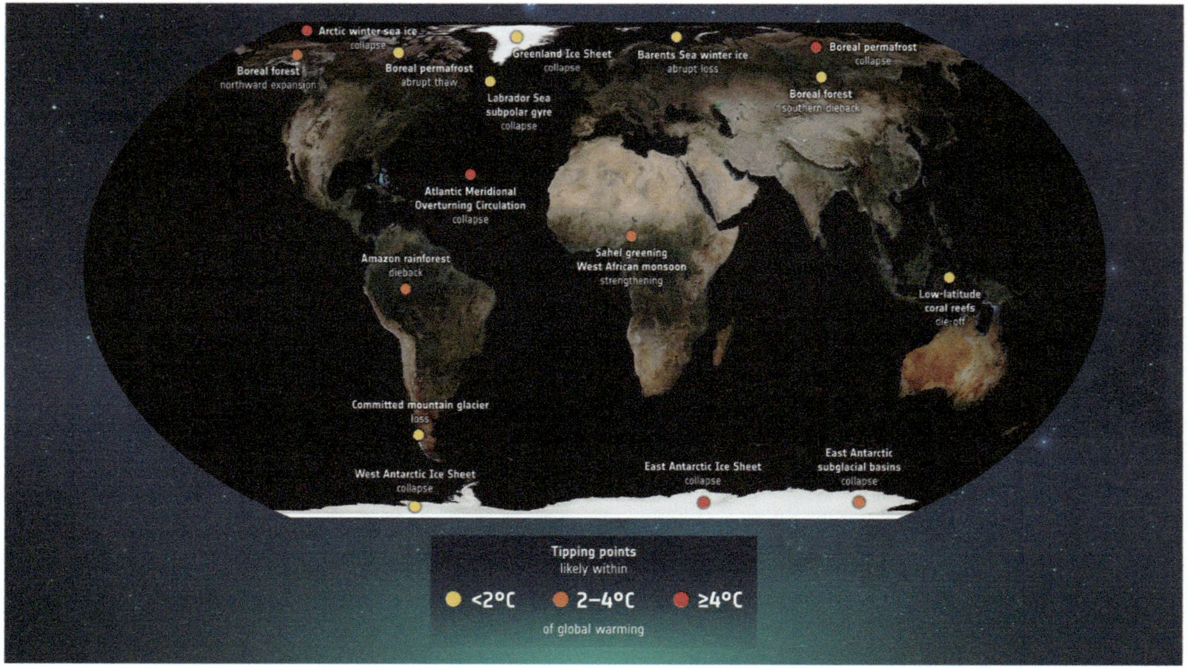

Figure 2.1 Climate 'tipping elements' and the levels of global warming at which they may be triggered. (Source: www.esa.int, copyright European Space Agency – ESA.)

significant Earth System change today?; how much change is required to declare the Holocene is at an end?; and when, precisely, did humans' combined activities instigate such change? Throughout, and as just noted, the global scale of change has been the major preoccupation, notwithstanding the varied regional- and metropolitan-scale drivers of such change operating across multiple decades.

So, in key parts of the geoscience community, the Anthropocene has been presented in a somewhat geographical way. The focus has been more temporal than spatial, and this is especially evident in debates about when, precisely, the Anthropocene began. This appears to have had a knock-on effect in parts of social science and the humanities, where the human aspects of the Anthropocene story have been explored by historians, sociologists, philosophers, and others. Many of these commentators have offered broad insights into things like the causes, human impacts, and moral significance of the Anthropocene. For instance, consider the book *Abundant Earth*, authored by American ecocentric philosopher Eileen Crist (2019). It deems the Anthropocene to be a resolutely 'ecocidal' event – one of mass destruction and death inflicted on the non-human world. It offers an impassioned defence of animals, insects, plants, trees, and whole ecosystems. However, it does so in a very general way and conveys little sense of nature's spatially varied yet continuous texture at, just below, and above the Earth's surface. Likewise, in some of the more public-facing literature about the Anthropocene that has appeared in recent years, the geography is rather underplayed. An example is environmental journalist Elizabeth Kolbert's best-selling book *Under A White Sky: The Nature of the Future* (2021). It's peppered with engaging stories about major environmental changes and proposed solutions from all around the world. But the stories, at base, are in service of a general argument that the changes are too big and too fast and that, together, we need to apply the metaphorical handbrake. There's no textured analysis of the geographies of cause, impact, and response when it comes to major planet-wide changes instigated by human action.

Having said all this, there are a number of more recent geoscience publications about the Anthropocene that are a little closer to the ground – more granular and less abstract, in other words – than some of the earlier work by Earth System Scientists and stratigraphers. Consider the following examples, noting their titles:

- Pastor, A. V., *et al.* (2022) 'Understanding the transgression of global and regional freshwater planetary boundaries', *Philosophical Transactions of the Royal Society A* 380, 2238: 20210294. This paper, as its title indicates, assesses global- and regional-scale boundaries.
- Jaureguiberry, P., *et al.* (2022) 'The direct drivers of recent global anthropogenic biodiversity loss', *Science Advances* 8 eabm9982. This paper identifies how the direct human drivers of ongoing biodiversity decline vary across the major continents of the Earth.

- Wang, S., *et al.* (2023) 'Mechanisms and impacts of Earth System tipping elements', *Reviews of Geophysics* 61, 1: e2021RG000757. This paper explores regional-scale tipping elements, such as the Amazon ecosystem, that can together tip the entire Earth System into a new regime if significantly perturbed.
- Liu, T., *et al.* (2023) 'Teleconnections among tipping elements in the Earth System', *Nature Climate Change* 13, 1: 67–74. This paper pays special attention to the Amazon and the Tibetan Plateau in a wider Earth System context. This and the Wang *et al.* paper are more granular than the Mackay *et al.* paper discussed in Box 2.1

These and similar publications begin to offer some insight into the geographies that are constitutive of the Anthropocene. We say 'begin' because the geographies are still quite abstract and macro; we say 'constitutive' because it's not that the Anthropocene *has* an evolving geography but, rather, that the Anthropocene is nothing more nor less *than* a geography. Its geography is part of its DNA; it defines the very nature of the phenomena. A few years ago, important research published by the British physical geographers Mark Maslin and Simon Lewis hinted strongly at this fact. They proposed a possible start date for the Anthropocene much earlier than that proposed by members of the AWG, who've adduced evidence that it began around 1950 (e.g. Zalasiewicz et al. 2015). Maslin and Lewis (2015) point to the colonisation of central and south America from the fifteenth and sixteenth centuries by the Portuguese and the Spanish. By 1610, there was a notable drop in atmospheric carbon dioxide levels, detectable to this day with the right dating techniques. Maslin and Lewis attribute this global process to a regional cause, namely, mass deaths among Indigenous Americans, largely caused by diseases imported by Iberian colonists. Because of the geographical separation of European and American ecologies, Indigenous peoples had no natural defences against the imported diseases. The huge population drop, occurring over several decades, subsequently reduced the amount of land and vegetation that was devoted to human uses. Accordingly, natural vegetation regrowth occurred and wild biodiversity bounced back, thus increasing biomass and sequestering CO_2 from the atmosphere. So it was, Maslin and Lewis proposed, that a particular set of changes in one part of the world – caused by its new links to another part thousands of kilometres away, courtesy of imperialism – had worldwide environmental effects (we'll say more about the Maslin and Lewis proposition in the next chapter).

Maslin and Lewis's research gestures towards a thicker conception of the Anthropocene's geographies than has been found in many – perhaps even most – of the key geoscience publications about our evolving post-Holocene planet. Because it couples global environmental change, regional environmental change, and momentous changes to specific societies, their research also signals the need to understand how particular human geographies get fashioned and what their wider impacts are. As we'll see in later chapters, a number of people within and beyond Geography are attending closely to these geographies as they impact, and are affected by, the planet's variegated and interconnected physical geography.

The hidden geographies of the Anthropocene: towards a thicker understanding

The Anthropocene, we stated above, is geographical through and through. Its geographies are multiform, intersecting and fluid, even on human time-scales, never mind planetary ones. Change seems to be the only constant: there is, variously, loss, restoration efforts, attempts at conservation, new material risks, intensified biophysical threats, novel ecosystems, reintroduced species, new infrastructures, unwelcome surprises, and so on – not just here or there but virtually *everywhere*. But while we've noted the insufficiency of thin depictions of the Anthropocene's geographies, we now need to explicate what a thick geography looks like, or at least what its key dimensions are. These dimensions will be highlighted in different ways in the rest of this book as we explore the work of numerous geographically-minded analysts.

To begin with, we note that the word 'geography' means Earth-writing or, more broadly, refers to any systematic attempt to represent phenomena as they're arrayed on the face of the planet. To see reality geographically, then, is all about perspective, not subject matter. For where specialist disciplines, such as Chemistry, Biology, and Sociology, isolate certain phenomena for detailed investigation – atoms, micro-organism, and family units, say – geographically-minded people examine the metaphorical tapestry of life – i.e. how things come together, and also often stand apart or become frayed, at scales ranging from the very local to the global. Geography thus spans the subject matter of numerous specialist fields. When Geography began to establish a foothold in European universities in the mid-to-late nineteenth century, it aspired to be a synthetic or 'bridging' discipline that examined the interconnected threads of both the social and the physical worlds. It had high ambitions: the goal was not merely to describe the world (a simply enormous empirical undertaking then and now) but to explain it, to pass judgment on it and to sometimes suggest ways to plan things like cities or exploit natural resources. This meant combining all sorts of information and insight usually confined to separate disciplines, such as economics and agronomy. No wonder the fortunes of Geography in Western universities have proven to be rather uneven in the last 170 years: the discipline's ambitions were perhaps *too high* from the get-go. David Livingstone (1992) regarded these ambitions as a rather risky 'experiment' that might founder. No wonder that from the mid-twentieth century, academic geographers began to specialise in areas such as geomorphology and urban social geography: this allowed depth of understanding, though at the expense of Geography's synthetic aspirations, which had – unfortunately – seen too many geographers skim the surface of the world they studied. Indeed, these aspirations often led to elaborate exercises in fact collecting and mapping, and little else. This is a key reason why human and physical geographers began to specialise from the 1950s onwards. They narrowed their individual vision so as to deepen understanding and improve the external reputation of Geography. In turn, this raised the spectre of internal fragmentation within Geography.

Regardless of this, nothing in the world operates on the head of a pin; everything has a geography, and the geography really matters to the things in question. For instance, the unequal global distribution of oil, gas, and coal reserves has had profound economic impacts on the world's 190-plus nation-states, not to mention geopolitical impacts. This brings us to those dimensions of the world that many geographers, but also others outside Geography, routinely seek to offer a thick understanding of. This understanding, we argue, should animate our sense of what the Anthropocene is, whichever narrative is favoured. Before listing these dimensions, we need to remember that changes to our anthropogenic Earth System simply can't be understood as 'averages' or aggregates – for instance, changes to the cryosphere or biosphere 'in general'. Instead, what's occurring is a mix of coincident and successive changes all over the planet in a myriad of specific locations – these comprise rippling or 'lumpy' time-space dynamics, if you will, in which causes, impacts, and responses are operating simultaneously. It's the accumulation of these micro-level changes that's having macro-scale consequences and, as we see with climate change, these larger consequences rebound differentially on all manner of places. The Anthropocene is a tracking back-and-forth, of cause-and-effect, impact and response, across multiple scales of space (and time). Its geographies are 'thick' not thin, as summarised here in Table 2.1.

To be more specific, examining the Anthropocene through a geographical lens has us seeing the following six important, overlapping dimensions of our Earthly reality:

- *Place and region*: Notwithstanding economic globalisation, mass human migrations, enormous geographic transfers of species, rapid transportation, and instantaneous communication technologies, the world remains intensely differentiated and distance still matters a lot. Despite what some

Table 2.1 Geographies 'thick' and 'thin'.

Thin geographies	Global scale perspective; lack of local and regional detail; generalisations about causes, physical impacts, and human responses; data aggregation in search of general trends
Thick geographies	Close attention to local, regional, and global scale interactions; a focus on the spatial origin and propagation of causal processes, on local and regional impacts, and on various ground-level responses to these impacts. A focus on how global patterns of human inequality, difference, and connectivity shape cooperative and non-cooperative actions to mitigate, steer, and adapt to planetary change

claim, we certainly haven't witnessed 'the end of geography' in this sense. Most people inhabit just one place during any given year or period of their lives, and the places where people are raised tend to be a locus of loyalty, affection, and self-identity. Where one lives matters in other ways too: for instance, inhabiting a zone where hurricanes and wildfires hit frequently is very different from living in the middle of London which is different in turn to living in outback Australia. Key questions arise. Which locations on the Earth, historically and today, are most responsible for driving global scale changes to the Earth System? Which locations are suffering the largest burdens of Earth System change? What sorts of biophysical change is unfolding in different places and what precise impacts is it having on different groups of people, even in the same place? Which regions and what industries will lead a 'sustainability transition', if one eventuates? What are the dynamics of power, responsibility, and liability among different places and regions? These are important questions.

- *Society and nature*: The Anthropocene proposition implies a new level of 'denaturalisation' of the Earth's surface. In other words, it indicates that the ontological division separating what's humanly influenced, or made, and what's non-human, is eroding at a larger scale than ever before. The 'post-natural' narrative we summarised in Chapter 1 highlights this. Yet the 'nature' that's on the retreat or being unsettled is differentiated: the pace and character of 'de-naturalisation' varies by place, region, and period. The receding horizon of 'nature' is uneven rather than a straight line: human impacts vary in type and intensity. Again, key questions arise. For instance, what 'post-natural' processes and phenomena are coming into existence in different parts of the world? – for instance, as the atmosphere warms, species will be forced to migrate and Holocene ecologies will be reshuffled. To what extent can and should different societies embrace rather than resist – through conservation and restoration efforts – an increasingly post-natural reality? What new threats will accompany the ever-growing human imprint on natural systems? How should different societies reduce their exposure to these threats? How might environmental injustices imposed on certain places be redressed? Can and should non-human entities – tigers, glaciers, lakes, or dolphins – be accorded greater rights and protections in the face of Anthropocenic change so great it might tip us into a 'hot house Earth' situation within a century or two? What does it even mean to talk about and to value 'nature' anymore, now that we live on a 'human planet'? Some societies never operated with the Western notion of 'nature' in the first place. That notion is not, confusingly, itself natural!
- *Long-distance ties, teleconnections, and enduring impacts*: The Anthropocene is very much about 'action-at-a-distance' or what you might call 'distantiated-cause-impact-and-response'. It's important to understand the geographical and temporal pathways that allow historical and present-day activities in certain places to 'travel' in space and time so as to have major

impacts elsewhere and elsewhen. There is a biophysical side of this, to do with material processes operating over thousands of kilometres and long spans of time. There's also a social side to this, to do with how people, their technologies, and their organisations move enormous quantities of material and goods around the planet, with associated waste streams. Currently, the World Trade Organisation is at the heart of this, along with a few other supranational organisations. One key reason to consider both sides is to anticipate, as best we can, the future impacts of post-Holocene change in all parts of the world. 'Hysteresis' describes a delayed impact when a large process set in motion today unfolds in the future, even if we were to immediately stop that which causes the process in the first place. Global sea level rise is an example: it will occur worldwide to various degrees, even if we stopped emitting any more GHGs into the global atmosphere tomorrow. Foreseeing the complex time-space choreography of cause, impact, and response is important work for researchers to undertake, and will have a bearing on public policy decisions. In recent years, concepts such as 'assemblages' and 'actor-networks' have been used by human geographers and others to get at the tapestry of connectivity. As the tapestry alters, different actors are able to have a large long-distance impact, such as the Chinese government orchestrating its 'belt and road' initiative in Asia.

- *Scales of cause, impact, and response*: As just intimated, the Anthropocene is all about scalar cross-connections. In Chapter 1, we mentioned the 'derangements of scale' that are now unfolding. Scale refers to the geographical and/or temporal extent of an event, phenomenon or process – for instance, the Atlantic Meridional Overturning Current is very large, whereas a local election may only involve, say, 30,000 people in an area of just 140 km^2. In the Anthropocene, the analysis of scale is both important and challenging. A certain collapse of scale seems now to be occurring; previous 'stratifications' of reality seem to be dissolving. How do local-scale occurrences, like power station emissions or civic waste streams, cumulatively cause changes at the regional, international, or global scale? And what sort of change are we talking about? Something incremental or more dramatic? So-called 'emergent behaviour' describes existing phenomena – say a grassland or a wetland – 'flipping' into new ways of operating because of an accumulation of small pressures occurring slowly over time. The natural 'layering' or hierarchical 'nesting' of reality during the Holocene epoch seems to be breaking down. And once a major change has been instigated at large spatial and temporal scales, how does it react back on the smaller scales at which peoples' lives are lived? How, if at all, can the world's business people, civic leaders, and statespeople cooperate across international borders to address the emerging problems of the Anthropocene?
- *Borders and boundaries*: This last question brings us to a fourth geographic dimension. In the Anthropocene, the natural divisions of vegetation, ice,

soils, and water on the face of the Earth have become more scrambled (older biophysical 'separations' and 'stratifications' have been eroded). But human administrative-political boundaries – for instance, between countries – continue to criss-cross the globe. As the 1987 United Nations Brundtland Report about 'sustainable development' famously said, 'the Earth is one but the world is not'. This raises key questions, which are hardly new but now assume intensified significance, about the role that political and ethnocultural divisions play in the dynamics of human co-operation, disagreement, unilateralism, coordinated action, compensation for environmental harm, and reparations for irrecoverable losses (such as Pacific Island states literally disappearing into the rising oceans). The changing natural and artificial divisions that pattern an otherwise continuous, integrated globe have a huge bearing on the character and quality of current and future lives. Since 1950 or so, the United Nations has been the leading body trying to diffuse conflict and forge cooperation among the world's peoples. As global environmental change becomes larger and faster, mechanisms will be needed to effectively tackle issues of joint action, of damages, and of injustices. These mechanisms will, among other things, be used to determine whether large-scale geoengineering measures are applied worldwide to keep the Earth hospitable for human life. If major environmental calamities befall certain countries due to Earth System change, then we might see serious stress on intergovernmental relations and even militarised attempts to grab land from other countries in the interests of environmental security. A new and hostile geopolitics may arise. How do political and administrative borders serve to prevent much-needed international cooperation? How do powerful states like the United States pursue their own domestic interests by refraining from serious transnational cooperation to reduce the human impact on the Earth?

- *Landscapes*: The world's continents form a patchwork of landscapes, some very large – like Canada's farmed prairies – and some much smaller, if still big relative to a single human being. The word 'landscape', in modern English, refers to the appearance of our lived environments – to the positioning, elevation, colour, size and texture of everything from high-rise apartment blocks to farm fences and fields to glacial valleys seen from on high. Landscapes are about aesthetics or, more accurately, kinaesthetics because they can engage *all* of our senses for good or ill if we are immersed in them (less so if we merely observe them from a point on high). Landscapes affect the quality of our lives, but also, as many social scientists and humanities scholars have shown, they can express inequalities in society, projecting the tastes and values of some but not others, and even excluding some people, both symbolically and literally. In this light, the unfolding Anthropocene raises important questions about the landscapes that will emerge by design and by force of 'natural' change. What kind of landscapes do people want? How much control can societies have over their proximate landscapes if Earth System 'tipping points' are crossed? Whose visions of landscape will or should prevail? We can pose similar questions about the near-shore 'sea-scapes' that are found world over. For instance, major moves towards renewable energy systems and 'carbon capture and storage' will profoundly reshape many rural and offshore landscapes worldwide.

Clearly, there are lots and lots of overlaps between the 'angles of geographic vision' on reality itemised above. Together, they encourage us to get at questions of location, specificity, connectivity, co-dependence, and asymmetrical influence in a world full of qualitative differences among things, both good and less good – i.e. differences of wealth, of vulnerability, of biodiversity, of socio-economic power, of military might, of adaptive capacity, of risks to life, and so on. Everything exists somewhere, but its character and fortunes are increasingly shaped by forces hailing from the recent human past and from elsewhere in the world today. This is the key lesson of geoscientists' Anthropocene propositions, even if – as we noted in the previous section – many geoscientists offer a thin sense of the geography involved. A thicker sense of geography clearly needs to go beyond description and encompass explanation, evaluation, prediction, foresight, proposals for policy and action, and more besides.

We should note too that the geoscience debates about when, precisely, the Anthropocene began make a big difference to which geographies are focussed on by analysts within and beyond Geography. For instance, we mentioned the research of Lewis and Maslin earlier, contrasting it to the Anthropocene Working Group's view that the Anthropocene started around 1950. In each case, very different geographies of societal action and implication, of cause and effect, of impact and response, come into view. Lewis and Maslin's work shines a bright light on European colonialism and its complicated but distinctive early patterning; meanwhile, the work of the Group, led for several years by Jan Zalasiewicz, is more focussed on escalating and expanding industrialisation in Europe and North America from around 1880 onwards, and especially from 1950. In sum, because the world's geographies have been changing for centuries, albeit at different rates, the 'slice of time' chosen to mark the Anthropocene's beginning dictates which geographies we see as historic initiators. This will become very clear in the next three chapters.

Forms of geographical knowledge: getting to grips with the Anthropocene

Clearly, the Anthropocene is a very grand concept; it signifies change to pretty much *everything* and is thus among the most encompassing geoscience ideas ever created. We've just explained the different ways the Anthropocene involves novel interfusions between the world's physical and human geography.

We now want to clarify what this means for the kinds of knowledge and insight that are required to make proper sense of the post-Holocene world that humans are creating. Fundamentally, the question is: moving forward, who should be granted the authority to speak about – to *speak for* – the Anthropocene and in what idioms? This question directs us to the issue of expertise that we touched on in Chapter 1. Which narratives, and whose voices, are most likely to be heard in the key realms of public understanding, political decision-making, and business behaviour? Who gets to speak for the Earth and to make cognitive and normative claims on pretty much all of humanity?

The first thing to say is that, while the Anthropocene proposition originated in geoscience, geoscientific knowledge is insufficient – albeit necessary – in order to understand it. There are three key reasons why. One is obvious: namely, geoscience brackets the human side of the Anthropocene story. Researchers like the late Will Steffen, Jan Zalasiewicz, and Frank Oldfield only cover one half of the story; this leaves space for human geographers, historians, philosophers, anthropologists, and others to fill the vacuum. These 'people disciplines' explore topics and issues that geoscience does not and cannot. A second reason is that geoscience's focus is *cognitive* – i.e. it's focussed on empirically verifiable descriptions, explanations, and predictions about the planet's biophysical systems and phenomena. Yet, as the previous section of this chapter has indicated, the Anthropocene raises profound questions of both ethics and aesthetics. These questions involve the articulation and defence of certain human values – such as rights, duties, liabilities, responsibilities, needs, and beauty. These values are not questions of science but, rather, involve the use of argument, logic, and persuasion to win the hearts and minds of people. They raise *normative* issues as much as cognitive ones, and trigger emotions as much as engaging the 'cool' faculty of human reason. The Anthropocene, in short, crosses the 'is-ought', fact-value divide that Western societies have long taken for granted. Fundamentally, it invites reflection on the right or best way for humans to occupy their one and only planet. It raises deep moral questions. While geoscience can help us understand what's happening in and to the world, it can't help us decide how to live on Earth.

The third reason why the Anthropocene can't be represented in purely geoscientific terms relates to technology and democracy. A number of geoscientists are very worried indeed about our planetary future. For instance, the Exeter University Earth System scientist Tim Lenton has warned repeatedly of the dangers of crossing tipping-points by triggering 'tipping elements' such as the polar ice caps. While they're entitled to speak out and sound the alarm, the risk is that scientific advice and technology substitute for proper democratic debate about what to do to prevent the worst outcomes. We mentioned the 'catastrophist' narrative in Chapter 1. One worry is that scientifically backed claims about the need to avoid a bleak environmental future will encourage some governments to employ risky geoengineering measures at a very large scale. As we're currently seeing with artificial intelligence (AI) worldwide, scientific inventions frequently outrun societal deliberation about how best to regulate them. Relatedly, it's important that the Anthropocene not default to being a science and technology issue in the minds of citizens, politicians, and others. It's as much an issue of what humans choose to value as it is an issue of understanding, and responding to, the fact of massive biophysical change. It's only when values and human goals are debated and agreed upon that science and technology find their proper place, which is in their service. We ought to design our science and technology to fit with the relationships, norms, institutions, and goals we have in mind, rather than vice versa.

To phrase all this a bit more formally, the Anthropocene needs to be understood and evaluated in a number of different registers, equally and together. It requires *descriptive, explanatory, and predictive knowledge*, which is a largely cognitive affair. It requires *practical knowledge* relating to policies and technologies that will produce desirable outcomes. It also requires *critical and evaluative knowledge*, which involves more than facts and evidence and is a largely normative affair. What sort of world do we want to inhabit, and what values will underpin it? What measures and means are required to achieve the ends we have in mind? Finally, the Anthropocene requires the exercise of *prudence and wisdom* because the stakes involved in deciding how to move forward in the Anthropocene are very high indeed.

The Ancient Greek philosophers distinguished between *episteme* (understanding), *techne* (effective practical action), and *phronesis* (sound or wise decision-making). Then, writing in the 1960s, the eminent German social philosopher Jürgen Habermas (1968) distinguished between *technical-instrumental knowledge*, *critical-emancipatory knowledge*, and *hermeneutic knowledge*. The latter involves grasping different social groups' 'life-worlds' or particular ways of making reality meaningful. The Anthropocene requires thinking and acting across *all* of these domains at a global scale and in the long term – which means also at the local scale and in the short-term because of the intimate scalar connections in play (see Box 2.2).

As we noted above, though without phrasing it in quite this way, a thick conception of the Anthropocene's geographies necessarily involves attempts at synthesis of this kind. We need to know: what's happening where and why; who currently is, and in the past was, responsible; what's likely to happen in different parts of the world; and what can and should be done, according to whose values. Geography as a discipline can make contributions here, but so too can geographically-minded researchers in other disciplines. The discourse of geography exceeds the discipline of that name, in both its thick and thin renderings. It requires efforts of inter-, multi-, cross-, and trans-disciplinary inquiry – or, in the case of Geography as a discipline, of *intra*-disciplinary cooperation among specialists in sub-fields like biogeography and cultural geography. To conclude this chapter, we now consider how a wide range of knowledge and insight about Anthropocenic geographies can be marshalled.

Box 2.2 Forms of understanding the Anthropocene

Episteme – knowledge, testable and revisable, of what we believe exists; cognitive understanding in the realms of description, explanation, and prediction.

Techne – practical knowledge, based on trial, error, experiment, and lived experience, of how to interact with and manipulate material reality in order to achieve certain goals, such as eating or flying. Jürgen Habermas called it 'technical-instrumental' knowledge focussed on deriving means to given ends.

Hermeneutics – knowledge of how other peoples and cultures, but also specific individuals, perceive reality and make it meaningful for themselves. Since perceptions vary widely today and historically, hermeneutic knowledge requires 'going native' or 'getting inside the mind' of others in a subjective sense.

Critical-emancipatory understanding – knowledge geared to realising justice and improved life conditions for those who are variously discriminated against, marginalised, silenced, or harmed by dominant social groups within and between the world's nations. This knowledge seeks to create a more egalitarian world where social power does not create large gaps between winners and losers.

Phronesis – sound decision-making, based on combining all manner of knowledge in specific contexts. Wisdom, foresight, and good judgement are all hallmarks of sound decision-making.

Beyond one world thinking: pluriverses, present and future

The Anthropocene clearly requires a 'wide band-width' approach and the combination of all manner of expertise if it's to be properly understood, and responded to effectively in practice. But this can easily create misapprehension about what, precisely, is required. To explain what we mean, we'll take the case of Geography the discipline and consider how its members can best shed light on the geographies of the Anthropocene. The choice is between 'one world' and 'multiple worlds' (pluriverse) approaches. Let us explain.

In some ways, the proclaimed advent of the Anthropocene may seem like manna from heaven for university-based geographers. For decades, leading scholars in the field have bemoaned the lack of unity among their peers. Specialisation has cut deep over the decades: sustained intra-disciplinary cooperation and robust synthetic work have seemingly been in short supply, at least in Anglophone Geography. So, the Anthropocene might appear as a welcome impetus to reunify the discipline and to represent, in a 'thick' way, the emerging and ever-altering geographies of the post-Holocene world (Taylor and O'Keefe 2021).

However, quite aside from any practical difficulties of getting lots of specialists in Geography to conduct far more joint (integrated, joined-up) research, there's a deeper issue. It's tempting to imagine there's just one world in which all humans live and that this world is amenable to accurate and truthful depiction by suitably qualified experts. Indeed, this belief seems so blindingly obvious to many as to be beyond question. The belief involves a mix of *ontological monism* – 'there's a single biophysical and social reality out there awaiting discovery' – and a matching *epistemological monism* – 'there are a set of relatively objective methods we can use to discover the nature of reality because they can generate evidence that, suitably interpreted, most people would find convincing'. Here, then, cooperation among geographers to thickly elucidate the geographies of the Anthropocene presumes the following: the metaphorical *tapestry* of the world can be represented truthfully if we unite specialists who each, accurately, study the key *threads* that are woven together to constitute our complex, differentiated yet unified reality. Or, to use a visual metaphor, the idea is that different geographers and others can, in time, assemble all the pieces of the three-dimensional jigsaw that is our phenomenal world. The jigsaw pieces span human and physical geography and specialists in each arena know a lot about each of them.

But, there are, in fact, serious problems with this approach to analysing the Anthropocene. One of us has pointed this out in a number of publications (e.g. Castree et al. 2014; Castree 2015).

First, it presumes that 'getting at' reality is, overwhelmingly, a question of *truth* – of disclosing evidence so that we can describe, explain and sometimes predict reality with considerable confidence. While, clearly, the pursuit of truth matters immensely – especially in our so-called 'post-truth' times – there are other key things in-play in the Anthropocene. As the previous section explained, there are also questions of *values* – of goals, of ethics, of aesthetics – that need to be tackled. While evidence can shape our answers to these questions, they're essentially about *arguments, wants, hopes, desires and commitments*. No 'neutral' body of independent evidence can settle these questions for us, as German sociologist Max Weber noted over a century ago. Instead, the arts of justification, logic and persuasion are involved to deliver things like rights, entitlements and, ultimately, justice.

Second, related to this, different societies possess, in a real sense, different perceptions of reality. As anthropologists like Phillipe Descola and Eduardo Viveiros de Castro have shown time-and-again, societies make their immediate world meaningful in a wide range of ways. They construct belief systems, categories, taboos and rituals that help them navigate through life's challenges, milestones and pleasures. Susan Crate (2021) provides a fairly recent example in her illuminating study of the Siberian Sakha people, *Once Upon the Permafrost: Knowing Culture and Climate Change in Siberia.* She elucidates the 'hermeneutic' element of our existence that Habermas and other commentators rightly regard as an ineliminable part of the human condition. The 'same' world can be apprehended in very different ways and these diverse ways matter to people and comprise their 'common-sense'. It's not that one way is 'right' and another one wrong – ask any person who's deeply religious, for example: they're unlikely to relinquish their beliefs just because other people may dissent strongly from those beliefs. Some, like anthropologist Arturo Escobar (2018), now suggest we take seriously the idea of 'pluriverses': i.e. different embodied modes of thought and practice that might, in some sense, be incommensurable. The 'ontological politics' of bringing them into productive dialogue requires tolerance, parity-of-esteem and open-mindedness to allow a 'world of many worlds' to flourish.

Thirdly, the variety of extant values and beliefs in the world speak to why we routinely disagree and actively criticise current arrangements, rethink the significance of past events and have rival views about what the future should look like (locally and beyond). Evaluation and critique, as we also noted in the previous section, are vital elements of Anthropocene sense-making. They're not adornments and appendages tacked-on to 'hard science'. In social science and the humanities – including human geography – there are a myriad of critical perspectives on causes, impacts and responses (for instance, Marxist, ecocentric or feminist). The Anthropocene is not a single super-complex problem to be 'solved', especially given the huge uncertainties about future Earth System behaviour. Instead, it's an evolving condition that will spark debate about right, proper and just courses of action at different geographical and temporal scales. This is why there are no fewer than five major Anthropocene narratives (see Chapter 1). This is at it should be, and Geography ought to reflect this in the way its practitioners go about their business.

Finally, the Anthropocene raises questions about how actions today will shape the fortunes of our descendants, of non-humans yet to exist, and of physical environments more broadly. By definition, representing the future involves deep *uncertainty* – informed guess work, logical extrapolation from the present, formal scenario building, back-casting and imaginative forward-projections are the only tools available, and none are even close to being foolproof. There can be no gold-plated statements about how the future will look or should turn out – locally, regionally or globally. By definition, the future will, in part, be actively shaped by what we want today and what we do tomorrow in light of what we say we want. While we're obviously not entirely free to act as we please, there's no 'one future' we can confidently anticipate, however clear in our heads we might be about the future we'd like to see eventuate. Instead, we must work towards creating futures that, in our view, our descendants would be happy to inherit.

Former climate scientist and now society-environment geographer Mike Hulme is one of those who, in light of these reasons, insists there's no one way to get to grips with the Anthropocene and its geographies. Despite his scientific training, including past involvement in the widely respected Intergovernmental Panel on Climate Change (IPCC), he refuses the notion that one or other form of knowledge offers privileged or sufficient insight relative to other forms. Science, facts and truth claims are not nearly enough, he argues. Hulme concludes a 2020 article as follows:

> There's no trick that will force a convergence of human values. The stories that people tell about themselves, their past, their futures, and their place on the planet, will continue to divide. Mobilizing some new 'solution science' resting on the putative cultural authority of science won't eradicate political conflict. We live on one Earth, but we imagine many futures and hence aren't susceptible to alignment of our actions toward securing a common single destination (311).

This stance invites us to understand that the Anthropocene – indeed, our world at *any* point in history – is best seen *kaleidoscopically*. There are many metaphorical jigsaws to be pieced together, and intentionally reconfigured, according to different values and goals. The Anthropocene isn't amenable to a singular view that can trump all others and command common assent. Instead, a range of legitimate perspectives exist and can, potentially, enter into processes of debate, mutual learning and, where possible, mutual accommodation. This is true across all the spatial and temporal scales we care to consider. The fact of those different Anthropocene narratives mention in Chapter 1 indicates the range of perspectives available – no one is 'right', though in reality they vie for our attention and hope to sway us. It's not so such that 'anything goes' and more that there's a growing tussle to see which versions of present and future reality will serve to 'co-produce' an Earth to-come. We do indeed live in 'one world' in the sense we are all interconnected – but our ways of living, our values and our hopes are varied and so an urgent debate is needed about how best to coexist and keep the planet habitable for all people.

Power geometries of knowledge and belief: geography's fruitful 'indiscipline'

This brings, finally, us to important questions about the relative power of bodies of knowledge and belief. Since the early 1970s, Anglophone Geography has become home to a striking diversity of cognitive and normative approaches to the study of the world's many geographical dimensions, including applied (e.g. policy relevant) ones. This is especially so in the fields of human and environmental geography. Some of these approaches began as minority positions, while others are only now beginning to make serious in-roads into the discipline – such as

Indigenous peoples' perspectives hailing from places like Australia. Overall, Anglophone Geography today is not only wide-ranging in its subject matter, in keeping with its history. It's also deeply heterodox and pluralistic: there are many available ways to 'join the dots' between, say, biogeography and urban social geography. A dazzling array of ontological, epistemological, methodological and normative possibilities present themselves to geographers who are willing to roam the variegated landscape of their discipline.

Linked to this, many geographers have become keenly interested in the geographies of geographical knowledge, information, and other forms of representation. Geographical understanding is *itself* geographical (just as knowledge of history is shaped by when a particular historian lives and works). Researchers such as Heike Jöns, David Livingstone, Charles Withers, Rebecca Lave, Mark Carey and Katherine McKittrick have inquired into how all forms of knowledge – even scientific knowledge – bear traces of the particular context in which they were, or are, produced, into how some forms of knowledge are able to 'travel' to other places and take-hold, to how some knowledges get overlooked or even denigrated, and into how 'hegemonic' knowledges emerge within and between countries, and come to shape behaviour on a wide front. Even thick geographical understandings are contingent upon who's creating them, where, how and with what ends in mind. The understandings are always situated ones, even when claiming some sort of universal relevance. In hindsight, for instance, it's clear just how shot-through with racial and colonial prejudices early work by academic geographers was from the mid-nineteenth century onwards (see Driver, 2000). More broadly, maps of all kinds have been shown to incorporate complex and contingent sets of values (Harley 2002; Monmonier 2018). As Bill Rankin phrased it in his fine book *After the Map*, 'Every cartographer must make choices about *what* to show and *how* to show it. Likewise, all maps make assumptions about how they will be *used*, and this affects their content' (2016, 7). Both historically and today, there are 'power-geometries' of knowing that criss-cross our world and which, at any one time, are rarely egalitarian in respect of whose knowledge counts the most.

In sum, so-called 'expert' insights provided by geographers and others into the Anthropocene necessarily incorporate certain norms and values that are, precisely, *not* universally held or recognised. These, in turn, influence how our changing world is depicted. For instance, as Hulme (2010) noted long ago, those who claim their knowledge of the world is truly global and stands above all the particularities of any one part of the globe are, in a real sense, projecting their epistemic power over others less able to match such claims. There's a somewhat presumptive claim on the attention of *all* peoples and cultures, one bordering on epistemological sovereignty. The expert-led IPCC has, for instance, been pre-eminent in making global statements about climate and tailored its work to the political project of avoiding certain levels of GHG emissions and certain atmospheric temperature levels. Useful as the Panel's scientific work has been, the risk – Hulme noted – is that its ability to average across all manner of places using vast amounts of data, models and global scenarios might imply that local level knowledges of those experiencing climate change are only relevant insofar as they 'speak' to climate science's big picture focus. Local knowledges of weather and environment have value at this level, and shouldn't be judged as secondary to the superordinate claims of certified climate scientists (see Rodrigues and Shepherd 2022; Sadai et al. 2022). These scientists don't stand above or outside the world any more than anyone else, even though their aims and methods differ from those of ordinary people on the ground (Mahony and Hulme 2018). Indeed, their knowledge is often anaemic in the sense we described earlier in this chapter.

In light of this, we can understand why the different Anthropocene narratives summarised in the last chapter can't be resolved in a single 'correct' one using evidence and logic alone. Instead, there are different, well considered ways of framing the perceived realities of the Anthropocene – as we said in Chapter 1, *Anthropocenes* not a single Anthropocene. These framings are necessarily synthetic ones – each one links different topics and forms of understanding because the Anthropocene is about everything-change and demands a synthetic sensibility. But no one narrative can offer the 'total picture', for reasons we've enumerated above. Instead, we have sometimes contradictory, sometimes complimentary big-picture representations of reality (e.g. see Macgregor and Houston 2018). These representations are richest when the geographies depicted in them are full-blooded, not pale and anaemic.

Anglophone Geography today is, in many respects, very well placed to furnish plural and sometimes dissonant Anthropocene narratives rich in geographical insight. Its practitioners can paint different pictures of Anthropocenic geographies because a great many of them acknowledge the problem of trying, somehow, to claim to present *the* total view. This reflects Geography's heterodoxy and, in particular, the prominence of many diverse perspectives attuned to social power, inequality and injustice – for instance, feminist political ecology, animal geography and so-called 'critical physical geography' (which acknowledges the direct implication of physical geographic science in societies that are full of competing political interests and dissonant values). As British geographers Jamie Lorimer and Timothy Hodgetts phrase it, 'Geography is a capacious field of inquiry and geographers populate a distinctly undisciplined discipline, lacking a canon of texts and figures and thus open to borrowing ideas and sharing them with others' (2024, 8).

What this means is that there's some appetite in Geography for novel fusions of theory, method, interpretation and practice in attempts to understand the fabric of life on Earth (see Box 2.3). This isn't to say all geographers are equally open minded, and nor is it to say that the majority are 'relativists' who believe that any one perspective on reality is just as valid as the next. On the contrary, many geographers continue to conduct very specialist and relatively narrow research, while more collaboration- and synthesis-minded geographers will typically advance their perspective over and against other ones. No one conducts research without some conviction that it offers insights as good as, if not *better*, than other research about the same topics and issues. The challenge for all those seeking to represent the Anthropocene and its geographies is to recognise and engage with the plurality of narratives in play. This necessarily introduces questions of morality, politics, power relations and the pragmatics of what's possible into the heart of the endeavour. Adapting Andrew Barry and Georgina Born's (2013) phrasing, it invites an 'agonistic' intra- and interdisciplinary mode of operating, one that seeks to actively shape that which it's aiming to depict in an honest, politically explicit fashion. Indeed, part of this fashioning will be working directly with communities, policy makers, public officials, NGOs and others to build a better Anthropocene than the one we might otherwise have coming at us. Research counts for more – is less purely academic – when its insights arise from collaborations with people outside universities.

Box 2.3 An experiment in knowing the world differently: understanding flood risk and management in northern England

Human and physical geographers are increasingly worker together to help communities address the challenges of environmental change. Often this means working at very local scales, and working closely with communities who are directly affected by various environmental challenges. The 'Environmental Competency Group' (ECG) model was developed by a group of geographers in the UK to bring together social scientists, physical scientists and residents of the flood-hit town of Pickering northern England. Between 1999 and 2007 Pickering suffered from regular flooding, with devastating economic impacts. However, funding for a £20m concrete flood wall to line the river in the town was rejected by the government, and things were at an impasse (Figure 2.2).

Conventionally, flood management has been dealt with by 'experts', often from government institutions, making decisions based on scientific flood modelling and engineering knowledge. The 'public' is often consulted, but rarely in a way which meaningfully affects solutions. Under the ECG model, by contrast, groups of academics, flooding experts and local residents met on equal terms. Everyone was deemed an 'expert' with important and valid knowledge to contribute to the problem. This included highly situated, 'local' knowledge about how water flowed through the catchment system – detail which had been lacking from the scientific models. ECGs met twice a month and did fieldwork together, interviewed affected residents, and discussed possible solutions to the

Figure 2.2 Pickering Beck flowing past Pickering Castle. (Source: Shutterstock.)

town's flooding problem (Lane et al. 2011). A new initiative, *Slowing the Flow*, was launched, which emphasised 'working with nature' to protect the town from flooding. 129 'leaky dams' made from woody debris were constructed, and water storage areas, heather obstructions and bits of new woodland were established. At a cost of £2m, this was considerably cheaper than a concrete wall and it seems to have worked. In 2015, when flooding ravaged Northern England again, Pickering stayed dry.

The ECG model re-envisions what 'environmental expertise' is, who holds it, and how it can be harnessed in the service of democratic decision-making. By democratising flood management in Pickering, and by extension democratising 'science' itself, new solutions were found which mainstream institutions and experts might have missed. The technique is now being used by geographers in other parts of the world (like coastal Louisiana), bringing together diverse 'experts' to find locally-appropriate solutions to seemingly intractable environmental problems (Barra, Hemmerling, and Baustian 2020).

Conclusion

The Anthropocene, in an important sense, doesn't exist outside the way it's depicted. Our increasingly post-Holocene reality is what literary critic and philosopher Timothy Morton (2013) has called a 'hyper-object'. It's too vastly distributed in space and time, relative to any one person, to be perceived directly. We thus rely on depictions of it, coming from geoscientists and others, in order to render it intelligible. 'No reality without representation' is an appropriate refrain here. We'll be moved to think and act differently about the Anthropocene according to which *ideas* about it come our way, which stories are told to us. Many of the ideas will come from expert communities – so-called 'epistemic communities', to which people like we two authors belong – across various disciplines and interdisciplinary fields. What Mike Hulme (2009) once said about the idea of climate change applies to ideas about the Anthropocene: these ideas will alter our social worlds, or not, depending on who has the power to affix content and significance to them so that we pay attention.

In this chapter, we've argued that all ideas about the Anthropocene need to foreground its complex geographies. There are different ways of doing this, we've argued, but there a set of geographical dimensions of reality that are of common relevance to different Anthropocene narratives. A thick rather than thin exploration of these dimensions is, we have argued, advisable. While there's a place for 'geography lite' depictions of the world, at the end of the day the Anthropocene – whatever narrative we use to depict it – is about the very stuff of the planet we inhabit in our own respective, but deeply interdependent, places. Geographers are well set to thicken geographical discourse about the Anthropocene, though many others are in a good position to contribute too, as later chapters in this book will make plain. The geographies of the Anthropocene amount to more than a vast catalogue of facts arrayed at different scales of Earthly space and time. Instead, they are dimensions of existence that require apprehension in descriptive, explanatory, predictive, evaluative and – in the end – political terms. Politics is about deliberation and then decision, underpinned by moral-ethical beliefs and justifications. And the Anthropocene will, one way or another, require serious action by institutions, communities and the world's c. 200 countries.

This chapter and the previous one have, we hope, served to set-up the rest of this book. In the next part of *The Anthropocene*, we delve in to the historical geographies of this slice of geological time. We'll examine arguments about when the Anthropocene might have begun, and show how debates about the Anthropocene are altering our understandings of the meaning and impacts of things like European colonialism and the industrial revolution. We aim to show that the historical stories we choose to tell about the Anthropocene have a significant bearing on how we perceive the nature of global environmental change, and how we might think about responding to it at different scales.

Note

1 Although the Holocene has not been officially declared over, by 'post-Holocene' we refer to the changes to the functioning of the Earth System that are increasingly being described as the Anthropocene 'event', for how they represent a departure from Holocene conditions – see Chapter 1.

References

Barra, M. P., S. A. Hemmerling, and M. M. Baustian. 2020. "A Model Controversy: Using Environmental Competency Groups to Inform Coastal Restoration Planning in Louisiana." *The Professional Geographer* 72 (4): 511–20.

Barry, A., and G. Born. 2013. 'Interdisciplinarity: Reconfigurations of the Social and Natural Sciences'. In *Interdisciplinarity*, edited by A. Barry and G. Born, 1–36. London: Routledge.

Castree, N., et al. 2014. "Changing the Intellectual Climate." *Nature Climate Change* 4 (9): 763–8.

Castree, N. 2015. "Changing the Anthropo(s)cene: Geographers, Global Environmental Change and the Politics of Knowledge." *Dialogues in Human Geography* 5 (3): 301–16.

Crate, S. 2021. *Once Upon the Permafrost: Knowing Culture and Climate Change in Siberia*. Tucson: University of Arizona Press.

Crist, E. 2019. *Abundant Earth*. Chicago: Chicago University Press.

Driver, F. 2000. *Geography Militant*. Oxford: Blackwell.

Escobar, A. 2018. *Designs for the Pluriverse*. Durham: Duke University Press.

Forrester, J. 1971. *World Dynamics*. Cambridge: Wright-Allen Press.

Habermas, J. 1968. *Knowledge and Human Interests*. Berlin: Suhrkamp Verlag.

Haeckel, E. 1868. *Natürliche Schöpfungsgeschichte*. Berlin: Georg Reimer.

Hamilton, C., and J. Grinevald. 2015. "Was the Anthropocene Anticipated?" *Anthropocene Review* 2 (1): 59–72.

Harley, J. B. 2002. *The New Nature of Maps: Essays in the History of Cartography*. Baltimore: JHU Press.

Hulme, M. 2009. *Why We Disagree About Climate Change*. Cambridge: Cambridge University Press.

Hulme, M. 2010. "Problems With Making and Governing Global Kinds of Knowledge." *Global Environmental Change* 20 (4): 558–64.

Hulme, M. 2020. "One Earth, Many Futures, No Destination." *One Earth* 2: 309–11.

Kolbert, E. 2021. *Under A White Sky: The Nature of the Future*. London: Bodley Head Books.

Lane, S. N., N Odoni, C. Landström, S. J. Whatmore, N. Ward, and S. Bradley. 2011. "Doing Flood Risk Science Differently: An Experiment in Radical Scientific Method." *Transactions of the Institute of British Geographers* 36 (1): 15–36.

Liu, T., et al. 2023. "Teleconnections Among Tipping Elements in the Earth System." *Nature Climate Change* 13 (1): 67–74.

Lorimer, J. and T. Hodgetts. 2024. *More Than Human*. London: Routledge.

Mahony, M., and M. Hulme. 2018. "Epistemic Geographies of Climate Change: Science, Space and Politics." *Progress in Human Geography* 42 (3): 395–424.

Maslin, M., and S. L. Lewis. 2015. "Defining the Anthropocene." *Nature* 519 (7542): 171–80.

McGregor, A., and D. Houston. 2018. "Cattle in the Anthropocene: Four Propositions." *Transactions of the Institute of British Geographers* 43 (1): 3–16.

Meadows, D., et al. 1972. *The Limits to Growth*. Falls Church: Potomac Associates Books.

Monmonier, M. 2018. *How to Lie with Maps*. Chicago: Chicago University Press.

Morton, T. 2013. *Hyper-Objects: Philosophy and Ecology After the End of the World*. Minneapolis: University of Minnesota Press.

Rankin, B. 2016. *After the Map*. Chicago: Chicago University Press.

Revkin, A. 1992. *Global Warming*. New York: Abbeville Press.

Rockstrom, J., et al. 2009. "A Safe Operating Space for Humanity." *Nature* 461 (7263): 472–75.

Rodrigues, R. R., and T. G. Shepherd. 2022. 'Small Is Beautiful: Climate-Change Science as If People Mattered'. *PNAS Nexus* 1 (1): pgac009.

Ruddiman, W. 2003. "The Anthropogenic Greenhouse Era Began Thousands of Years Ago." *Climatic Change* 61 (3): 261–293.

Sadai, S., et al. 2022. 'The Paris Agreement and Climate Justice: Inequitable Impacts of Sea Level Rise Associated With Temperature Targets'. *Earth's Future* 10 (2): e2022EF002940.

Steffen, W., et al. 2011. "The Anthropocene: Conceptual and Historical Perspectives." *Philosophical Transactions of the Royal Society A* 369: 842–67.

Steffen, W., et al. 2015. "Planetary Boundaries: Guiding Human Development on a Changing." *Science* 347 (6223): 1259855.

Stoppani, A. 1873. *Corso Di Geologia*. Milan: Bernardoni and Brigola.

Taylor, P. J, and P. O'Keefe. 2021. "In Praise of Geography as a Field of Study for the Climate Emergency." *The Geographical Journal* 187 (4): 394–401.

Zalasiewicz, J., et al. 2015. "When Did the Anthropocene Begin? A Mid-Twentieth Century Boundary Level Is Stratigraphically Optimal." *Quaternary International* 383 (2): 196–203.

Part II

Geo-histories

3 The Colonial Anthropocene

Empires and Planetary Transformation

Introduction

Some 480 years before Paul Crutzen had his Anthropocene 'Eureka!' moment at a scientific conference (see Chapter 1), the city of Cuernavaca was known as Cuauhnahuac and was a rich, densely populated urban centre, a key economic and political stronghold of the Aztec Empire. It was therefore a key target of the invading Spanish, who marched on the city in early 1521, before moving on to the Aztec capital, Tenochtitlan. The infamous conquistador Hernán Cortés himself joined the attack, and the city was pillaged and burned before the invading army moved on. Cortés returned to the city in 1523, founding churches, setting up a sugar plantation, and commissioning a palace to be built – in a classically colonial move – on the site of an Aztec tribute collection centre in the centre of the city now known, to the conquerors at least, as Cuernavaca (Haskett 2005).

What, you might well ask, has this distant moment of contact and conquest got to do with an atmospheric chemist and his colleagues announcing humanity's arrival in the Anthropocene? For a growing number of scholars, the pre-industrial period of European expansion, particularly the colonisation of the Americas, is the key turning point that set the world on the path towards the Anthropocene. This wasn't a period of high greenhouse gas emissions or of grand perturbations of global biogeochemical cycles. It was, however, a period in which flora and fauna began, with human help, to circulate the planet and to create novel ecosystems, in which people developed new ways of organising and exploiting human and nonhuman life, often violently, in the pursuit of profit, and in which the effects of anthropogenic transformations of landscapes and environments can be seen on ever-increasing scales.

In this chapter, we'll explore and evaluate arguments that read the Anthropocene as fundamentally a product of European colonialism. Significantly, most of these arguments bring together evidence and propositions from both Physical and Human Geography, as well as ideas from a range of other disciplines which contend that the geographies of the Anthropocene – particularly its spatial inequalities and injustices – are firmly rooted in the geographies of

DOI: 10.4324/9781003162469-5

colonial expansion, exploitation and violence. We contend that the development of what we called in Chapter 1 the 'Colonial Anthropocene' narrative has been one of the key contributions that Geographers have made to Anthropocene debates, and that that contribution is founded in large part upon Geography's unique disciplinary diversity, bridging the physical and human sciences.

For scholars concerned with the 'colonial present' – rather than just the colonial past (cf. Gregory 2004) – and its patterns of environmental exploitation and injustice, this version of the Anthropocene has proven particularly compelling. Throughout the chapter, we'll show how selecting a colonial onset point for the Anthropocene is not solely about evaluating scientific data, but is also about asserting 'a critical project' concerned with understanding and transforming the colonial origins and legacies of environmental injustice; a project which, in Davis and Todd's (2017, 763) description, 'understands that the ecocidal logics that now govern our world are not inevitable or "human nature", but the result of a series of decisions that have their origins and reverberations in colonization'.

Historical geographies of empires and environments

Historical geographers have had a lot to say about the spatial dynamics of empires and colonialism over the last few decades (Driver 2013), and 'have perhaps had more impact within the field of colonial, imperial and postcolonial studies than in any other branch of broader historical studies' (Lester 2015, 120). Geographical analyses of systems of imperial rule, the growth of colonial cities, networks of transport and trade, patterns of human mobility and social change, and much more have helped refine understandings of the nature of empires – such as the British, French, and Dutch – and their significance in the making of the modern world (Butlin 2009; Pawson 1990). Historical geographers have also pioneered the study of the role that geographical knowledges themselves played in imperial endeavours – expressed in maps of colonial territories, for example which enabled exploration, settlement, and resource exploitation (Godlewska & Smith 1994), or in the reports of travellers, explorers, and 'militant Geographers' sent to survey new lands and to claim them for European powers (Driver 2000; McEwan 2000). Robin Butlin 2009 survey *Geographies of Empire* takes stock of several decades of scholarship and conceptual advances by geographers and spatial thinkers in other disciplines, reinforcing the idea that it's now difficult to comprehend the history of European imperialism without a spatial language of networks, connections, landscapes, transfers, and flows.

But what about the relationship between empires and environmental change? What about the 'ecocidal logics' which are seemingly pushing us deeper into the Anthropocene, and which some propose have their roots in the practices and ideologies of colonialism (Davis and Todd 2017)? Butlin's chapter on what he rather benignly calls 'environmental interactions' documents how British, French, and Dutch empires laid waste to large tracts of forest, transformed water courses, and changed the face of ecosystems by sending flora and fauna around the world as part of experiments in the 'acclimatisation' of different species to different climates, usually with an eye on economic profitability. Yet Butlin also documents how many colonial administrators became deeply concerned by the environmental changes wrought by colonial settlement, agriculture, and trade, and essentially became early 'environmentalists'. But in sum, Butlin's account of the geographies of imperial environmental interactions remains predominantly tied to *local* changes of ecosystems and landscapes, and to the political struggles over how those changes could be justified or ameliorated.

More recently, the 'environmental interactions' of European colonialism have been re-interpreted on a more global scale. In 2015, two physical geographers and Earth System scientists at University College London, Simon Lewis and Mark Maslin, published an analysis which argued that the onset of the Anthropocene could be precisely dated to 1610, a year in which atmospheric CO_2 – as measured in ice cores from Antarctica – reached a curious low-point. Lewis and Maslin argued that this dip in atmospheric CO_2 was anthropogenic – human in origin – and could be linked directly to the European colonisation of the Americas. This, they contended, was the first stirrings of humanity as a force capable of altering the functioning of the Earth System as a whole (Lewis and Maslin 2015).

The planetary impacts of the 'Columbian exchange'

To understand how that was possible, we need to understand something of the human geography of the Americas in the period before Europeans arrived. The first groups to settle in the Americas probably arrived from east Asia and northern Eurasia between 20,000 and 15,000 years ago. Around 10,000 years ago, many of the descendants of these groups developed arable farming techniques and a sedentary way of life; populations grew and became denser, and extensive land-use change occurred.

Estimating the human population of parts of the Americas before European contact is nonetheless difficult. Inferences can be made from archaeological findings such as buildings and pottery, and from traces of agricultural activity such as charcoal and tilled soils. Generalisations can then be made from one place to another, based on the characteristics of climate and ecology and assumptions about how many people a landscape could theoretically have supported. Archival records, such as censuses and tax records kept by later colonial powers, can also be helpful to the historical demographer.

Koch et al. (2019) bring together a vast range of estimates of the historical demography of North, South, and Central America and the Caribbean and, using advanced statistical modelling, offer new estimates of pre-Columbian populations. They estimate that by 1492, around 60.5 million people lived in the Americas. This is less than the 70–88 million thought to have called Europe home at around that time, and the approximately 100 million in China and Mongolia. These European and Asian populations were also denser than those of the Americas, although not by much. The picture that is emerging from archaeological studies, historical anthropology, and demography is of pre-Columbian societies in the Americas that were large, complex, highly urbanised, and technologically sophisticated. Pre-Columbian American environments were far from the pristine wildernesses often imagined; complex irrigation systems fed farms, terraces were used to turn hillsides into useful agricultural land, and afforestation and controlled burning were used to manage landscapes and ecosystems (Dull et al. 2010). Koch et al. (2019) therefore estimate that some 61.9 million hectares of the Americas were given over to 'anthropogenic land use'. This was a markedly populated environment.

Such evidence might be used to support the arguments of the likes of Ruddiman et al. (2020) and Ellis et al. (2013) that the Anthropocene began with the emergence of settled agricultural societies, and the implications that farming had for forest ecosystems (Roberts, Boivin, and Kaplan 2018), several 1000 years ago. This is when people started to really transform the face of the Earth not only on the Eurasian continent but in the Americas too, and with implications not only for Earth surface processes, but for atmospheric greenhouse gases as well.[1] But for Maslin and Lewis (2015) and Koch et al. (2019), it's what happened next in the Americas that is crucial.

The Europeans who followed Columbus to the Americas came to kill, conquer, enslave, proselytise, and exploit the material riches of the continent (see the various implements symbolically wielded by the men crowding behind Columbus in Figure 3.1). The violence that followed was, as a growing number of historians and geographers argue, central to the emergence of a new global economic system. But for physical geographers, what mattered most were the biogeographical changes that ensued. As travel and trade between Eurasia and the Americas accelerated, plant and animal species began to circulate the globe on an unprecedented scale, in a process dubbed the 'Columbian Exchange' (Crosby 1972). Trade routes created a 'new Pangea', referencing the unified supercontinent which began to break apart around 200 million years ago. Across the Atlantic Ocean from the Americas came new agricultural staples like potatoes, tomatoes, and peppers. From Eurasia came rice, coffee, pigs, and horses. But it was the diseases that the conquerors and settlers brought with them that would prove so devastating. Diseases like measles, smallpox, the bubonic plague, influenza, malaria, typhus, and cholera crossed the Atlantic for the first time; diseases which can mostly be traced to domesticated farm animals in Europe.

European populations had built up relatively high levels of immunity after centuries of close co-existence with agricultural animals, but populations in the Americas had no such immunity (Crosby 2004). Furthermore, American

Figure 3.1 Christopher Columbus landing on the island of Guanahane (San Salvador, Bahamas) on October 12, 1492, as depicted in a painting by Dioscoro Puebla, 1892. (Source: Shutterstock.)

communities did not have the same entanglement with domesticated animals as Europeans, and as such had no 'zoonotic' diseases to send back the other way. Instead, wave after wave of epidemics surged through populations all over the American continent, causing unprecedented loss of life.

Koch et al. (2019) synthesise estimates of mortality in the Americas in the decades following European contact and conclude that around 90% of the pre-contact population succumbed. Warfare, enslavement, and famine following the collapse of indigenous agricultural systems were significant, but it was the disease epidemics that accounted for most of the devastation (Figure 3.2). In the Caribbean, it's possible that the population collapsed by as much as 99% by the year 1570. Such epidemics are known as 'virgin soil epidemics', as they involve populations coming into contact with new diseases against which they have minimal immunity. But most recorded virgin soil epidemics (such as the fourteenth-century 'Black Death' in Europe) exhibit a mortality rate of around 30%. They were largely caused by a single pathogen and lasted up to a decade. In the post-1492 Americas by contract, successive waves of different pathogens set off waves of depopulation that stretched over several decades, even centuries. Indigenous populations, often forced by colonial policies into intensive labour regimes and dense new settlements where diseases could spread easily (Moore 2017a), had little protection against what has since been dubbed the 'Great Dying'.

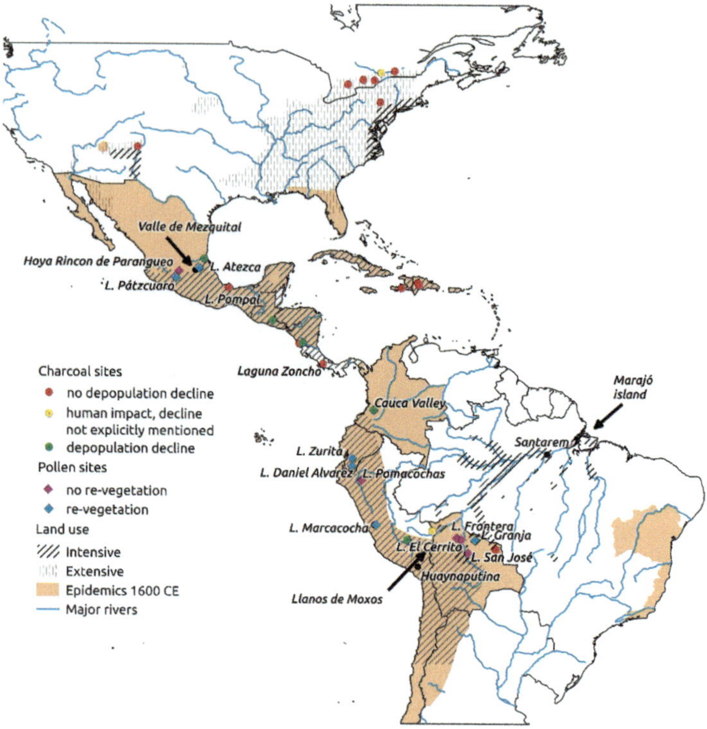

Figure 3.2 Map showing pre-Columbian land use and regions known to have been affected by disease outbreaks by 1600. Charcoal and pollen deposits allow a detailed reconstruction of historical land use change. (Source: Koch et al. 2019, 16)

This decimation of large agricultural (and in some cases urban) societies had significant consequences for the land. As societies and agricultural systems collapsed, huge amounts of land experienced 'secondary vegetation succession', meaning that once-farmed land saw the gradual return of prior plant species – most significantly trees – with forest structure returning close to pre-clearance states within around 100 years. Koch et al. (2019) estimate that 55.8 million hectares of deforested agricultural land recovered to something approaching a 'natural' state, following the Great Dying. And as is well known, as trees grow, they capture and store carbon dioxide, with significant implications for the climate. Around 7.4 petagrams of carbon may have been removed from the atmosphere as this land was abandoned, which Dull et al. (2010) and Koch et al. (2019) argue was a crucial contributor to the observed drop in atmospheric carbon dioxide levels of 7–10 ppm between c. 1500 and the early 1600s. In turn, this drop in atmospheric CO_2 contributed to a significant period of global cooling (sometimes known colloquially as the Little Ice Age) – the only one in the last two millennia (Koch et al. 2019). Combined with observations of the globalisation of flora, fauna, and pathogens (the 'Columbian Exchange'), these climatic observations lead Lewis and Maslin (2015) to argue that the year 1610, the year that atmospheric CO_2 reached its nadir, should be considered the onset of the Anthropocene. This was the first time, they argue, that humanity had had a recognisably global impact on the functioning of the Earth System.

In many ways, the argument of Lewis and Maslin (2015) that the Anthropocene began in 1610 might seem a surprising one. The Anthropocene is often associated with global climate change, and therefore with recent global warming. But Lewis and Maslin identify a moment of global *cooling* as a key threshold in humanity's interference with the Earth System. And while we might think about the Anthropocene as concerning the steady expansion of humans at the expense of 'natural' landscapes and ecosystems, this moment of global cooling was a product of a hemispheric 'recovery' of natural systems as almost entire societies were swept away by disease, warfare, and enslavement. Among Anthropocene scholars the proposal has met some criticism – for example, that the 1610 climatic event identified by Lewis and Maslin (2015) exists within the range of natural climatic variability; and even if anthropogenic (a point disputed by Rubino, Etheridge, and Trudinger 2016 and Bush, Nascimento, and Åkesson 2021), it pales in comparison to what came later with the Industrial Revolution and the Great Acceleration, when measured on the scale of human alteration of the global atmosphere and other systems. It has also been argued that Lewis and Maslin's (2015) proposal rests too much on an ecological and biogeographical argument about the globalisation of flora, fauna and pathogens, whereas the core arguments about the Anthropocene concern the altered functioning of the planet's more foundational biogeochemical systems and cycles (Hamilton 2015) – i.e. it's not just about human transformations of ecosystems on the planet's surface, but about the transformed exchanges of matter and energy between the atmosphere, biosphere, lithosphere, hydrosphere, and so on – 'the intrinsic biophysical processes that regulate the stability of the Earth System' (Steffen, Richardson, and Rockström 2015, 9).

So, while Lewis and Maslin's proposal has not convinced all Anthropocene geologists and Earth System scientists (see Zalasiewicz, Waters, and Barnosky 2015), it has aligned with a version of the Anthropocene which is popular among human geographers, historians, anthropologists, and others. This is partly because, in their own words, what they argue is that:

> In narrative terms, the Anthropocene began with widespread colonialism and slavery: it is a story of how people treat the environment and how people treat each other.
>
> (Lewis and Maslin 2018, 13)

This is an argument using concepts which have traditionally been the preserve of human geographers and other social scientists and humanities scholars: a story of labour relations, empires, and human violence. As such, it has appealed to people who commonly work with such concepts to construct their accounts of history. This is the fifth narrative we identified in Chapter 1 – the 'Colonial Anthropocene' – and it resonates too with many adherents of the 'eco-Marxist' narrative. For scholars working with these narratives of the Anthropocene, the Columbian Exchange and what followed did not just have huge environmental consequences, but also set in train a whole host of social, cultural, and political transformations that put the world on course for the multiple crises of the present.

World-ecology and the Capitalocene

The year 1492, when Christopher Columbus made landfall in what came to be called Hispaniola, marked the start not only of radical biogeographical transformations, but also what historical geographer Jason W. Moore calls a 'geocultural shift' which was fundamental to the rise of capitalism as a planetary force (Moore 2019, 53). Within both standard and critical narratives of the historical development of the Anthropocene, 'capitalism' tends to appear on the scene, fully formed, at some point during the British Industrial Revolution. Moore insists on a longer history and a more nuanced geography, pointing us towards the socioeconomic transformations that were occurring across Europe from the late medieval period onwards and, crucially, to the geographies of European expansion from the fifteenth century onwards, of which the 'Columbian Exchange' was a central component (Moore 2018).

Plagues and epidemics are a recurring feature of Anthropocene histories and geographies. As we mentioned above, the fourteenth-century 'Black Death' in Europe had devastating effects on the population (albeit not as devastating as the later epidemics in the Americas). But it also set in train an 'epochal crisis of feudal civilization' (Moore 2015, 181), as wages rose, workers demanded a bigger slice of agricultural profits, and European feudal elites faced limitations on their ability to reproduce their power and wealth. Centuries of agricultural expansion and intensification had cemented a regime of lord-peasant relations whereby seigneurial lords – the landowning, militarily-active aristocracy – benefited from the agricultural surpluses produced by tenant farmers. But the struggle to control these surpluses meant that little was reinvested in increasing agricultural productivity. Only geographical expansion could resolve the tensions. Forests were cleared, but soils were quickly exhausted by arable farming with minimal inputs to protect soil fertility. A geographical 'limit to growth' had been reached. But there was nothing 'natural' about that limit: overall agricultural productivity *could* have been enhanced without further geographical expansion by, for example, allowing peasants to switch to growing kitchen produce rather than just marketable cereals. But the class relations of feudal society – the system of extracting a surplus – prevented that from happening (Moore 2015).

'Simply put', Moore concludes, 'the lord-peasant relation was fundamentally antagonistic to long-run ecological sustainability' (Moore 2002, 304). From around 1300, agricultural yields across western, northern, and much of central and eastern Europe began to decline, helped on their way by colder weather, as populations continued to grow. A growing, under-nourished population, with increasingly interconnected towns and commercial centres, was therefore highly vulnerable to famine and disease. Famine became normalised as a regular occurrence in the early fourteenth century (having been largely absent for two centuries or more), but the population crash of the Black Death began to tip the scales in favour of the (surviving) peasants. By the mid-fifteenth century, rents extracted by feudal lords in England, Germany, and Italy were 40% lower than they had been at the time of the Black Death, while wages for labourers were as much as 300% higher (Moore 2002, 307). Numerous 'peasant revolts' united rural and, in some instances, urban labourers, against not only the rent-extracting aristocracy but also the emerging bureaucracy of tax-extracting and war-making states.

There thus emerged in opposition a new, informal coalition of seigneurial landowners, states, and urban merchants and financiers who shared a new interest in the geographical expansion of production and exchange. Landowners struggled to recover their surplus extraction from the land and turned to less labour-intensive livestock farming, particularly sheep, which tied them into an emerging world market for wool and cloth. They thus had an interest in expanding that market. For states, declining revenues could only be offset so much by conquering other, smaller European countries. Other frontiers of conquest therefore became attractive. And while higher borrowing costs to finance those wars of conquest buoyed the emerging urban capitalists, they too suffered from contracting European markets (as seigneurial revenues declined) and troubled trading networks across the Eurasian continent. So, when the city of Venice managed to displace Genoese capital and military force from the eastern Mediterranean by the later fourteenth century, the latter started to look to the opposite side of the compass, and the apparent investment opportunities that lay to the west. As such, Genoese finance houses became the bankers to the Castilian (i.e. Spanish) and Portuguese crowns, financing the exploration, conquest and settlement of islands in the eastern Atlantic, the search for an alternative route to the 'Indies' (i.e. southern and eastern Asia), and later the conquest of the 'New World' on the other side of the ocean (Moore 2002).

The subsequent period of European expansion was an early example of what geographer David Harvey calls a 'spatial fix' – a geographical reorganisation (usually of an expansive kind) designed to manage, at least for a while, the crisis tendencies inherent in capitalist regimes of wealth accumulation

(Harvey 1981). The biogeographical 'new Pangea' would, in the eyes of that new coalition of lords, kings, and bankers,

> serve as a virtually limitless storehouse of Cheap labor, food, energy, and raw materials. It's here, in the Atlantic zone of modern Pangea, that capitalism and today's planetary crisis originated.
>
> (Moore 2019, 53–54)

Moore capitalises the word 'Cheap' as for him it's a key concept: '*Cheap* is the opposite of a bargain – cheapening is a set of strategies to control a wider web of life' (Patel and Moore 2018, 3):

> Cheap is a strategy, a practice, a violence that mobilizes all kinds of work – human and animal, botanical and geological, with as little compensation as possible. We use *cheap* to talk about the process through which capitalism transmutes these un-denominated relationships of life making into circuits of production and consumption, in which these relations come to have as low a price as possible.
>
> (Patel and Moore 2018, 22)

The European expansion into the Atlantic was a search for new, 'cheap' and 'uncommodified natures whose work [could] be appropriated for free or low cost through violence, cultural domination, and markets' (Eichen 2020; Moore 2019, 54). As such, Patel and Moore insist that capitalism is far more than just an 'economic' system, and that the post-feudal development of capitalism was about much more than the spread of systems of economic production and exchange. It was a 'geocultural' revolution, at the heart of which were transformations in the way in which natural and social worlds were conceived, categorised, understood, and exploited.

Knowing and rationalising natures

Moore's work is concerned with understanding and explaining the historical development of capitalism and its relationship to non-human natures. A chief argument of his, building on earlier theorists, is that we can't understand the development of capitalism's economic systems without attending to what we might think of as their political and cultural infrastructures: modes of power and political organisation, and modes of knowledge, perception, and rationality (Moore 2017b). Moore's work thus attends, as he puts it, to the 'fundamental co-production of Earth-moving, idea-making, and power-creating across the geographical layers of human experience' (Moore 2015, 3), echoing Marxist geographer Neil Smith's idea of the 'tight dialectic' between environmental change and environmental knowledges (see Prologue). Put simply, the spread of early capitalism required new ways of seeing and understanding the world.

There's some risk here in generalising about complex processes of cultural change which took place across many decades and centuries, in many parts of the world, in the period when Europeans began establishing colonial empires. Jason Moore arguably also underplays the significant role that religion played in driving colonial expansion, with many early European colonisers aiming to harvest the souls of converted 'heathens' often as much as they wished to harvest new commodities (Lachenicht, Henneton, and Lignereux 2016), or using theological arguments to justify racialised oppression and exploitation (Wynter 2003). Religious persecution within Europe between the fifteenth and eighteenth centuries also drove colonial expansion and settlement (Lachenicht 2017). But Patel and Moore nonetheless offer us some useful and intriguing insights into emerging European mindsets with Columbus himself – an exemplar of early capitalism's new 'way of seeing'. They characterise Columbus as 'from the outset an assessor with a keen sense of cheapness and power, able to cast his eye on nature and be frustrated that he couldn't instantly see money' (Patel and Moore 2018, 50). For example, in the diary of his first voyage to the Americas, there are passages such as this:

> I can never tire my eyes in looking at such lovely vegetation, so different from ours. I believe there are many herbs and many trees that are worth much in Europe for dyes and for medicines but I do not know them, and this causes me great sorrow.
>
> (Columbus, quoted in Patel and Moore 2018, 50)

Poor Christopher! Not satisfied by the aesthetic pleasure of encountering such beautiful American natures, he wallows instead in the frustration and sorrow of not being able to extract all their apparent economic potential:

> There are a thousand other kinds of fruits, which it is impossible for me to write about, and all must be profitable.
>
> (Columbus, quoted in Patel and Moore 2018, 138)

Viewing the 'New World' in this way, as a pool of potentially infinite riches, was a cause of great excitement in Europe (Aráoz et al. 2026). It is arguably no coincidence that around this time Europeans were starting to think of themselves as thoroughly separate from nature, above it, and with a divine right to exploit it. While medieval philosophies had placed humans 'inside' creation, new ways of thinking about 'Man' [sic] and 'Nature' were appearing which cleaved apart humanity and the rest of the living world. Later on, such ideas would find their expression in formulations like Rene Descartes' mind-body dualism, which emphasised the inherent separateness of the rational, thinking human mind from its fleshy container and, by extension, the rest of the material world. But while all of the living, nonhuman world was being placed in a single new master-category of 'Nature', not all of humanity was invited into the category of 'Man' (Alvares 2008). The early modern period also saw the emergence of a new binary code: that of 'civilized' and 'savage'. Essentially, the master-category of 'Man' meant precisely that: male, but also 'civilized'; therefore, the supposedly 'savage' peoples encountered on colonial frontiers, from Ireland to the Caribbean, were relegated to the category of 'Nature'. As such, they too were cast as a readily exploitable pool of *cheap* work and energy:

> The binary code of Civilization and Savagery constitutes a pivotal operating system for modernity, one premised on dispossessing human beings of their humanity. Such dispossession — which occurred not once but many times over — was the fate meted out to indigenous peoples, to the Irish, to virtually all women, to African slaves, to colonial peoples around the world. It's this capitalist geoculture that reproduces an extraordinary cheapening of life and work, essential to every great world economic boom but also violent, degrading, and self-exhausting.
>
> (Moore 2019, 53)

So while European explorers and conquerors were often dependent on local knowledge to help navigate terrain and identify valuable resources (Hofman et al. 2018), the colonial frontier was primarily experienced by Indigenous peoples as a space of violence, dislocation, and dispossession. The Portuguese and Spanish expansion into the Atlantic quickly moved from exploration to settlement, with places like Madeira and the Canary Islands, which took Spanish forces three decades to wrest from indigenous control, seeing the first stirrings of another new kind of economy: the plantation system. This was a system that exemplified a new economy defined by regular 'booms', but which also illuminates Moore's remark about 'self-exhaustion'.

When Christopher Columbus first made it to the Americas in 1492, the island of Madeira was in the middle of a boom in sugar production. From the mid-fifteenth century, the island had witnessed an unprecedented convergence of (Portuguese) state power, (Italian and Flemish) financial investors, and Canarian and – for the first time – African slaves to power a 'commodity revolution' that embodied the emerging spirit of capitalism. This was the world's first cash-crop bonanza, with sugar grown on a new plantation model where both nature (in the form of crops, weeds, and pests) and humans (slaves and indentured labourers) could be rigidly and violently controlled. By 1506, 2000 tons of sugar flowed annually from this small island to Lisbon, Antwerp, Genoa, and other cities which were starting to grow rich on Atlantic trade. But by 1530, production had contracted by some 90%. Jason Moore argues that what happened on Madeira heralded an 'epochal shift in human history', defined by 'new flows of capital, power, and nature', and that it was indicative of what capitalism would become: 'the pursuit of endless financial accumulation meant the endless conquest of nature' (Moore 2009, 346). Madeira's forests, a key source of fuel for sugar processing, were quickly exhausted, soils lost their nutrients and stability, the sugar monocultures became increasingly vulnerable to pests and diseases. From this increasingly exhausted land, the sugar frontier moved on to Sao Tome, to Brazil, and later to the Caribbean, where similar dynamics of boom and bust took place (Eichen 2020). Moore argues that this early capitalism, driven by a new urge for expansion and accumulation and forged through commodity frontiers such as that of sugar, normalised regional socio-ecological crises as a new 'way of life', and set the world economy on a path that would lead to the environmental crises of the present (Box 3.1). As such, he and many like-minded scholars have argued that the Anthropocene would be better known as the 'Capitalocene'.

Box 3.1 Situating the Anthropocene: the frontier and the plantation

We've been throwing around a lot of inherently spatial and geographical terms in this discussion, so let's take a closer look at two crucial ones.

The restless frontier

While a frontier conventionally 'marks a limit', either to the extension of political power or, in more historical usage, to the domain of 'civilization' as opposed to unknown, uncouth or unsettled lands (Gregory et al. 2009, 264), Jason Moore encourages us to think of frontier spaces as crucial to the historical geography of capitalism: a 'peculiar system', which has 'always depended on frontiers of Cheap Natures — uncommodified natures whose work can be appropriated for free or low cost through violence, cultural domination, and markets' (Moore 2019, 54). These frontiers were spaces where a bonanza of 'free gifts' (Burkett 1999) could be had, in the form of uncommodified land, labour, and materials. Commodity frontiers followed the geography of land that was relatively easy to appropriate (often by force), of labour that was readily forced or coerced, and of resources (like trees or silver) that could be taken without any apparent need to compensate anyone. A further feature of commodity frontiers in this period is their rapid exhaustion – like the Madeira example above, cycles of boom and bust happened over mere decades, whereas previous cycles of socio-ecological change, in Europe at least, had stretched over centuries. The sugar frontier, like others which preceded it (e.g. timber) and followed it (e.g. tobacco, coffee), was almost constantly in search of a new spatial fix to crises of accumulation, and embodied 'early capitalism's profound geographical restlessness' (Moore 2010, 93).

The plantation

While processes of turning 'cheap' natures into highly profitable commodities exhibited a restless geography, the plantation was all about keeping things – and people – in place, and eking as much out of them as possible.

The term 'plantation' initially meant the process of colonial settlement itself, with settlers 'planted' in a new place, with the intention of growing their community and 'civilizing' their indigenous neighbours. The 'Plantation of Ulster' from 1609 onwards is an early example of this meaning of the term. This was an organised colonisation of the largely Gaelic, Catholic and rebellious Irish province of Ulster by English-speaking Protestants from northern England and southern Scotland, personally directed by King James I.

Over time, however, 'plantation' came to mean a site for the cultivation of agricultural commodities, and it quickly became associated with those sites using slave labour. After early experiments on Madeira and the Canary Islands, such 'plantations' spread around the Atlantic world, most notably the Caribbean islands and the eastern fringes of the American continent. We can understand plantations as spaces designed to rationalise and render more efficient the production of certain crops; they were spaces where nature was cultivated and disciplined with the aim of maximising profit (Miller 2025). In contrast to the unruly richness of natural ecosystems, and even to the relative diversity of peasant smallholdings with their crop rotations and kitchen gardens, plantations were (and still are) 'green deserts' (World Rainforest Movement 2007), where vast amounts of land, labour and capital are given over to the production of a single cash crop, such as sugar, tobacco or, to pick a pressing contemporary example, palm oil.

Plantations were a form of 'abstract space': rationally ordered, bureaucratically governed, standardised, and amenable to control and processes of commodification. Historical geographer Jim Duncan (2002) took the notion of abstract space from philosopher Henri Lefebvre to examine the functioning and life of coffee plantations in nineteenth-century Ceylon, modern-day Sri Lanka. Duncan noted how standardised plantation designs followed a spatial model geared towards constant surveillance of both plants and workers. And although Duncan's case study featured indentured rather than enslaved workers, his arguments about how workers were treated as mere economic resources to be maximally exploited resonate with earlier plantations on the sugar frontier, where enslaved Africans toiled under the shadow of severe disciplinary regimes, sexual abuse, and extreme violence.

Plantations were therefore spaces designed for the cultivation and disciplining of both human and nonhuman natures (Eichen 2020). Duncan (2002) dubs them 'laboratories of modernity', for how they pre-figured later instances of abstract space designed to control life – such as prisons or army barracks – and to maximise profits – such as factories and even supermarkets and shopping malls (Woods 2017; Miller 2025). Below, we explore arguments about how the plantation can be considered one of the defining features and perhaps even causal elements of the Anthropocene, to the extent that this new epoch might be better dubbed the 'Plantationocene'.

Transforming land and labour

> Capitalism's uniqueness is found in the historical geography of endless accumulation.
>
> (Moore 2022, 124)

While the management of regular socio-ecological crisis became a 'way of life' for expanding European empires and globe-trotting early capitalists, for millions of Indigenous and enslaved peoples, the centuries after the Columbian Exchange brought extreme suffering, caused by the aforementioned epidemics, and by transformations of land and labour (Whyte, Talley, and Gibson 2019).

Take Potosí, in the Andean mountains of the Spanish Viceroyalty of Peru. A frontier town where labourers worked in the harshest, most poisonous conditions to extract from dangerous mines the silver that would form the material basis of a new, globalising economy. By the 1570s, Potosí was the 'epicenter of a commodity revolution that reorganised Peru's peoples and landscapes to serve capital and empire' (Moore 2010, 58). At its peak, Potosí's population rivalled that of Madrid, and its silver would form the material basis of new, globally circulating currencies for centuries to come.

Unlike silver frontiers further north in present-day Mexico, in Potosí the geographical concentration of the precious metal meant that centralised control over production was easier. While Indigenous communities had their own means of extracting silver and traded extensively with Spanish colonists, these were gradually displaced by new, proto-industrialised modes of production (Figure 3.3). And despite falling Indigenous populations, the colonial state

Figure 3.3 'A section of a silver mine in Potosi, in Bolivia, and the manner of working it'. Engraved for *The Universal Magazine of Knowledge and Pleasure* ca. 1750.

was able to generate a steady supply of labour by instigating taxes and rules, which meant that a certain proportion of the male population, from hundreds of miles around, had to be given over to essentially slave labour in the mines. Local populations were stripped of their lands as Indigenous agricultural systems gave way to extensive wheat farming, and rapid urbanisation occurred as 1.5 million people were shepherded into Spanish-style towns.

At Potosí, new techniques of using mercury to extract silver from crushed ore increased production massively, albeit at the expense of widespread pollution and rapid deforestation for fuel wood – by 1714, the frontier of deforestation to feed Potosí's fuel needs had reached Paraguay. The implications of this deforestation for ecosystem functioning were not lost on people at the time. For example, one anonymous observer wrote in 1603 that:

> …today, because of all the work done on the mountain, there is no sign that it ever had a forest, [but] when it was discovered [sic] it was fully covered with trees … today not even weeds grow on the mountain, not even in the most fertile soils where trees could have grown. This is the most frightening, because now the mountain is covered with loose gravel, with little or no fertile land'
>
> (quoted in Moore 2010, 72)

Potosí represented the wider trends of deforestation and landscape transformation which defined the European march through the Americas (Aráoz et al. 2026; Eichen 2020), and which heralded a worldwide process of colonial 'terraforming' and expanding extractive infrastructures which continues into the present (Woods 2014; LaDuke and Cowen 2020). For Jason Moore, this process exemplified early capitalism as a system which depended on cheapening both nonhuman and human life, and a "sociopathic disregard for their care" (Moore 2017b, 197). Worker mortality was high, especially as mines went deeper in search of more ore, but the steady supply of labour guaranteed by the colonial system meant that this posed no immediate threat to the profitability of the mines. Nonetheless, going deeper increased production costs, so mine owners began to ignore colonial prohibitions in order to eke out more from their workers with production quotas. Rules about rest periods, shift lengths, and the number of times a worker could carry material to the surface were ignored. Workers were kept in the mines from Monday to Saturday, might carry two dozen loads of 25kg to the surface per day, rest periods were disregarded, diseases flourished in the regularly-flooded mine shafts, and tunnel collapses became more frequent, perhaps due to the rising costs of timber for supports. One mine owner, Luis Capoche, describes the pits as a 'harsh executioner of Indians…their lives are made miserable by the fear of death' (quoted in Moore 2010, 77). At the surface, silicosis and tuberculosis compounded the misery, while mercury poisoning afflicted labouring bodies and regional ecosystems, cascading through food chains back into human bodies, where its effects on fertility are thought to have stunted the demographic recovery from the diseases and other forms of mortality that had followed the colonisers.

Eventually, social and ecological exhaustion meant that the silver frontier would move on – by the eighteenth century, New Spain (present-day Mexico) was the epicentre of global silver production, producing a commodity whose trade placed Europeans at the centre of economic networks which linked Central and South America with expanding markets for silver currency in East Asia (Pomeranz 2021). But the example of Potosí is instructive: like Madeira's sugar frontier, it illustrates the tendency of early capitalism towards relatively rapid socio-ecological crisis, it illuminates how this early colonialism involved radical re-organisations of space (e.g. in relation to land ownership and urbanisation), and it highlights how an emerging economic world-system (see Box 3.2) rested on a sharp division of labour between a raw material-producing colonial periphery, and imperial metropoles where wealth and profit accumulated (Wallerstein 2004).

This division of labour had developed over the preceding 200 years of colonisation and trade under a guiding ideology which became known to its critics as 'mercantilism' (Williams 1944, 2020). This embodied a theory that the wealth of the world was finite, and that it was the task of nations to capture as much of that wealth as possible. Ideally, that wealth should be held in gold or silver, and the impetus to trade and conquest thus became the acquisition of raw materials from regions of plenty, so that they may be turned into goods for trade with regions where such goods were scarce. Mercantilism 'idealized extracting raw materials from colonial locations, adding value by treating or transforming them in the home country, and finding or founding markets to unload the finished goods' (Hahn 2020, 36).

Gold and silver themselves were the ultimate raw materials, which could simply be hoarded. Alternatively, nations would seek to shape their balance of trade with other nations by selling them manufactured goods in exchange for silver and gold. The role of colonies was simple – to supply to the 'mother country', and to the mother country alone, raw materials for processing and onward trade, and to provide a market for those processed and manufactured goods. Mercantilism got its clearest articulation in the seventeenth- and eighteenth-century French policy *Le Principe de l'Exclusif*, which meant that French colonies in places like the Caribbean could only sell goods to France, on French ships, and could only import from the metropole too – they weren't even allowed to trade with neighbouring island colonies in the region (Williams 2020). Mercantilist ideas informed bans on any industrial production within colonies, such as the prohibitions that French and British governments upheld on sugar processing in the Caribbean; the work of adding value to the raw material was to be done at home, in the refineries of London, Paris, and other proto-industrial cities. Through the political structures of colonialism, the economic ideology and practice of mercantilism birthed a world-system defined by new 'core-periphery' relationships.

Box 3.2 From world-system to world-ecology

Jason Moore's work is inspired by Immanuel Wallerstein's 'world-systems analysis'. This is an approach to the study of social change which views 'society' not as bounded by national borders, but as a component and function of different modes of economic production. Wallerstein identified the historical movement from societies organised around reciprocal and familial economic relationships towards societies, such as those in feudal Europe, organised around tributary systems whereby a large class of agricultural workers gives their surplus production to elites. The capitalist mode of production, while also class-based, is defined by ceaseless accumulation, the normalisation of private property, the setting of prices by market forces, and, crucially, the emergence of a world-economy which further transcends geographical borders (Gregory et al. 2009; Wallerstein 2004). Within this capitalist world-economy, which Wallerstein dates to the late fifteenth century onset of European colonialism and economic mercantilism, a clear division of labour exists between 'periphery' areas, with low incomes and low-profit production of largely raw materials, and 'core' areas, with higher incomes and higher profit margins, often based on taking low-value materials from the periphery and turning them into high-value goods. As such, Wallerstein's analysis is a variant of dependency theory, which highlights how resources tend to flow from poor countries to rich ones, to the benefit of the latter.

While Wallerstein certainly didn't ignore nature or the environment in his analysis, Moore suggests that this approach sees capitalism as too detached from the material world. He therefore suggests that we understand capitalism as a world-ecology. Capitalism, Moore suggests, 'is not an economic system, it is not a social system; it is a way of organizing nature' (Moore 2015, 2). It 'did not emerge as a set of relations between humans, but rather through the nexus of human interaction with the rest of nature' (Moore 2010, 59), or what Moore likes to call the 'web of life'.

So, rather than being something external to 'nature' which then acts upon it, capitalism needs to be understood as a system which 'produces' different natures in different times and places, from the denuded ecosystems of Madeira or Potosí to the global nature of the 'New Pangea', or indeed the stressed global ecologies and climates of today. But by trying to understand how different natures are produced through the interactions of humans and nonhumans within the web of life, the world-ecology perspective also emphasises 'life-making alternatives forged in resistance' to imperialism and capitalism (Moore 2017c, 180). Capitalism can't be said to have generated a new world-ecology unimpeded and without resistance – a point we'll return to throughout this book.

For human geographer Kathryn Yusoff, examples like Potosí are illustrative of how industrial mining might be considered the real 'natal moment', the beginning, of the Anthropocene (Yusoff 2018, 6, 2024). Yusoff, like Moore, is interested in how the material transformations wrought by colonialism and capitalism also involved *intellectual* transformations in ways of seeing and categorising the world. The rapid expansion of mining and metallurgy in the early modern period was when humans really started leaving their mark in the rocks, and early colonial ways of seeing the world as consisting of so many standing stocks of exploitable materials (such as silver or gold) were directly parallel to the racialised relegation of certain humans to a category of, essentially, inhuman matter:

> The racial categorization of Blackness shares its natality with mining in the New World... That is, racialization belongs to a material categorization of the division of matter (corporeal and mineralogical) into active

> and inert. Extractable matter must be both passive (awaiting extraction and possessing of properties) and able to be activated through the mastery of white men. Historically, both slaves and gold have to be epistemically made through the recognition and extraction of their inhuman properties.
>
> (Yusoff 2018, 2–3)

As Katherine McKittrick puts it, building on the work of Sylvia Wynter, slavery and the New World mining and plantation frontiers:

> transformed – or more specifically converted – the enslaved into units of labor. The enslaved perceptively and conceptually became homogeneous units of labor, planted in the New World not to inhabit (people, settle) the land but to mechanically produce monocrops and fuel the economic system.
>
> (McKittrick 2021, 154)

As mining and plantation frontiers spread across the 'New World', the west coast of Africa was increasingly and similarly seen as a standing stock of inhuman matter with useful properties – 'energy, reproducibility, and transformation' (McKittrick 2021, 70). So much so that the 'Gold Coast' in West Africa was nicknamed 'the Mine' by Portuguese traders for its seemingly bounteous reserves of both gold and slaves, a name which stuck for several centuries (McKittrick 2021). Yusoff's overarching argument is therefore that the standard narrative of the Anthropocene, which paints humanity as a singular actor acting on the Earth System, erases these historical processes of racialisation and exploitation which were the engines of capitalist growth and which define 'an epoch of imperial world building that is not yet at its end' (Yusoff 2018, 1).

It's estimated that around 12 million people were dispatched from Africa in Atlantic slave ships, with around 10 million surviving the journey to the Americas and the Caribbean between the fifteenth and nineteenth centuries (Inikori and Engerman 1992). On arrival, most slaves were sold to the owners of the plantations, which were rapidly remaking the physical and human geographies of the European colonies. The significance of the plantation system in the early development of capitalism, and in the emergence of new, usually brutal relationships between humans and with the rest of nature, is such that some have argued that we might do better to think of the Anthropocene not just as the Capitalocene, but as the 'Plantationocene'.

The Plantationocene

The plantation is a space of homogeneity, of environmental modernisation, and control (see Figure 3.4). For anthropologists like Donna Haraway and Anna Tsing, both early-modern and modern plantations are spaces where

Figure 3.4 Jamaica: A sugar plantation. Coloured aquatint by P. Fumagalli, ca. 1821. To the top left is the Master's House, dominating the scene, with slave houses below and buildings for sugar production on the right. (Source: Wellcome Collection.)

new modes of control over nonhuman natures were developed. Anthropologists have even used the language of 'slavery' to discuss how plants and microbes were forcibly enrolled into new modes of production, with the application of fast-developing agricultural sciences, technologies, and chemicals to ensure ever-rising profitability from the growth of single 'cash crops' (Haraway et al. 2015). Such observations have identified the plantation as ground-zero for new, 'modern' relationship between humans and nonhuman natures which, thanks to the plantation's characteristics of sameness, scalability and reproducibility (Duncan 2002; Tsing 2015), has proliferated since the period of early European colonisation to become a dominant landscape feature across the world, and a key feature of various supply chains: food, textiles, paper, wood, and so on (Kröger 2014). As such, the argument goes, the term 'Plantationocene' might offer a better, more precise descriptor than 'Anthropocene' of what defines the new, dominant relationship between humans and the rest of nature (Barua 2023).

But while plantations undoubtedly introduced radically new forms of socioecological relations, and perhaps also provided the model for later carbon-intensive production systems like steam-powered factories, analyses of the Plantationocene have only recently started to attend to the *human* labour, lives, and power structures which characterised plantations in early colonial and industrial periods (Eichen 2020; Wolford 2021). Geographer Janae Davis and colleagues argue that the Plantationocene proposition, like the 'standard narrative' of the Anthropocene (Chapter 1), obscures histories and geographies of human inequalities. In response, they synthesise work by African American geographers on the plantation as a space of racialised violence (McKittrick 2011; Woods 2017) and argue that the true significance of the plantation lies in the 'interrelated – *yet distinct*' violences metered out to nonhumans, dispossessed indigenous communities, and enslaved human workers (Davis et al. 2019, 7, original emphasis).

Davis and colleagues draw particular attention to the 'plot', or the small gardens which slaves were allowed to keep in the margins of the plantation. Donna Haraway (2015, 162) observes that such spaces provided 'refuges for biodiverse plants, animals, fungi, and soils' amid the wider 'green desert' of the plantation. But Davis et al. point out that they did so not just on account of being gardens, but because they were spaces produced by a distinctive 'socioecological ethos' (Davis et al. 2019, 8). This is an ethos that can be traced to West African tenets of 'good use':

> Contrary to the idea of 'rational use' motivating colonial expansion which promoted the seizure, enclosure, privatization, intense cultivation, and commodification of land as a means of European accumulation… 'good use' principles assigned moral, spiritual, and relational meanings to land that assured its more-than-economic value. Land was sacred, as evident in observations of African women ingesting soil before being forced aboard slave ships. While in the Americas, indigenous Africans and African Americans adapted good use tenets and practices to cultivate new socio-ecological realities.
>
> (Davis et al. 2019, 8)

The plantation slavery system rested on the violent displacement and dislocation of Africans from home, from familial connections, and from local, autochthonous cultures. But while the literally dehumanising plantation system sought to reduce Black life to little more than quantified bodily labour, Davis et al. (2019, 7) cast the cultivation of the plot as a means of fashioning 'alternative ways of being while confronting, refusing, and resisting racial violence'. This was part of a rich history of 'radical practices of black humanization – marronages, mutinies, funerals, carnivals, dramas, visual arts, fictions, poems, fights, dances, the blues…' – all geopolitical responses to the racism and objectification of plantation slavery, and to the emerging 'geoculture' of capitalism (McKittrick 2021, 158–59). Like these practices, cultivating the plot 'served as a creative source of hope, vision and perseverance' (Davis et al. 2019, 8).

In this light, the Plantationocene is not just a story of the spread of an ecologically-homogenising system of economic production. It is also a story of human violence, and of resistance, of different modes of relating to nonhuman natures which perhaps hold a key to navigating the Anthropocene today (Curley and Smith 2024; Ferdinand 2022)). Davis et al. (2019) draw a direct link between these plantation practices, post-slavery practices of economic survival in the urban United States, and late-twentieth-century environmental justice movements, which brought together African American, Indigenous, and other minority activists together to campaign for conjoined social, economic, and environmental justice. In the declaration of the First National People of Color Environmental Leadership Summit (1991), we can hear echoes, Davis et al. suggest, of the alternative socioecological ethos of the plantation plot:

> WE, THE PEOPLE OF COLOR, gathered together at this multinational People of Color Environmental Leadership Summit ... do hereby re-establish our spiritual interdependence to the sacredness of our Mother Earth; to respect and celebrate each of our cultures, languages and beliefs about the natural world and our roles in healing ourselves; to insure environmental justice; to promote economic alternatives which would contribute to the development of environmentally safe livelihoods; and, to secure our political, economic and cultural liberation that has been denied for over 500 years of colonization and oppression, resulting in the poisoning of our communities and land and the genocide of our peoples.

Geographies of knowledge and resistance

This articulation of an alternative socioecological ethos was the product of hundreds of years of indigenous and marginalised communities living through the social and ecological upheavals which possibly await much of the rest of humanity in the Anthropocene (Whyte 2018). Those upheavals were frequently met with violent struggle, revolt, rebellion, and 'marronage' (see below) – all political practices by which autonomy and alternative lifeworlds to those of agrarian and plantation capitalism were sought and enacted. But the upending of indigenous relationships to land and nature was not always a product of a straightforward imposition of extractive capitalist logics onto colonised lands; as colonial states grew in power and complexity, the destruction of indigenous life-worlds was frequently committed in the name of an early kind of environmentalism.

In some parts of European colonial empires, settlers sought to deliberately transform environments and climates to make them more hospitable. Early European settlers in New England, for example, reckoned that the region's relatively extreme climate (in comparison to old England, for instance) could be tamed by the kind of land domestication and deforestation which had transformed European landscapes over preceding centuries. Elsewhere, deforestation was seen as something much less desirable. Emerging scientific understandings of vegetation-atmosphere relationships led colonial thinkers in places like modern-day South Africa, India, and New Zealand to blame erratic weather – particularly droughts – on recent deforestation committed both by settlers and by Indigenous communities. During the eighteenth century, many tropical island colonies of France and Great Britain were made into laboratories for testing competing theories of regional climate change, often concerning the relationship between deforestation and reduced rainfall. Island colonies became sites for testing the effectiveness of different policy measures such as instituting forest reserves and slowing the expansion of things like sugar plantations. Richard Grove contends that such measures were influenced by contemporary environmental management practices in East Asia (Grove 1995), but through their circulation around imperial networks, colonial actors started to treat colonial environments and natures as homogenous and amenable to similar types of control (Moulton 2023).

Whether or not they were effective as climate-protection devices, forest reserves became a key strategy of colonial environmental management, which itself became 'a precise measure of the effectiveness and prestige of the colonial state' (Grove 1995, 127). This complicates any view of colonial states as simple handmaidens of the capitalist destruction of nature. It is perhaps too charitable though to see such colonial institutions as 'protectors' of the nonhuman world against the ravages of capitalism's practices of primitive accumulation; rather, we might understand them as actors who recognised capitalism's tendency towards regional socioecological crisis, and who sought to ensure and fix in place the longer-term productivity of local commodity production, as against the relentless expansion and exhaustion of commodity frontiers. They embody what the political philosopher Hannah Arendt identified as a core driving feature of colonialism – the tension between capitalism's geographical restlessness and core belief in limitlessness, and the nation-state's sometimes contradictory tendency towards localism – trying to fix things in place amid a world of increasing circulation and flux (Charbonnier 2020).

While in places like Mauritius, environmental degradation pitted colonists against each other in a battle between environmental preservation and the expansion of local commodity frontiers, colonial forest policy soon became another means of securing colonial interests and marginalising local communities. In nineteenth-century British India, for example, theories about forests and climate protection were joined by the desire to secure timber supplies as Britain's domestic reserves started to dry up. Colonial 'environmentalists', as Gregory Barton dubs them, looked down on local forest management practices and positioned themselves as enlightened protectors of nature and its bounties (Barton 2002). In instituting forest reserves, communities were systematically denied access to the ecosystems from which they drew their livelihoods, in a model which was soon replicated in places like Australia and the United States. This was an early instantiation of 'fortress conservation' – 'protecting' nature by putting an exclusionary fence around it – which has disenfranchised and displaced Indigenous communities all over the world, variously in the name of resource husbandry, game hunting, or 'wilderness protection' (Büscher 2016; see Chapter 7).

Marroon ecologies

Recent research in Geography has nonetheless helped to complicate our understandings of colonial land and conservation politics, and has centred the agency of marginalised communities in transforming, maintaining, and protecting ecologies. Studies of the historical geographies of 'marronage', for example add to our knowledge of the strategies of resistance to plantation capitalism and of the cultivation of alternative lifeworlds. 'Marronage' is a term for the strategies of 'maroons', a collective term for fugitive slaves who escaped from the plantation system in the Caribbean, eastern North and South America, and in parts of the Indian Ocean, and established new settlements and forms of social organisation, often in consort with Indigenous communities. Celeste Winston describes marronage as 'Black fugitive placemaking' that resists colonialism and capitalism (Winston 2021, 2186). Danielle Purifoy likewise considers marronage to be:

> an Afro-diasporic spatial freedom practice that rejects structures of unfreedom (particularly slavery) that have been imposed on Afro-diasporic peoples across at least the last four centuries. Integral to these practices is the interdependence of Afro-diasporic communities and the landscapes they inhabit.
>
> (Purifoy 2023, 2)

Marronage 'is both a fugitive movement from bondage and the replication of an alternative world(view)' (Wright 2020, 1135–36). It frequently involved flight into 'marginal' landscapes – forests, mountains, swamps – that had escaped or resisted colonial transformation owing to their inherently 'unruly' natures (Vickers 2023). Such landscapes evaded the scrutiny of colonial surveyors and resisted the ecocidal rationalisation of the plantation (Yusoff et al. 2025). Unruly natures became accomplices to the unruly maroons and their strategy of building alternative worlds beyond the reach of colonial power. In spaces deemed unsuitable for cultivation and development, fugitives could step beyond the reach of the authorities and the plantocracy and develop new, collective, and communal means of subsistence and wealth building (Ferdinand 2022). In places like the interior mountain ranges of Jamaica (Figure 3.5), and the swamps of Virginia and North Carolina, new forms of agriculture were practiced which prioritised agroecological diversity and long-term sustainability, as well as new forms of social organisation.

Alex Moulton (2023) shows how in the Jamaican highlands the colonial authorities begrudgingly acknowledged that maroon societies had effectively protected forests from destruction. Nonetheless, the colonial authorities, keen to pin deforestation on the encroachment of Black peasants rather than the white plantocracy, ignored the extent to which the maroons had developed and deployed complex forms of botanical and agroecological knowledge to

Figure 3.5 'Trelawney town, the chief residence of the maroons', from '*The history, civil and commercial, of the British Colonies in the West Indies*' by Bryan Edwards (1801). Named by the colonial authorities in Jamaica after Governor Edward Trelawny, the town was known to its residents as Cudjoe's Town, after the leader of the Maroon forces who fought the British in the 1730s before signing a treaty with them in 1739.

produce new natures which supported human health and wellbeing as well as maintaining the biodiversity of the forests. Scholars like Andreas Malm (2018) have pointed to these historical geographies of marronage as providing lessons for ongoing struggles against racial oppression and ecological destruction. In the claustrophobic world of the Anthropocene, research shows that spaces can still be created where people and natures can live according to different logics, and where alternative worlds might be built according to radically different worldviews – something we will explore further in Chapter 10.

Conclusion: colonialism, capitalism, and the historical geography of the Anthropocene

While it's long been acknowledged by historical geographers that European colonialism in the centuries after 1492 had some significant environmental implications, they've tended to be regarded as fairly local-to-regional in nature. The debate about the onset and nature of the Anthropocene, across Physical and Human Geography and beyond, has cast the environmental impacts of European colonialism in a new light. In this chapter, we've looked at how physical geographers have pointed to the 'Columbian Exchange' as key to the emergence of new global ecologies, as well as setting in train a series of events in the Americas that might have been responsible for a significant bit of global anthropogenic climate change. The argument that the beginning of the Anthropocene should be dated to the onset of European colonialism accords with that of Jason Moore, who similarly sees the years after 1492 as epoch-making in their significance. However, Moore isn't pointing to ice cores and pollen records to make his case. Rather, he makes a historical argument that the transition from feudalism to capitalism in Europe was driven by regional socioecological crises, and that the transition to capitalism was fundamentally dependent on the enactment of new, exploitative relationships to the 'web of life' in distant parts of the globe.

Moore's work, like that of world-systems analysts (e.g. Bunker and Ciccantell 2005), places great emphasis on global interconnections, flows of matter, people, and capital, and on big 'systems' (like capitalism) that structure how people behave and the decisions they make. Nonetheless, this approach does not ignore the local; indeed, Moore stresses that the historical emergence of capitalism as a world-ecology 'must be historicized' (i.e. understood and explained in relation to a wider historical context) 'and situated geographically' (Moore 2017b, 197). He therefore contends that the 'Capitalocene… insists on a deep historical perspective. It also insists that we all become geographers' (Moore 2017b, 197). We're inclined to agree!

The work of Moore and others now offers us a relatively fine-grained picture of how the development of early colonialism and capitalism unfolded through and between particular nodal places, like Madeira and Potosí, as they became plugged into and re-shaped emerging global networks of investment and production. Moore's use of the notion of 'commodity frontier' highlights the pace by which capitalist modes of production transformed particular places and then quickly moved on, helping to emphasise how, while the transformation of landscapes by activities like mining wasn't new, there was something fundamentally new about this restless geography. Moore's conclusion is that to understand our current era of planetary crisis as the 'Anthropocene' is a conceit – there is no *anthropos* to be held responsible, and such a framing just reproduces a dualist notion of separate 'Man' [sic] and 'Nature' which Moore (and many others!) would argue is at the root of our problems anyway.

Instead, Moore proposes that things like climate change be understood as *capitalogenic* rather than anthropogenic, and that the Anthropocene should be understood as the Capitalocene, *the age of capital* (Moore 2019). To invoke capital in this way 'isn't about blame [for individuals]; it's about identifying the system that has devastated life on this planet', and to say 'capitalogenic' is to invoke not just a system of economic production, 'but the power and violence that has made endless accumulation possible' (Moore 2017c, 179).

Identifying this system as the problem therefore implies that, if the worst effects of the Anthropocene are to be avoided, it is the system that has to change.

Moore certainly isn't alone in offering what we labelled in Chapter 1 as an Eco-Marxist narrative of the Anthropocene. But among Marxist-inspired theorists of capitalism and nature, he is relatively unique in identifying early European empire-building (rather than industrialisation) as the key trigger (or 'detonator' – Tsing et al. 2020) of the Anthropocene. As such, Moore's work forms part of a broader conversation in Geography and related disciplines about the nature of capitalism – how it was forged not in the factories of northern England nor the banking houses of London, Amsterdam, or Genoa, but in spaces like the slave ship and the plantation. As such, his work informs the narrative of the Colonial Anthropocene (Chapter 1) as well as claims that we might better think of this new geological epoch or event as the 'Plantationocene'. This is closely related to the work of theorists of 'racial capitalism', who have examined how capitalism and racism are co-constitutive of each other,[2] and of how they combine to create stark environmental justices (Pulido 2017; Saldanha 2020).

Such arguments pose a fundamental challenge to the standard Anthropocene account of humanity acting as a single, undifferentiated force, and emphasise that 'the Anthropocene is a geohistorical event, the stratigraphic signature of which cannot be easily decoupled from the histories of race and racism, capitalism, and European imperialism' (Baldwin and Erickson 2020, 4). These critiques of the 'standard narrative' (Chapter 1) make clear that 'the Anthropocene is not a neutral concept' – it can either obscure the histories, geographies, and legacies of race and racism, or it can occasion public debate on those geohistories and their role in the past, present, and future of environmental change (Baldwin and Erickson 2020, 5; Luke 2020). The work explored in this chapter is a significant part of that growing and vital public debate, and we think it's significant that this is a space where Human and Physical Geographers are increasingly talking to each other about the colonial origins of the Anthropocene and environmental change (Knitter, Augustin, and Biniyaz 2019), and where a greater diversity of voices is to be found than among the official adjudicators of the Anthropocene's stratigraphic status (Todd 2015). But this isn't just about talk and debate, nor about simply getting the history of the Anthropocene 'right' (Thomas 2022). Identifying the origins of the Anthropocene in the emergence of European colonialism and racial capitalism has important political implications too: it directs our attention to how achieving 'environmental justice' is inseparable from achieving social and racial justice (see Chapter 6), and to how building a good or just Anthropocene might mean building a decolonial Anthropocene (Curley and Smith 2024; Davis and Todd 2017).

When we look at the 'thick geographies' of the Colonial Anthropocene, in places like mines and plantations, we can see not only power and domination but also forms of resistance, creativity, and even hope. In the next chapter, we will look at further evidence in support of the idea that this new epoch of ours might best be known as the Plantationocene or even the 'Racial Capitalocene' (Vergès 2017), given how important plantation economics and slavery were to the emergence of the Industrial Revolution. But we will also continue tracing some of the geographies of resistance through which this world-ecology was challenged and resisted; geographies which might inform those concerned with resisting the ongoing material transformations of the Anthropocene today.

Study questions

1 Why does Jason W. Moore (2019, 53–54) claim that it is 'in the Atlantic zone of modern Pangea, that capitalism and today's planetary crisis originated'?
2 What role did disease play in the onset and early consequences of European colonialism?
3 What lessons might the historical geographies of *marronage* hold for strategies of responding to environmental change in the present?

Notes

1 William Ruddiman's controversial 'early anthropogenic' or 'early Anthropocene' (Ruddiman, Crucifix, and Oldfield 2011) hypothesis states that 'millenia ago farming began to transform landscapes sufficiently to emit greenhouse gases and extend the natural warmth of the current interglaciation that had been initiated by orbital variations' (Ruddiman 2017, 4).
2 The short film *Geographies of Racial Capitalism with Ruth Wilson Gilmore*, produced by the Antipode Fountation, provides an excellent introduction to the concept and its use in geographical research and activism – see Antipodeonline (2020).

References

Alvares, C. 2008. "New World Slavery: Redefining the Human." Annali d'Italianistica 26: 131–53.

antipodeonline, dir. 2020. "Geographies of Racial Capitalism With Ruth Wilson Gilmore." *An Antipode Foundation Film* 16: 18.

Aráoz, H. M. 2026. 'The Conquistal Mineralocene: Potosí, the Earth Eaters, and the World's Machination'. In *Contesting the Anthropocene*, edited by Kalthmeier, Olaf, et al. London: Routledge.

Baldwin, A., and B. Erickson. 2020. "Introduction: Whiteness, Coloniality, and the Anthropocene." *Environment and Planning D: Society and Space* 38 (1): 3–11.

Barton, G. A. 2002. *Empire Forestry and the Origins of Environmentalism*. Cambridge: Cambridge University Press.

Barua, M. 2023. "Plantationocene: A Vegetal Geography." *Annals of the American Association of Geographers* 113 (1): 13–29.

Bunker, S. G., and P. S. Ciccantell. 2005. *Globalization and the Race for Resources*. Baltimore: JHU Press.

Burkett, P. 1999. "Nature's 'Free Gifts' and the Ecological Significance of Value." *Capital & Class* 23 (2): 89–110.

Büscher, B. 2016. "Reassessing Fortress Conservation? New Media and the Politics of Distinction in Kruger National Park." *Annals of the American Association of Geographers* 106 (1): 114–29.

Bush, M. B., M. N. Nascimento, C. M. Åkesson, et al. 2021. "Widespread Reforestation before European Influence on Amazonia." *Science* 372 (6541): 484–87.

Butlin, R. A. 2009. *Geographies of Empire: European Empires and Colonies c. 1880–1960*. Cambridge: Cambridge University Press.

Charbonnier, P. 2020. '"Where Is Your Freedom Now?" How the Moderns Became Ubiquitous'. In *Critical Zones: The Science and Politics of Landing on Earth*. Cambridge: MIT Press.

Crosby, A. W. 1972. *The Columbian Exchange: Biological and Cultural Consequences of 1492*. Westport: Greenwood Publishing Group.

Crosby, A. W. 2004. *Ecological Imperialism: The Biological Expansion of Europe, 900–1900*. Cambridge: Cambridge University Press.

Curley, A., and S. Smith. 2024. "The Cene Scene: Who Gets to Theorize Global Time and How Do We Center Indigenous and Black Futurities?" *Environment and Planning E* 7 (1): 166–88.

Davis, H., and Z. Todd. 2017. "On the Importance of a Date, or Decolonizing the Anthropocene." *Acme* 16 (4): 761–80.

Davis, J., A. A. Moulton, L. Van Sant, and B. Williams. 2019. "Anthropocene, Capitalocene, … Plantationocene?: A Manifesto for Ecological Justice in an Age of Global Crises." *Geography Compass* 13 (5): e12438.

Driver, F. 2000. *Geography Militant: Cultures of Exploration and Empire*. London: Wiley.

Driver, F. 2013. 'Research in Historical Geography and in the History and Philosophy of Geography in the UK, 2001–2011: An Overview'. *Journal of Historical Geography* 42 (October): 203–11.

Dull, R. A., R. J. Nevle, W. I. Woods, D. K. Bird, S. Avnery, and W. M. Denevan. 2010. "The Columbian Encounter and the Little Ice Age: Abrupt Land Use Change, Fire, and Greenhouse Forcing." *Annals of the Association of American Geographers* 100 (4): 755–71.

Duncan, J. S. 2002. "Embodying Colonialism? Domination and Resistence in Nineteenth-Century Ceylonese Coffee Plantations." *Journal of Historical Geography* 28 (3): 317–38.

Eichen, J. R. 2020. "Cheapness and (Labor-)Power: The Role of Early Modern Brazilian Sugar Plantations in the Racializing Capitalocene." *Environment and Planning D: Society and Space* 38 (1): 35–52.

Ellis, E. C., J. O. Kaplan, D. Q. Fuller, S. Vavrus, K. Klein Goldewijk, and P. H. Verburg. 2013. "Used Planet: A Global History." *Proceedings of the National Academy of Sciences* 110 (20): 7978–85.

Ferdinand, M. 2022. Decolonial Ecology. Cambridge: Polity.

FNPOCELS. 1991. 'Principles of Environmental Justice'. First National People of Color Environmental Leadership Summit.

Godlewska, A., and N. Smith. 1994. *Geography and Empire*. Oxford: Blackwell.

Gregory, D. 2004. *The Colonial Present: Afghanistan. Palestine. Iraq*. Oxford: Blackwell.

Gregory, D., R. Johnston, G. Pratt, M. Watts, and S. Whatmore. 2009. *The Dictionary of Human Geography*. Hoboken: John Wiley & Sons, Incorporated.

Grove, R. H. 1995. *Green Imperialism: Colonial Expansion, Tropical Island Edens and the Origins of Environmentalism, 1600–1860*. Vol. 3. Cambridge: Cambridge University Press.

Hahn, B. 2020. *Technology in the Industrial Revolution*. Cambridge: Cambridge University Press.

Hamilton, C. 2015. "Getting the Anthropocene so Wrong." *The Anthropocene Review* 2 (2): 102–7.

Haraway, D. 2015. "Anthropocene, Capitalocene, Plantationocene, Chthulucene: Making Kin." *Environmental Humanities* 6: 159–65.

Haraway, D., Noboru, I. Gilbert Scott, K. Olwig, A. L. Tsing, and N. Bubandt. 2015. "Anthropologists Are Talking – About the Anthropocene." *Ethnos* 1844 (November): 1–30.

Harvey, D. 1981. "The Spatial Fix – Hegel, Von Thunen, and Marx." *Antipode* 13 (3): 1–12.

Haskett, R. S. 2005. *Visions of Paradise: Primordial Titles and Mesoamerican History in Cuernavaca*. Norman: University of Oklahoma Press.

Hofman, C. L., J. Ulloa Hung, E. H. Malatesta, J. S. Jean, T. Sonnemann, and M. Hoogland. 2018. "Indigenous Caribbean Perspectives: Archaeologies and Legacies of the First Colonised Region in the New World." *Antiquity* 92 (361): 200–216.

Inikori, J. E., and S. L. Engerman. 1992. *The Atlantic Slave Trade: Effects on Economies, Societies and Peoples in Africa, the Americas, and Europe*. Durham: Duke University Press.

Knitter, D., K. Augustin, E. Biniyaz, et al. 2019. "Geography and the Anthropocene: Critical Approaches Needed." *Progress in Physical Geography: Earth and Environment* 43 (3): 451–61.

Koch, A., C. Brierley, M. M. Maslin, and S. L. Lewis. 2019. "Earth System Impacts of the European Arrival and Great Dying in the Americas After 1492." *Quaternary Science Reviews* 207 13–36.

Kröger, M. 2014. "The Political Economy of Global Tree Plantation Expansion: A Review." *Journal of Peasant Studies* 41 (2): 235–61.

Lachenicht, S., L. Henneton, and Y. Lignereux. 2016. "Spiritual Geopolitics in the Early Modern Imperial Age. An Introduction." *Itinerario* 40 (2): 181–87.

Lachenicht, S. 2017. "Refugees and Refugee Protection in the Early Modern Period." *Journal of Refugee Studies* 30 (2): 261–81.

LaDuke, W., and D. Cowen. 2020. "Beyond Wiindigo Infrastructure." *South Atlantic Quarterly* 119 (2): 243–68.

Lester, A. 2015. "Commentary: New Directions for Historical Geographies of Colonialism." *New Zealand Geographer* 71 (3): 120–23.

Lewis, S. L, and M. A. Maslin. 2015. "Defining the Anthropocene." *Nature* 519 (7542): 171–80.

Lewis, S. L., and M. A. Maslin. 2018. *The Human Planet: How We Created the Anthropocene*. London: Penguin.

Luke, T. W. 2020. "Tracing Race, Ethnicity, and Civilization in the Anthropocene." *Environment and Planning D: Society and Space* 38 (1): 129–46.

Malm, A. 2018. "In Wildness Is the Liberation of the World: On Maroon Ecology and Partisan Nature." *Historical Materialism* 26 (3): 3–37.

McEwan, C. 2000. *Gender, Geography and Empire: Victorian Women Travellers in West Africa*. Aldershot: Ashgate.

McKittrick, K. 2011. "On Plantations, Prisons, and a Black Sense of Place." *Social & Cultural Geography* 12 (8): 947–63.

McKittrick, K. 2013. "Plantation Futures." *Small Axe: A Caribbean Journal of Criticism* 17 (3): 1–15.

McKittrick, K. 2021. *Dear Science and Other Stories*. Durham: Duke University Press.

Miller, Nicholas. 2025. *Plantation Knowledge: Agricultural Colonization, Exploitation, and Exchange Since 1500*. Albany: SUNY Press.

Moore, J. 2010. "This Lofty Mountain of Silver Could Conquer the Whole World": Potosí and the Political Ecology of Underdevelopment, 1545–1800." *Journal of Philosophical Economics* IV (1): 58–103.

Moore, J. 2019. "The Capitalocene and Planetary Justice." *Maize* 6 (July): 49–54.

Moore, J. W. 2002. "The Crisis of Feudalism: An Environmental History." *Organization & Environment* 15 (3): 301–22.

Moore, J. W. 2009. "Madeira, Sugar, and the Conquest of Nature in the "First" Sixteenth Century: Part I: From "Island of Timber" to Sugar Revolution, 1420–1506." *Review (Fernand Braudel Center)* 32 (4): 345–90.

Moore, J. W. 2015. *Capitalism in the Web of Life: Ecology and the Accumulation of Capital*. London: Verso.

Moore, J. W. 2017a. "The Capitalocene, Part I: On the Nature and Origins of Our Ecological Crisis." *The Journal of Peasant Studies* 44 (3): 594–630.

Moore, J. W. 2017b. "Confronting the Popular Anthropocene: Toward an Ecology of Hope." *New Geographies* 9: 194–99.

Moore, J. W. 2017c. "World Accumulation and Planetary Life, or, Why Capitalism Will Not Survive Until the "Last Tree Is Cut." *IPPR Progressive Review* 24 (3): 175–202.

Moore, J. W. 2018. "The Capitalocene Part II: Accumulation by Appropriation and the Centrality of Unpaid Work/Energy." *Journal of Peasant Studies* 45 (2): 237–79.

Moore, J. W. 2022. "Anthropocene, Capitalocene & the Flight from World History: Dialectical Universalism & the Geographies of Class Power in the Capitalist World-Ecology, 1492–2022." *Nordia Geographical Publications* 51 (2): 2.

Moulton, A. A. 2023. "Towards the Arboreal Side-Effects of Marronage: Black Geographies and Ecologies of the Jamaican Forest." *Environment and Planning E: Nature and Space* 6 (1): 3–23.

Patel, R., and J. W. Moore. 2018. *A History of the World in Seven Cheap Things: A Guide to Capitalism, Nature, and the Future of the Planet*. London: Verso.

Pawson, E. 1990. 'British Expansion Overseas c. 1730–1914'. In *Historical Geography of England and Wales*, 521–44. London: Elsevier.

Pomeranz, K. 2021. *The Great Divergence: China, Europe, and the Making of the Modern World Economy*. Princeton: Princeton University Press.

Pulido, L. 2017. "Geographies of Race and Ethnicity II: Environmental Racism, Racial Capitalism and State-Sanctioned Violence." *Progress in Human Geography* 41 (4): 524–33.

Purifoy, D. 2023. "Black Towns and (Legal) Marronage." *Annals of the American Association of Geographers* 113 (7): 1–16.

Roberts, P., N. Boivin, and J. O. Kaplan. 2018. "Finding the Anthropocene in Tropical Forests." *Anthropocene* 23 (September): 5–16.

Rubino, M., D. M. Etheridge, C. M. Trudinger, et al. 2016. "Low Atmospheric CO2 Levels During the Little Ice Age Due to Cooling-Induced Terrestrial Uptake." *Nature Geoscience* 9 (9): 9.

Ruddiman, W. F., M. C. Crucifix, and F. A. Oldfield. 2011. "The Early-Anthropocene Hypothesis." *Holocene* 21 (5): 713–887.

Ruddiman, W. F., F. He, S. J. Vavrus, and J. E. Kutzbach. 2020. "The Early Anthropogenic Hypothesis: A Review." *Quaternary Science Reviews* 240 (July): 106386.

Ruddiman, W. 2017. "Geographic Evidence of the Early Anthropogenic Hypothesis." *Anthropocene* 20 (December): 4–14.

Saldanha, A. 2020. "A Date With Destiny: Racial Capitalism and the Beginnings of the Anthropocene." *Environment and Planning D: Society and Space* 38 (1): 12–34.

Steffen, W., K. Richardson, J. Rockström, et al. 2015. "Planetary Boundaries: Guiding Human Development on a Changing Planet." *Science* 347 (6223): 1259855.

Thomas, J. A., ed. 2022. *Altered Earth: Getting the Anthropocene Right*. Cambridge: Cambridge University Press.

Todd, Z. 2015. 'Indigenizing the Anthropocene'. In *Art in the Anthropocene: Encounters Among Aesthetics, Politics, Environments and Epistemologies*, edited by Heather Davis and Etienne Turpin, 241–54. London: Open Humanites Press.

Tsing, A., J. Deger, S. Keleman Alder, and F. Zhou. 2020. *Feral Atlas: The More-Than-Human Anthropocene*. Stanford: Stanford University Press.

Tsing, A. L. 2015. *The Mushroom at the End of the World: On the Possibility of Life in Capitalist Ruins*. Princeton: Princeton University Press.

Vergès, F. 2017. 'Racial Capitalocene'. In *Futures of Black Radicalism*, edited by Gaye Theresa Johnson and Alex Lubin, 72–82. London; New York: Verso.

Vickers, M. P. 2023. "On Swampification: Black Ecologies, Moral Geographies, and Racialized Swampland Destruction." *Annals of the American Association of Geographers* 113 (0): 1674–81.

Wallerstein, I. M. 2004. *World-Systems Analysis: An Introduction*. With Library Genesis. Durham: Duke University Press.

Whyte, K., J. L. Talley, and J. D. Gibson. 2019. "Indigenous Mobility Traditions, Colonialism, and the Anthropocene." *Mobilities* 14 (3): 319–35.

Whyte, K. P. 2018. "Indigenous Science (Fiction) for the Anthropocene: Ancestral Dystopias and Fantasies of Climate Change Crises." *Environment and Planning E: Nature and Space* 1 (1–2): 224–42.

Williams, E. 1944. *Capitalism and Slavery*. London: Penguin UK.

Williams, E. 2020. *From Columbus to Castro: The History of the Caribbean*. Morrisville: Lulu Press, Inc.

Winston, C. 2021. "Maroon Geographies." *Annals of the American Association of Geographers* 111 (7): 1–15.

Wolford, Wendy. 2021. "The Plantationocene: A Lusotropical Contribution to the Theory." *Annals of the American Association of Geographers* 111 (6): 1622–39.

Woods, C. 2017. *Development Arrested: The Blues and Plantation Power in the Mississippi Delta*. London: Verso.

Woods, D. 2014. "Scale Critique for the Anthropocene." *The Minnesota Review* 2014 (83): 133–42.

World Rainforest Movement. 2007. *Brazil: Why Landless Women Are Opposed to Monoculture Eucalyptus Plantations*. Montevideo: World Rainforest Movement.

Wright, W. J. 2020. "The Morphology of Marronage." *Annals of the American Association of Geographers* 110 (4): 1134–49.

Wynter, S. 2003. "Unsettling the Coloniality of Being/Power/Truth/Freedom: Towards the Human, After Man, Its Overrepresentation–An Argument." *CR: The New Centennial Review* 3 (3): 257–337.

Yusoff, K. 2018. *A Billion Black Anthropocenes or None*. Minneapolis: University of Minnesota Press.

Yusoff, K. 2024. *Geologic Life: Inhuman Intimacies and the Geophysics of Race*. Durham: Duke University Press.

Yusoff, K., S. Mutegi, K. Karanja, and O. Hopkins. 2025. "Planetary Vernaculars." *E-Flux*: https://www.e-flux.com/architecture/insurgent-geologies/667529/planetary-vernaculars

Zalasiewicz, J., C. N. Waters, A. D. Barnosky, et al. 2015. "Colonization of the Americas, "Little Ice Age" Climate, and Bombproduced Carbon: Their Role in Defining the Anthropocene." *Anthropocene Review* 2 (2): 117–27.

4 The Industrial Anthropocene

Introduction

In the previous chapter, we explored arguments that the Anthropocene began with the European colonisation of the Americas. These arguments take two forms: (1) that the demographic and environmental impacts of that colonisation were so great, that they represent the first instance of humanity's actions having global environmental consequences; and (2) that the emergence of a new capitalist 'world-ecology' through colonialism marked the first stirrings of a new way of using 'nature', and of a new way of organising life, society and economies through which human impacts on the environment would intensify over the ensuing centuries. The first argument is largely biophysical, and includes a proposed place in the geological record for the 'Golden Spike' marking the Anthropocene's onset (1610 – see Lewis and Maslin 2015). The second argument is more socio-economic and is concerned less with identifying precise thresholds and more with identifying precursors to what came later, and with extending backwards our historical sense of when, where, and how the 'modern world' started to emerge (Saldanha 2020).

While the second argument seems to have convinced many Human Geographers, the first argument doesn't seem to have succeeded in convincing the keepers of the geological record-books. In his paper 'Getting the Anthropocene so wrong', Clive Hamilton argues that Lewis and Maslin completely misunderstood the point of the Anthropocene. He argues that it's not about spotting sporadic anthropogenic impacts on the global environment (like the 1610 dip in atmospheric CO_2) but about identifying the start of *sustained* transformations of the Earth System, which take it beyond the range of natural variability. For Hamilton, the Industrial Revolution is a much stronger candidate, as it 'marked the beginning of the burning of coal on an industrial scale, which had a discernible effect on the functioning of the Earth System', and which will continue to do so long into the future (Hamilton 2015, 104).

DOI: 10.4324/9781003162469-6

The Industrial Revolution has long loomed large in discussions of the Anthropocene. In the first paragraph of Paul Crutzen's formative 'Geology of Mankind' article, he argued that

> The Anthropocene could be said to have started in the latter part of the eighteenth century, when analyses of air trapped in polar ice showed the beginning of growing global concentrations of carbon dioxide and methane. This date also happens to coincide with James Watt's design of the steam engine in 1784.
>
> (Crutzen 2002, 23)

The invention of the steam engine has commonly been taken as a key inflection point, with the burning of fossil fuels (initially coal) representing a radically new phase of the 'mastery of fire by our ancestors' (Steffen, Crutzen, and McNeill 2007, 614). For Earth System Scientists, the embrace of fossil fuels helped humanity move beyond the 'bottleneck' of reliance on the energy available in plants, wind, and water. Fossil fuels heralded 'an era of far looser constraints upon energy supply, upon human numbers, and upon the global economy', and 'opened an era of intensified and ever-mounting human influence on the Earth System' (Steffen, Crutzen, and McNeill 2007, 616). For Will Steffen and colleagues, atmospheric CO_2 concentrations were the key marker of the move into the Anthropocene, and they proposed that the dawn of the Industrial Revolution represented the start of the Anthropocene's first stage, with a second stage of intensified impacts beginning in the mid-twentieth century (Steffen, Crutzen, and McNeill 2007; see Chapter 5).

This Earth System Science emphasis on human perturbations of the global atmospheric system can be contrasted to those hewing closer to the Physical Geography tradition of identifying human impacts on landscapes and ecosystems and their variations across time and space (Ellis 2017). In such conversations, the Industrial Revolution has been conversely identified as the 'Late Anthropocene' (Glikson 2013), a culmination of millennia of human transformations of physical environments whose origins can be traced to the rise of agriculture or to early humans working out how to set fire to things (Davis, Hatch, and Hoare 2025; Goudie 2020).

So, whether the Industrial Revolution is conceived of as a 'big bang' moment when everything changed, and humanity started cooking the climate with its new fiery machines, or conversely as the outcome of a slow, steady process of getting to grips with the potentials of combustion, there's little doubt that it represents a period of profound remaking of human-environment relations. However, around 2010, the Industrial Revolution began to fall out of favour as a candidate for the formal onset of the Anthropocene. Why? On the one hand, the physical geography of the Industrial Revolution's environmental (and geological) impacts – recorded for example in fly ash deposits from coal burning – were found by stratigraphers to be too 'diachronous', i.e.

varying greatly in age by place. That tells a story of how fossil-fuelled industrialism started in a very small corner of the world and then spread gradually, with ecological and geomorphological impacts following in its wake (Goudie and Viles 2016). Even atmospheric CO_2 concentrations show a relatively gradual uptick as the Industrial Revolution got going (see Figure 5.1). This was not quite the globally synchronous signal that stratigraphers needed to see to declare a new epoch. On the other hand, human geographers like Jason Moore (2022) worried that both Anthropocene scientists and some historians tended to 'fetishize' the Industrial Revolution – to see it as a spontaneous outburst of innovation and invention, rather than understanding it as part of deep-seated world-historical transformations stretching over centuries.

As a result, physical scientists have pulled the onset of the Anthropocene into more recent times, specifically the mid-twentieth century (McCarthy et al. 2025); meanwhile, many historians, geographers, and other social scientists have pushed it further back in time, seeing the Industrial Revolution as inextricable from the kinds of processes we examined in Chapter 3. So why do we still need to talk about the Industrial Revolution?

The social, demographic, ecological, and Earth System changes that defined the Industrial Revolution, and which it set in train, are undeniable, even if their residues in the rocks don't quite satisfy the stringent stratigraphical requirements of a new geological epoch. And with the Anthropocene increasingly being thought about as an 'event' rather than an 'epoch', and therefore as a *process* unfolding unevenly across space and time (Lähde 2025), the period of the Industrial Revolution remains central to understanding the shape of the present. The historical geographies of the Industrial Revolution can give us a clearer picture of patterns and processes of change, of chains of causation and, crucially, can inform conversations about historical responsibilities and their significance for ensuring just and fair responses to the Anthropocene in the present (Lewis and Maslin 2015).

So, in this chapter, we ask: how and why did the Industrial Revolution happen? Was it driven by clever inventions, like the steam engine, or were there other geographical transformations at work? This points towards a rather well-worn, geographical 'why-of-where?' question beloved by television documentaries and school textbooks: why did the Industrial Revolution happen in Britain? Commonly, answers to this question speak of important local factors, and of 'geographical advantages' that the British Isles enjoyed over economic rivals: abundant coal in exploitable reserves near population centres; agricultural surpluses to feed a growing population; well-developed markets, high wages, and legal regimes to support commercial expansion (Allen 2011). However, in the discussion below, we'll show how these factors aren't enough to explain the geography of the Industrial Revolution. The debate about the onset of the Anthropocene has prompted a re-appraisal of how and why the move to fossil fuels happened. It's also helped re-cast the Industrial Revolution as being a thoroughly *global* story – not in the sense that it happened everywhere at once, but rather that explaining the onset of the Industrial Revolution requires understanding global patterns of trade, commodity flows, and divisions of labour. All these global relationships were, in turn, shaped by the enduring and evolving colonial relationships whose origins we explored in the previous chapter.

It's also important to acknowledge at the outset that the Industrial Revolution and the shift to fossil fuels were not the same thing. The industrialisation of manufacturing began to happen long before steam engines entered the scene, and for a long period in the early nineteenth century, coal and renewable power sources like water jostled for position within the industrial manufacturing economy of northern England. Explaining the rise of a fossil fuel-based energy regime is therefore a related but slightly different task to explaining the onset of industrialisation. Here, energy historians (e.g. Malm 2016a) have done important work in showing how the widespread adoption of coal was not driven solely by economic calculations or by the convenient availability of coal-burning machines; rather, to understand the emergence of a 'fossil economy', we need to understand the politics of the factory, and of factory cities such as Manchester in the UK, and the role that coal and steam played in global conflicts between workers and bosses, labour and capital.

The chapter therefore proceeds as follows: first, we'll explore the global connections which made the Industrial Revolution possible, picking up where we left off in the previous chapter. We'll then investigate how the human 'mastery of fire' embodied in steam engines was also about mastering a restive and growing working class. But we'll also argue that focusing on machines and fuel sources, while highly relevant to tracking the geological impacts of human activities, can also obscure the wider cultural transformations that were going on in this period, for example in the emergence of a new consumerism which was both a product and a driver of industrialisation. Another cultural-political phenomenon is of great relevance too: the worldwide resistance that industrialisation and the fossil economy faced, from disenfranchised artisans, starving rural labourers, and proto-environmentalists.

The overall picture which we wish to paint is one of an industrial Anthropocene which was highly diachronous, to borrow a geological term: it exhibits great geographical and historical texture, shaped by geographies of colonialism, social unrest, political resistance, and more. Attending to these geographies, we'll conclude, is vital for thinking about the Industrial Revolution's environmental legacies and about the distribution of responsibility for addressing them in the present. The Industrial Revolution and its environmental consequences were not the inevitable outcomes of the sudden appearance of steam engines in northern England towards the end of the eighteenth century. Rather, they were products of political and economic choices made by powerful actors, which could have been made differently. In facing up to the Anthropocene today, knowing these histories and geographies can help us all to recognise that the future of the Anthropocene is also not defined by inevitability; it too is a matter of human choice.

The global geographies of the Industrial Revolution

One day in 1784, an American ship deposited eight bags of cotton on the Liverpool dockside that so vexed the customs officials that they were immediately seized for quarantine. Surely, the officials reasoned, the cotton must have come from the West Indies, from Britain's colonial plantations in the Caribbean Sea, and therefore any trans-shipment by a vessel from America – where little cotton was being grown at the time – on their journey to the Mother Country would have been illegal, according to the mercantilist trading laws that governed trade between metropole and colony (see Chapter 3). By the latter part of the eighteenth-century Liverpool had become a prominent importer of cotton from the West Indies, the chief source of raw material for the growing spinning and weaving industries in northern England (Hall 2001). The customs officials could not believe that 'so large a quantity of cotton as eight bags could have been raised in the country' now known as the United States of America, and thus the bags languished in quarantine while their origin was confirmed, before languishing a while longer in a warehouse while a buyer could be found for their slightly sub-standard contents (Ellison 1886, 82). But these humble eight bags of second-rate cotton heralded an important shift by which, over the next few decades, 'industrialization in Britain and cotton slavery in the United States grew up together' (Hahn 2020, 105).

This process of industrialisation was a relatively long and slow one, which started long before the period we tend to dub the 'Industrial Revolution'. This was most notable in the textile trade in northern England. Historian Barbara Hahn defines industrialisation as consisting of three elements:

1 the mechanisation of tasks which were previously done by hand;
2 the geographical separation of production from consumption – 'the people who use goods are not the ones who made them', and production processes no longer taking place in the home; and
3 the regularisation and standardisation of the flow of physical objects, both in the form of raw materials and finished products: raw materials of consistent quality flow into factories at a steady pace, and mass-produced products come out the other end (Hahn 2020, 4).

Each of these three elements of industrialisation represents profound geographical transformations. Mechanisation saw a shift from artisanal craftwork to the repetitive work and machine-minding of the factory, bringing workers into a new relationship both with technology and with their bosses, and generating new experiences of space and time in the setting of the factory. The separation of production and consumption relegated the economic importance of work performed in the home, spawned new urban forms and industrial landscapes, and helped create new 'cultures of consumption' through which people related to the material world in new ways. And the regularisation and standardisation of material flows created new global divisions of labour, transformed trade networks and commodity flows, and remade environments and societies across the globe (Riello 2015).

Let's take this latter argument as our starting point for understanding the global geographies of the Industrial Revolution, before we zoom into these transformations of the spaces and practices of work and consumption.

The empire of cotton

The British East India Company (EIC) was initially founded to monopolise trade in spices – in fact, the company's first dividends to investors were paid in pepper – but by the seventeenth century, the EIC was heavily involved in the textiles trade, bringing intricately woven and patterned cloths from India to markets in Europe, Africa, and the New World. Ready buyers were available in Britain, but until the mid-eighteenth century, the brightly coloured fabrics were favoured for home furnishing and decoration rather than for wear. More receptive markets were available among North American and Caribbean colonists, however, as well as among slave traders along the West African coast. The Portuguese had been trading Indian cloth for African slaves since the 1580s, and were soon joined by Dutch, French, and British traders who were sponsored by their governments or organised into chartered companies like the EIC (Hahn 2020, 42). By the middle of the eighteenth century, it's estimated that the markets that rimmed the Atlantic Ocean accounted for somewhere between 80 and 94% of Britain's cotton exports, while cloth sold in Africa represented around 68% of Britain's exports across the eighteenth century. Much of that cloth was British-made wool, but around 40% of it was Indian-made cotton (Hahn 2020; Inikori 2002).

During this same period, the textile-producing industries of northern England had been maturing, integrating new machines into the production process and increasing the quality and consistency of finished wool, linen, and cotton-blend fabrics. The growing taste for colourful Indian fabrics therefore posed a threat to England's clothiers, who successfully lobbied parliament for a series of prohibitions against the wearing of imported, printed cloth. Plain fabrics could nonetheless still be imported from India, helping the English printing and dying industries to mature, while cloth manufacturers sought out new techniques of production, marketing, and distribution to compete with the imported cloth.

One such strategy was the spinning and weaving of cotton in Britain itself, using raw cotton imported from elsewhere. This is where individuals like Richard Arkwright have been given a prominent place in the story of the Industrial Revolution, as inventors of new machines and systems of production which made manufacturing processes more efficient (Arkwright is credited with building the first modern factory in Derbyshire in 1771). But while devices like Arkwright's new water frame for spinning yarn were no doubt crucial for the mechanisation and industrialisation of textile manufacturing, we must be careful not to ascribe too much agency to machines in processes of socioeconomic change. Barbara Hahn argues that Arkwright was not so much the 'heroic inventor' as a 'heterogeneous engineer' (Law 1989), who cleverly tinkered not only with machines but also with patents, rules, and regulations in order to remake the world around him such that his investments in his new machines would prove profitable. Most notably, Arkwright successfully lobbied parliament for the repeal of the Calico Act, which in 1701 had stimulated British printing by banning the importation of printed Indian cloth, but which in 1721 had also banned any printed, all-cotton cloth, even that made in Britain, from sale at home. The 1774 repeal meant that the production of all-cotton cloth was now both legal and a good economic prospect. The material was lighter than linen and fustian (a cotton and linen blend) and took bright colours well. Consumers loved it, as did the government, for whom it represented a decisive new stream of tax revenue (Riello 2015). Subsequent import tariffs on Indian-made cloths gave British cotton a crucial competitive advantage, and Arkwright profited handsomely both from the sale of his yarn and cloth and from licensing his machines and manufacturing process to others keen to invest capital into this new mode of production. The age of industrial capitalism was underway, born of the protectionist trade policies of mercantilism.

The repeal of the Calico Act coincided with the American War of Independence, a war whose outcome would fundamentally shift the geography of the cotton industry (Quark 2008). But it wasn't until the 1800s that the independent United States fully embraced cotton, which explains the dockside confusion over the origin of those eight bags of cotton in 1784. But the steady southern and westward expansion of the United States – the purchase of Louisiana Territory from France in 1803, the ceding of Spanish Florida and Gulf Coast territories in 1813 and 1819, and the later annexation of Texas in 1845 – opened a new restless frontier of plantation agriculture. Pre-independence planting had focused on crops like tobacco and rice, but the growing demand for raw cotton in the textile mills of northern England and, later, the northern states of the United States meant that plantation owners spied an opportunity to produce a durable, high-yielding cash crop with a global market, aided in no small part by Eli Whitney's 1793 breakthrough cotton gin, which provided huge boosts to productivity (Moore 2023) (Figure 4.1).

The United States also had the legal regime in place to support a system of production that would keep input costs low and profits high – chattel slavery. As the slave states of the Mississippi valley entered the Union, slaveholders were positively tripping over each other to seize land and to send groups of slaves to clear old-growth forests and cypress swamps and then plant cotton.

Figure 4.1 Eli Whitney's hand-powered cotton gin – 'gin' is a derivation of 'engine'. The machine greatly sped up the separation of cotton seeds from fibre. With the machines, plantation owners sought to grow bigger crops and therefore acquired greater numbers of slaves, helping to strengthen the institution of slavery in the American South in the early decades of the nineteenth century. (Source: Wikimedia Commons.)

After 10 or 15 years of planting, the soil would become exhausted, and plantations would move on (Riello 2015). Here, then, is another example of the kind of restive commodity frontier explored in Chapter 3, defined by the violent appropriation of both land and labour, the rapid exhaustion of resources (McNeill 2013), and a 'profound geographical restlessness' (Moore 2010, 93). 'King Cotton' quickly became crucial to the U.S. economy, and by 1860 cotton formed 60% of the country's exports, with most of it going to Britain, whose imports of raw cotton had increased one-hundred-fold since the 1770s.

While the plantation system had developed 'in mercantilist times' (Hahn 2020, 106), and evolved out of earlier household and feudal models of agricultural production, it was also a key geographical underpinning of industrialisation and modernity (Mintz 1985) – hence why some prefer to think of the

Anthropocene as the 'Plantationocene' (see Chapter 3). As Barbara Hahn puts it, 'the large scale of production, the division of labour, and the regularisation of tasks and techniques made plantation agriculture a precursor to industrial manufacturing' (Hahn 2020, 43). In this reading, plantations provided a model for the spatial organisation of work which would later find expression in the factories of northern England. The intense *profitability* of the slave trade and New World plantation agriculture has also been cited as a key driver of both industrialisation and financialisation in England (Berg and Hudson 2023). The historian and first Prime Minister of independent Trinidad and Tobago, Eric Williams, argued in his book *Capitalism and Slavery* that the financial assets generated by slave-trading and colonial plantations formed a key part of the capital which was invested in early industrialisation in England (Williams 1944] 2022).[1]

Some slave-traders enjoyed profit margins of 50% (Inikori 2002), but historians have struggled to come to any agreement on how significant these profits were for the new wave of investments being made in industrial mechanisation. It is possible to trace up to 'forty per cent of British commercial and industrial investment after 1750' to these slavery and plantation profits (Bonneuil and Fressoz 2016, 232), while recent calculations have suggested that the trading of slaves and the products of the American plantations comprised about 11% of British GDP by the early nineteenth century (Rönnbäck 2018). Contrary voices have argued that there were plenty of other sources of capital available to those wishing to invest in factories and machinery in northern England, and that early mechanisation did not require much capital investment anyway (Burnard and Riello 2020; O'Brien 1982).

Ghosts in the machine

Profits from slavery may therefore have been significant as sources of new capital investment, but not necessarily decisive. The economic historian Kenneth Pomeranz, in his influential thesis on the 'Great Divergence' between Europe and East and South Asia from 1800 onwards, argues that a more decisively influential factor was how colonial plantations allowed Britain to escape the ecological limits of the British Isles (Pomeranz 2021). The tripling of Britain's population, from around three to nine million between 1500 and 1800, coincided with a huge increase in domestic agricultural productivity thanks to new farming systems and technologies (Bonneuil and Fressoz 2016; Schandl and Krausmann 2007). But it also coincided with the accumulation of millions of what Pomeranz calls 'ghost acres' – those vast swathes of colonised land growing commodities for export to the mother country. The notion of 'ghost acres' was first introduced by geographer and biologist Georg Borgström in his 1965 book *The Hungry Planet*, a neo-Malthusian analysis of the world food trade and the dangerous dependence of many countries on food imports (Borgström 1965). Pomeranz developed the concept in his historical analysis to show that by 1831, plantation-grown sugar consumed in the UK represented the calorific equivalent of 2.6 million acres of British cereal land, while cotton from North America generated the textile equivalent of 23 million acres of sheep pasture, a figure which 'surpasses Britain's total crop and pasture land combined'. With North American timber added into the mix, we arrive at a total of around 25–30 million ghost acres which were feeding land-intensive raw materials into Britain, largely in exchange for non-land-intensive industrial products like cotton cloth (Pomeranz 2021, 276).

Alf Hornborg (2006) likewise analysed the labour time embodied in imports like raw cotton, alongside the embodied land. For example, the 1850 import of 223,623 tons of American raw cotton brought with it over 2.7 million acres of embodied land and over 616 million hours of labour, most of it by slaves. He argues therefore that British industrialisation did not just result in 'time-space compression' through new forms of transport and global patterns of trade (Harvey 1996), but functioned through 'time-space appropriation', as industrial efficiency gains in one place (northern England) were made at the expense of (human) time and (physical) space in the peripheries of the world-economy.

Geographers Jim Clifford and Stéphane Castonguay (2022) expand the ghost acres concept to account not just for the land surface used to produce commodities, but also the ecological changes inflicted upon – in their case – the timber frontier in nineteenth-century Canada. They find that between 1827 and 1900, replacing the timber imports to Britain solely from the St Laurence River valley in Canada 'would have required fully harvesting all the available forests in the British Isles' (Clifford and Castonguay 2022, 134). But beyond such abstract descriptions of 'extraterritorial extraction of natural resources', they also analyse the ecological and social impacts of the export of vast amounts of timber to Great Britain (Clifford and Castonguay 2022, 137). They detail the ecological transformations wrought upon forested areas as demand for different tree species waxed and waned (e.g. as demand shifted over time from shipbuilding to railway timbers and then to housebuilding), and document how the advancing timber frontier was central to the steady infringement of the sovereign lands of Indigenous peoples. In so doing, Clifford and Castonguay draw analytical attention away from the metropole and towards the 'colonial peripheries' and, with echoes of Jason Moore's work on the ecological crises of early colonial commodity frontiers (Chapter 3), offer a way of understanding how 'local' environmental changes resulted from industrialisation and commercialisation in distant corners of the world.

It is to these global dynamics of the world-ecology of British industrialism that environmental historians Bonneuil and Fressoz (2016, 234) refer when they conclude that the industrial revolution in Britain would have been 'physically impossible' without the British Empire, and without the later American system of slave-plantation agriculture. Pomeranz's study compared England and western Europe with Japan and the Yangtze delta region in China – regions which until the eighteenth century had comparable levels of population density, agricultural productivity, and commercial markets supported by mature financial institutions (Parthasarathi 2014). For Pomeranz, 'the exploitation of the New World, and of the Africans taken there to work...did more to differentiate western Europe from other Old World cores than any of the supposed advantages' that Europe held over its Asian rivals (Pomeranz 2021, 282–83). If a gradually industrialising Britain and western Europe were experiencing a socioecological bottleneck around 1800 (Steffen, Crutzen, and McNeill 2007), overcoming it was aided as much by colonialism as it was by the turn to fossil fuels. The resources, food, and capital which flowed into Britain didn't lead to the invention of the steam engine, but they did create the conditions by which emerging technologies of mechanisation were able to rapidly transform production processes in sectors like the textile industry.

Viewed this way, the Industrial Revolution can be understood not as a 'result of individual genius or institutions', but as an 'epiphenomenal effect of changing patterns of world trade' (Hahn 2020, 107). These changing patterns included the deepening of economic divisions between the global economic core and its peripheries. For example, through the post-Columbus depopulation of the Caribbean and the subsequent repopulation of the region with slaves, the region became:

> a perversely large market for imports [such as food and clothing for both slaves and settlers] and a source of land-intensive exports. In fact, it became the first periphery to assume a now familiar "Third World" profile: that of a large importer of both capital goods [including slaves] and manufactured goods for daily use, with exports that kept falling in price as production became more efficient, capital intensive, and widespread.
>
> (Pomeranz 2021, 267)

Back across the Atlantic, mechanisation in northern England was initially pursued as a response to Indian imports, while industrial expansion was fuelled by the steady supply of slave-grown Caribbean and then U.S. cotton. Later, when the U.S. Civil War of the 1860s put a brake on cotton exports, and once the British Raj had been established in India, a coercive transformation of parts of the Indian economy was pursued, from a position of proto-industrial cloth production to a focus on exporting raw cotton and importing finished cloth from Lancashire (Riello 2015). These shifting geographies of global trade, and of circulating capital, cloth, and enslaved humans, are crucial to understanding the 'why-of-where?' question of the Industrial Revolution. They show an emerging industrial capitalism embodying a new world-ecology which expanded and deepened global divisions of labour, and produced new global inequalities of resource extraction, with all its environmental consequences. And a great deal of this happened without a whiff of coal smoke. But come the early part of the nineteenth century, that was about to change too.

Fossil capital: power and the people

For Pomeranz, two things made the industrial revolution happen in Britain and caused the 'Great Divergence' from South and East Asia: the bounties of the New World, and coal. Britain had an advantage here in that its coal deposits were located conveniently close to the places where mechanisation in the textile industry had started to get underway; in the Yangtze Delta, by contrast, the proto-industrial and commercial centres were located some distance from exploitable coal reserves. The knowledge of British miners and geologists in the eighteenth and early nineteenth centuries was also such that they didn't fully realise just how much coal there was to exploit, so there wasn't the same orgy of consumption followed by rapid exhaustion as there had been with other commodities during the age of colonial expansion and mercantile proto-capitalism (Pomeranz 2021). But while we tend to associate the industrial revolution with the burning of coal, it didn't *start* with coal. As mentioned above, Arkwright's first factory was water-powered, as were most of the textile mills which followed. Indeed, it wasn't until the 1830s, 60 years after Arkwright opened his first factory, that the majority of textile mills had turned to steam power produced by coal-burning steam engines (Lawton and Pooley 2021). Understanding this move to fossil fuels, which was hugely consequential for the British and then for the global economy, and is one of the 'detonators' of the Anthropocene (Tsing et al. 2020), requires digging into the politics of mechanisation and industrialisation as they played out in the factories and streets of northern England.

If we go back to Arkwright's first factory in Cromford, Derybshire in the 1770s, we find a skilled workforce operating cutting-edge machinery, working round the clock to fulfil a global order book. Profit margins were slight but investing in machinery promised production at such a scale, and with such standardisation of product, that much money was to be made. Three hundred similar mills had appeared across northern Britain by 1800. They were key nodes in an emerging industrial capitalism, and a world-ecology predicated on the exploitation of distant ecologies and unfree labouring bodies. Arkwright's factory was also 'a key moment on the road to the environmental problems of ever-rising production and consumption we see in the twenty-first century' (Lewis and Maslin 2018, 195). Nonetheless, these factories might pass some twenty-first-century 'sustainability' checks, powered as they were by renewable energy. Mills were located next to fast-running streams, where the bountiful energy of flowing water could be expropriated for free. They invite an exercise in counterfactual imagination – could the industrial revolution have continued on the basis of renewable energy, generating wealth and raising living standards (at least for populations in the economic core regions) but without the environmental costs of which we are now so painfully aware?

Most analysts would agree that the scale of subsequent industrial expansion, from the 1830s onwards, was only truly possible with the insertion of fossil energy into the equation. Nonetheless, that wasn't apparent to factory owners at the time, like Richard Arkwright. While he was rolling out his factory model, and licensing others to do the same, coal-fired steam engines were being developed, and finding application in things like pumping water out of mines and grinding corn. Why did it take so many decades for the textile industry to adopt this supposedly transformative, epochal technology? Or, viewed another way, why would textile manufacturers ever abandon their free and abundant waterpower in favour of expensive, heavy, and dirty coal?

Industrialisation, as defined by Barbara Hahn (see above), 'took place well before the steam engine was economically important' (Hahn 2020, 99). But industrialisation based on renewable energy like waterpower had its problems. River flows could be intermittent – a big problem when you had fine margins and a global order book. This could lead to long hours for factory workers – 69-hour working weeks were common, more when the water stopped flowing (Lewis and Maslin 2018). The regularity of steam power, which could be turned on and off at will, was a big draw, as was the geographical freedom it offered new entrants to the textile industry – new factories need not be sited by fast-running water, just anywhere coal could be accessed relatively cheaply, something which was helped by Britain's fast-expanding canal network (Lawton and Pooley 2021). Nonetheless, coal was still relatively expensive in the early decades of the nineteenth century, and steam engines were inefficient – their efficiency could be as low as 4% when comparing energy inputs to outputs, while waterwheels exhibited efficiency of around 85% (Hahn 2020). So at this point in time, there was nothing inherently superior about steam power, and even James Watt himself was sceptical about its application in the textile industry (Gregory 1990). Steam had its advantages and disadvantages, but its gradual adoption cannot be explained solely by economic rationality or technological advancement. Much more was at play.

In 1824, trade unions were legalised by the UK Parliament. The following year saw a serious financial crisis, as exuberant lending by banks and speculative investments by private individuals (including in government bonds and land grants from the fictional country of Poyais in Central America, an invention of the Scottish soldier and con-man Gregor MacGregor) saw the economic boomtime of the Napoleonic Wars come to end, with the ironic rescue of the Bank of England by the Banque de France. While the causes of the crisis were external to the cotton industry, the increasing reliance of the industry on long-term capital investments (e.g. in factories and machinery) made it vulnerable to financial regression. While textile manufacturers during the 'Arkwright era' saw profits of up to 50%, they now plunged below 5%, and bankruptcies swept the industry. Drought hit the country in 1826, leading some mill owners, next to desiccated streams, to turn to steam power. Prices of domestically-produced food rocketed, while the Corn Laws – a mercantilist protection of domestic agriculture – maintained steep tariffs on imported grain, meaning bread and other food staples took a growing chunk out of workers' wages.

Cotton spinners in Lancashire walked out on strike, demanding higher pay to put bread on the table. Previous strikes had been met with factory owners simply hiring new workers, including migrants from Ireland. This time, however, the owners turned to machines to replace the striking workers. Richard H. Roberts' 'self-acting mule' would be a crucial element of the deepening mechanisation of the cotton industry, and of the deepening reliance on coal (see Figure 4.2). Later known as the 'Iron Man', Roberts' invention allowed large spinning mules to move automatically, thus mechanising one of the last

Figure 4.2 A self-acting mule of the type devised by Richard Roberts. The overhead belts transfer the power from the steam engine. The new design meant the mules could increasingly be minded by women on lower wages than male employees, while even cheaper children performed the dangerous work of sweeping up underneath. (Source: Wikimedia Commons.)

factory tasks requiring brute-force bodily strength. Designed to work with steam engines, over the next decade, Roberts' invention was used to replace striking workers from Preston to Glasgow, where hundreds lost their jobs, to be replaced by fewer operatives trained solely to operate the new automated and larger machines. By the 1840s, 'mule spinners' wages were half what they had been when they revolted in 1826', but the automatic mules required 60% more energy than previous models (Hahn 2020, 144). Waterfalls were still providing cheaper energy than coal in 1849, but expensive coal was increasingly attractive over a large, restive workforce, and steam-powered mechanisation raced through the cotton industry (Malm 2016a). As Hahn (2020, 150) summarises it:

> Steam power provided by the fossil fuel, coal, became one of the defining characteristics of the industrial age and the capitalist system partly because a few thousand mule spinners in Lancashire wanted higher wages. Bigger capital investments and higher operating costs replaced truculent labour.

So, the move to fossil fuels was as much about struggles between labour and capital as it was about efficiency or optimising operating costs (Malm 2013). Conventional Marxist analyses have long argued that the greater the amount of 'fixed capital' (e.g. investments sunk into machines and factory buildings), the more capitalists sought to take control of production processes by designing factories and buying machines which would dictate the behaviour of the workers. The move to fossil fuel-powered machines arguably fits this analysis. But the Iron Man self-acting mule *could* be operated with waterpower, and often was; there were 'still choices available in the industrial system' (Hahn 2020, 145), available paths that weren't followed (e.g. Malm 2014); nothing was inevitable or pre-determined by the technology itself. Nonetheless, in the ensuing decade, the historian and critical political ecologist Andreas Malm identifies the seemingly inexorable rise of a 'fossil economy', an economy 'most simply defined as one of self-sustaining growth predicated on the growing consumption of fossil fuels and therefore generating a sustained growth in CO_2 emissions' (Angus 2016; Malm 2016b, 222); an economic system within which we still seem to find ourselves.

The mid-nineteenth-century rise of the railways and the accelerating penetration of steam power into different sectors of the industrial economy meant that coal production and consumption in Great Britain rose rapidly during the nineteenth century. From 3 million tons in 1700, by 1850 Britain's coal output was 50 million tons and was growing at around 6% per year. By 1879, steam engines supplied 97% of the power used in the cotton industry, and 85% of that in the woollen industry (Gregory 1990, 362, 371).

Figure 4.3 shows how the mix of energy sources changed from the mid-sixteenth century to the middle of the nineteenth century in England and Wales. Energy consumption expanded 13-fold, approximately doubling every 50 years, with coal the main source of growth. The growth in coal production was highest between 1830 and 1854, as British capital, Malm (2016a, 249)

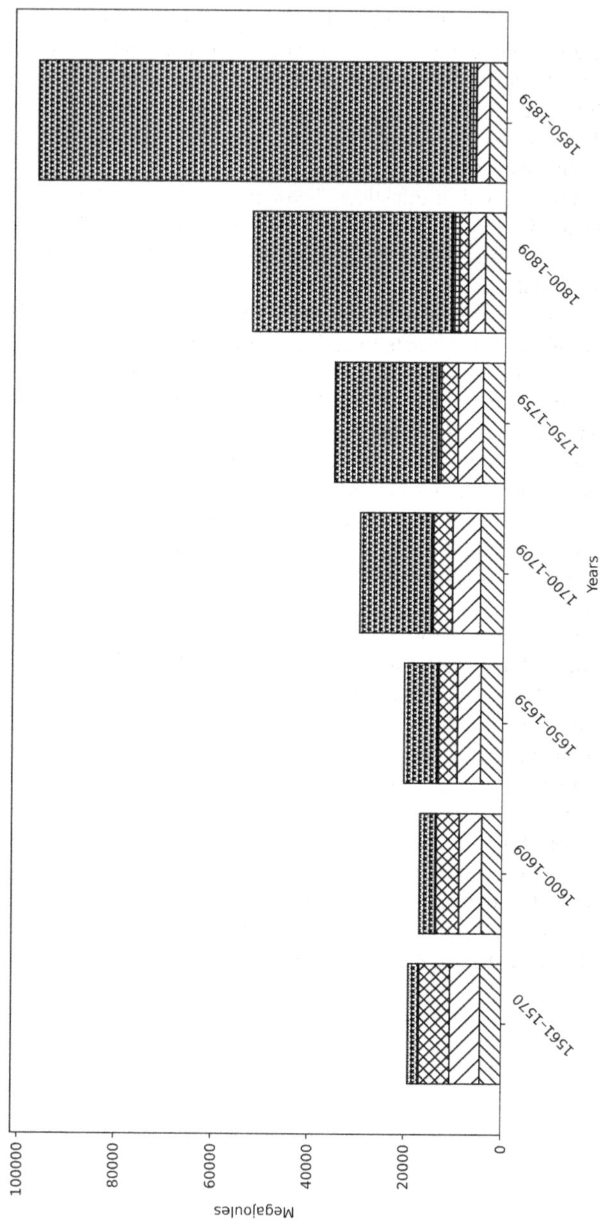

Figure 4.3 Annual energy consumption per person in England in Wales. (Adapted from Wrigley (2013, 6–7).)

argues, 'wrested itself out of the crisis' of the 1820s 'through an unprecedented mobilisation of energy' from fossil fuels. By the end of the 1830s, Great Britain was responsible for over 80% of global CO_2 emissions. By this point, China, still then the world's largest economy, was responsible for around 1% of CO_2 emissions (Driscoll 2021; Malm 2016a).

But while we might imagine growing coal consumption being primarily driven by the rise of steam-powered factories, industrial blast furnaces, and eventually railways, Wrigley (2013) attributes much of the pre-nineteenth-century English rise to growing domestic consumption. To fully understand the shift to fossil fuels, we therefore need to shift our gaze from the space of the factory to the domestic sphere, (Box 4.1) and the changing forms of labour and material consumption which were equally significant for the rise of the fossil economy.

Box 4.1 Situating the Anthropocene: at home in the city

Despite the rapid spread of the factory model of production, the geographical separation of work and home was never wholly complete. Homes remained important sites for spinning and handloom weaving throughout the nineteenth century, although such domestic workers largely lost their previously proud status as independent producers and took up the tasks to supplement meagre household incomes, for example when factory jobs for grown men began to disappear (Hahn 2020, 159). The home also remained an important site of coal consumption, and population growth was therefore a significant driver in the increase in coal use over the eighteenth century. While coal is often discursively associated with men doing manly things with big fires (a discourse which itself was a product of nineteenth-century pro-industrial propagandising), it is important to remember that a significant proportion of Britain's early coal boom was a function of increasingly gendered domestic tasks; the gendered historical geographies of coal consumption are yet to be fully explored. In 1816, domestic heating consumed over half of Britain's coal. Huge amounts of coal would continue to be consumed in this way, but as a proportion of overall consumption, it dropped steadily, with power-hungry industry and transport taking over. By the middle of the nineteenth century, coal consumption growth had been fully decoupled from population growth and was driven instead by economic expansion, thus representing the moment, for Andreas Malm, when the fossil economy was fully in swing (Malm 2016a, 249).

Much of the population growth that drove the early expansion of coal consumption was concentrated in cities. Manchester, in northwest

England – 'Cottonopolis' as it came to be known – can be described as the 'world's first industrial landscape', and forms many people's view of the quintessential industrial city, 'with its monumental, repetitive, red brick mills, and its spewing smokestacks and filthy tenements' (Hahn 2020, 90). Manchester had been a textile town for hundreds of years before steam engines and power looms came along. Its position at the centre of a gaggle of textile towns, built alongside the streams which flowed westward out of the Pennine hills, made Manchester a centre of both manufacturing and merchant activity. Over the eighteenth century, a coherent regional economy emerged, with different towns specialising in different aspects of the textile (particularly cotton) manufacturing process, with Manchester operating as a centre both for manufacturing and for finishing and packing (Gregory 1990). From there, products from all over Lancashire and the north of England were exported via the canal network and the port of Liverpool, from where raw American cotton was obtained, and the cycle started again. Like other industrial cities which came after it (see Cronon 1991), Manchester was a city enmeshed within, and for a significant period at the centre of, regional and global networks of trade and commerce, and global systems of unequal ecological and economic exchange (Hornborg 2012; Moore 2023). In Manchester's case, the exploited bodies of slaves and the exhausted soils of the American south were exchanged, through the application of proletarianised labour and energy (first from water, then from coal), for the finished manufactures of the textile industry, which were in turn re-exported all over the world.

Industrial cities like Manchester grew up close to coal seams, but not because their industries needed coal – that would come later. Instead, the crucial point was that their residents needed coal for cooking and heating, and coal's high transport costs meant it was vastly cheaper to buy the nearer you were to the mines (Lawton and Pooley 2021). That meant that wages in industrialising, coal-adjacent cities could be kept lower. And while owners of early factories situated by fast-running, rural streams commonly provided housing, schools, and healthcare facilities for their new workforces, in cities these costs of the 'social reproduction' could be ignored (Gandy 2018; Harvey 1996). As the Romantic writer Sir Walter Scott put it in 1820 (quoted Malm 2016a, 152):

> the manufactures are transferred to great towns, where a man may assemble five hundred workmen one week and dismiss them the next, without having any further … solicitude about their future fate than if they were so many old shuttles.

Workers grew accustomed to these precarious conditions, but also regularly railed against them through strikes, protests, and marches – the industrial city became the stage for the playing out of the class antagonisms of capital and labour. Contemporary critics, particularly of a romantic bent, railed against the alienation from nature that these cities represented. For example, where workers once toiled according to the rhythms of the seasons or the passing of day into night, now they were disciplined by the clock and by 'factory time', by the rhythms of the machine, of shift-work, and of the rigours of productivity measured in units of time (Glennie and Thrift 2009; Thompson 2017). The problems of nineteenth-century cities were conceived of in disarmingly modern, 'environmental' terms, for example in Marx's identification of a 'metabolic rift' between city and countryside whereby the city gobbled up the resources of the countryside without replenishing things like its soil fertility (Gandy 2018). Early environmentalists and socialists sought to heal the rift by promoting 'homeostasis', a kind of 'circular economy', between city and country, society and soil, by conceiving of cities and their challenging abundance of human excrement as 'fertiliser factories' for their agrarian hinterlands (Bonneuil and Fressoz 2016, 187).

Others – not least urban citizens – were vexed by the effluents pouring out of the factories themselves – the smoke, the soot, the dirty water. The French traveller and political theorist Alexis de Tocqueville described the scene in Manchester in the early 1830s:

> A sort of black smoke covers the city. The sun seen through it is a disc without rays. Under this half daylight 300,000 human beings are ceaselessly at work…From this foul drain the greatest stream of human industry flows out to fertilize the whole world.
>
> (Tocqueville 1835)

For industrialists and city leaders, this smoke and 'foul drain' was but a symbol of industriousness and progress, a thing to be celebrated. But a vibrant civil society movement sprang up against it. A committee of inquiry was established in 1819, and in 1842, the Manchester Association for the Prevention of Smoke was established. Four years later, the city opened its first parks, only for their trees to be killed by what would soon become known as 'acid rain'. Mancunians suffered Britain's highest rates of respiratory diseases and rickets, and in 1899, 'the majority of the Mancunians who volunteered for the second Boer War were rejected on account of their defective constitution' (Bonneuil and Fressoz 2016, 263). The proliferation of 'smoke abatement' societies towards the end of the century nonetheless speaks of the vibrancy of civil society moves against urban air pollution, and shows how the industrial city – in Europe and North America – was becoming a site of environmental struggle in which both working- and middle-class citizens were active (Montrie 2018).

Consuming the world

The significance of the Industrial Revolution for the Anthropocene lies not only in the move to the widespread burning of fossil fuels. For geographers like Lewis and Maslin (2018), it lies also in the spread of new levels of material consumption. The Revolution was driven as much by changing patterns of *consumption* as by changing means of *production* (which is obvious really – there wasn't much point making all that new stuff if there was no one to buy it!). New technologies and means of organising production co-evolved with the development of new markets for manufactured goods and agricultural products, and often these were closely interrelated: 'Machinery made in Glasgow supported sugar refining in the West Indies and sugar and syrup consumption in the UK, just as the tools, guns and chains produced in the English Midlands supported the capture, transport and control of slave labour upon which sugar production depended' (de Groot 2006, 173; Satia 2019). Earlier in this chapter, we explored the macro-geographical effects of the production of things like sugar in the colonies in helping Britain to escape its own local ecological limitations. But the consumption of things like food, drink, and clothing is geographically significant across a range of spatial scales. For example, cultural historians and cultural geographers have shown how changing eating habits are revealing of things like changing gender relations in the private spaces of the home (Blunt and Dowling 2022), as well as how acts of consumption in more public spaces are central to forming cultural identities: 'our sense of who we are, and how we present ourselves to others' (Horton and Kraftl 2014, 68).

Industrial cities were sites of consumption as well as production. Evidence from sources like the diaries and autobiographies of factory workers show that, despite the often appalling working conditions and the struggle for decent wages and steady jobs, the emerging class of industrial workers also revelled in the new found social and political freedoms brought by wage labour, and in the occasional amount of disposable income which allowed them to participate in new forms of material consumption (Griffin 2013). And it was London that was arguably the epicentre of an emerging consumer economy. While cities like Manchester and Leeds specialised in particular products, London's industry was highly heterogeneous and organised around smaller, more artisanal units of production. But the city was nonetheless a hub (arguably *the* hub) of global trade and the nerve centre for a rapidly evolving consumer society. It was in London that intelligence on shifting fashions and consumer preferences could be gained, and where the latest wares could be put on display in shops catering for all classes of consumer (Gregory 1990).

By the beginning of the nineteenth century, a new 'consumer culture' was already well in formation, with plantation-grown commodities like sugar, tea, and tobacco having moved from a status of exotic luxury to articles of widespread use (consumption by >25% of the adult population is generally taken as the key threshold). From the early nineteenth century, such items began to appear in the new breed of 'corner-shop', an otherwise humble purveyor of essential groceries which also functioned as a portal for the exotic products of a globe-spanning empire. A new diet of sweetness and stimulation (sugar, sweetened tea and coffee, tobacco) began to mesh with the new world of industrial production. A break for a sweetened cup of tea or a pipe gave welcome respite from the monotony of factory labour, while at home a relatively easy and calorific meal of tea, bread and butter, perhaps with jam or syrup, filled the 'calorie gap' of the labouring poor and fitted in with the complicated rhythms of male, female and child labour (Mintz 1985); so much so that in industrial northern England 'tea' became synonymous with a meal eaten at the end of the working day.

New spaces and cultures of consumption – from the working-class dinner table to the new breed of refined, middle-class 'tearooms' – linked British consumers materially to the networks of trade and ecological exchange of the British Empire and beyond. Tea, sugar, and tobacco were soon joined on the shelves of corner shops and department stores by coffee from Brazil and Kenya, flour from Canada, butter from New Zealand, and wool from Australia and South Africa. Through the work of advertisers and trade bodies, consumption of such goods was also made culturally meaningful as a means of supporting, and enjoying the fruits of, the imperial enterprise. Advertisers drew on imagery of exotic landscapes and racial stereotypes to make their products stand out, and in so doing perpetuated an 'imaginative geography' (Gregory 1995) of an empire populated by peasant farmers and servants labouring gratefully in the service of their colonisers, or of pioneering settlers spreading industry and civilisation to distant corners of the globe (Hund et al. 2013; Ramamurthy 2017). Colonial exhibitions (Figure 4.4) were an important site for popularising and domesticating the empire – selling its wares for domestic consumption, and fostering an idea of the wider empire as a (white, British) 'home' (Blunt and Dowling 2022) – and indeed the origins of the tea room can be traced to the refreshments stalls at such exhibitions. One commentator remarked how 'weary afternoon visitors' to the Colonial and Indian Exhibition of 1886 'enjoy their tea all the more because it is served to them by white-robed Sinhalese' (quoted in de Groot 2006, 184).

Figure 4.4 Jamaica at the colonial and Indian Exhibition, London 1886. (Source: Wikimedia Commons.)

Colonial exhibitions made a spectacle of the world-ecology of empire and industrial capitalism (Hassam 2017), just at the moment that its global reach was expanding further. Through this imperial expansion, fossil-fuelled capitalism became an increasingly global phenomenon.

The planetary mine: coal and empire

In the early 1830s, Thomas Bacon, an accomplished artist and Lieutenant in the Bengal Horse Artillery, was travelling around north-eastern India when he encountered the ruins of Rajmahal, a former seat of Mughal power over Bengal (Figure 4.5). In his travel writings, he takes a certain romantic or even gothic delight – as was common at the time – in the crumbling grandeur and the implacable tropical nature, as vines and trees reasserted themselves over this abandoned landscape of royal power.[2] But some hope for preservation of the *Sungi-dallán*, a grand marble hall and 'one the handsomest chambers', lay in its apparent recent conversion 'into a coal hole for the supply of Government steamers plying up and down the Ganges' (Bacon 1837, 265).

This marble hall was lucky (in Bacon's view) to have been enrolled into a growing infrastructure of coal-powered mobility which was helping to project British power across the Indian subcontinent, well before the formal

Figure 4.5 'Ruin at Rajmahal'. (Adapted from Bacon (1837), *First impressions and studies from nature in Hindostan*, p. 267.)

establishment of the British Raj, or direct crown rule, in 1858. The new technology of steamships enabled the rapid movement of colonial officials, troops and settlers, they allowed tax revenues to be transported quickly and safely from rural hinterlands to seats of government, and they allowed raw materials like cotton, silk, rice, indigo, opium and tea to be efficiently extracted from the interior to coastal ports, from whence they could be shipped to England or to new markets – often opened-up to imports through the related technology of 'gunboat diplomacy' – in places like China (e.g. Driscoll 2021).

For Irish engineer and steamboat proselytiser John Bourne, the new technology had turned the subcontinent's rivers into 'great steam highways for bringing cotton and other products of the interior to the coast, and for transmitting English manufactures in return' (quoted in Malm 2016b, 224). As such, they could complete and accelerate the kind of mercantilist circuit of trade which had re-shaped the economies and ecologies of other world regions, such as the Caribbean: raw materials out, manufactured goods in. As Malm (2016b) argues, this enthusiasm for steamboats summoned into being the industrial-scale mining of coal in India. From the 1830s onwards, British officials, explorers, and soldiers all participated enthusiastically in the search for coal. A Coal Committee coordinated the search, and geologists found steady employment in reading the rocks and the landscape and speculating about where coal might be found. Officials often encountered local communities who were perfectly aware of the coal reserves which lay beneath their feet but who had largely opted – much to the incredulity of the British – not to exploit them. Numerous coal fields were discovered and opened up to mining, run by both British and Indian entrepreneurs, some of which are still being exploited to this day (Chatterjee 2019). In some cases, the relative absence of an indigenous mining industry meant that the locals had to be essentially forced down the pits, using techniques of coercion reminiscent of Potosí (see Chapter 3). British investors buying up coal fields would often purchase much of the surrounding countryside too, so that the peasantry could be compelled to spend some of their time serving in the mines, on the pain of expulsion from their land (Malm 2016b).

Coal motivated and fuelled further colonial expansion across the region. In 1837, a British missionary happened upon coal in Borneo, with the most promising reserves found on the island of Labuan. Frantic exploration, lobbying, and diplomacy led to the British taking possession of the island in 1846, having seen off competition from the United States (Shulman 2019). The annexation had been backed by figures from the Royal Navy as well as from the Glasgow and Manchester Chambers of Commerce, all of whom spied the potential of:

> perfect[ing] the chain of posts that connects, by means of steam navigation, Southampton with Victoria [Harbour] in Hong Kong ... We already have bound British North America, the West Indies, Ceylon, and India, to the dominant country by steam navigation. We are now riveting China by the same bounds, and are about to open out the Indian Archipelago to commerce, civilisation, and Christianity, by the like agency... Thus in a very few years we may expect to see the world fairly *belted* by the steam navy to England.
>
> (*Singapore Free Press*, 1845, quoted in Malm 2016b, 229)

But as in India, British prospectors reportedly found the locals to be largely indifferent to the coal reserves and resistant to working in the new mines. In the face of what the British saw as the 'natural indolence' of the natives of the region, the colonisers turned again to forms of unfree labour, importing thousands of indentured Chinese and Indian workers to do the work that the residents of Labuan were seemingly incapable of doing. The ability and willingness to work with coal became, for the empire-builders, an index of a people's degree of civilisation: at the bottom were the 'savages' who merely used the material for trinkets and toys, preferring the energy sources of their forebears; atop the hierarchy the British saw themselves, from which vantage point they scanned the planet for more coal reserves, striking black gold in Trinidad, Tasmania, Vancouver, Natal and New Zealand. A horizon of infinite economic growth and great political power opened up, the fossil fuel increasingly seen as 'the key protagonist of national progress' and the shaper of national destiny (Coen and Jonsson 2022, 412). The infrastructure and ideology of the new fossil economy, 'utterly absent from all but a tiny corner of the canvas of history' thus made its way across the surface of the Earth as the nineteenth century wore on, and as British – and then French and American – empires expanded to cover almost the entire terrestrial globe (Malm 2016b, 231).

The hunger for coal was accompanied by hunger for other raw materials, to sate an accelerating economic metabolism in the industrialising world. Around five-sixths of British outward investment between 1815 and 1880 went into building an 'informal empire' of mines, plantations, railways, and so on that lay outside of formal British territories. As such, tin and rubber from Malaya, copper from the Andes and the Congo basin, guano from South America, and timber, wheat, and phosphates from North America poured into Britain and Europe, where many would be turned into manufactured products and re-circulated around the globe (e.g. Clifford and Castonguay 2022). But if cheap, bulky raw materials filled ships on the way into Britain's ports, smaller, high-value manufactured goods might only half-fill those ships on the way out. Coal helped solve this problem, and while Britain was scouring the globe for new coal seams, it also became, in the second half of the nineteenth century, a major coal exporter (Figure 4.6), fuelling steam trains and trams in Egypt, the navies of European allies, and fuelling hearths, street lights, and furnaces from Cuba to China (Barak 2020).

All this was aided by a new financial capitalism which, through new means like limited liability companies and an international monetary system based on sterling, drove outward investments into things like railways, mines, gas lighting, and plantations (Pawson 1990). This created what Bonneuil and

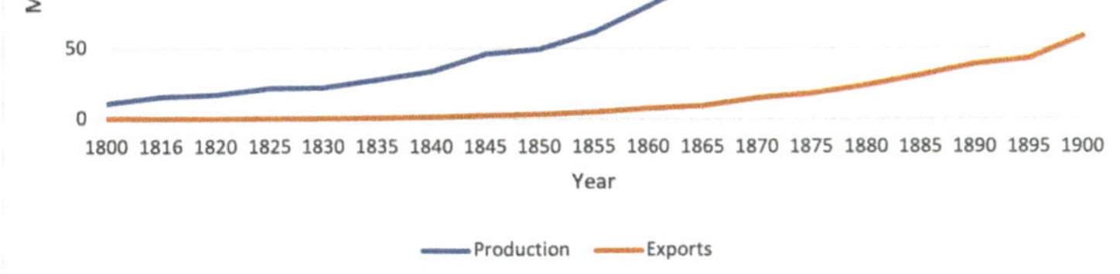

Figure 4.6 British coal production and exports, 1800–1900. (Based on data from Cook (2005, 208.)

Fressoz (2016, 119) describe as an 'unprecedented historical phenomenon: raw material was shipped to the other end of the world to be transformed and sometimes re-exported to its country of origin'. While we might look to earlier periods of things like sugar production to query how unprecedented this really was, nonetheless, the *scale* of the new movements of the nineteenth century certainly was unprecedented, and crucially was:

> embodied in technologies that were high emitters of CO_2, reorganizing flows of matter, energy and goods on a world scale. Transcontinental canals, railways, steam ships, docks, grain silos and telegraph lines constructed a second nature of planetary dimensions, penetrating to the interior of the peripheral countries and mooring them to the world economy.

This 'second nature' can be thought of as a new industrial world-ecology, 'which projected fossil capitalism onto the whole world' (Bonneuil and Fressoz 2016, 238).

Box 4.2 Avoiding teleology in the historical geography of the Anthropocene

These arguments arise from a 'new social and environmental history of energy attuned to questions of class, gender, race, and empire', which is revealing 'new kinds of dependency, exploitation, and resistance' (Coen and Jonsson 2022, 416). Much of this history has been written from a stance of opposition to the 'standard narrative' of the Anthropocene, which tends to treat humanity as a single actor, pushing the Earth System into a new state through actions which express something innate about 'human nature' (Davies 2016). This narrative often draws a direct link between the 'mastery of fire by our ancestors', which put humanity 'firmly on the long path towards the Anthropocene', and the industrial revolution's turn to fossil fuel (Steffen, Crutzen, and McNeill 2007, 614). Human geographers have likewise drawn such links, seeing the industrial revolution as an expression of the foundational and perhaps defining human skill of handling and manipulating fire (Clark and Yusoff 2014), with 'the recent propensity to tap into sedimented and fossilised biomass' being the latest of a long line of human augmentations of the planet's own 'pyrophytic tendencies' (Clark 2012, 269). These arguments originate from a variant of the 'End of Nature' narrative (Chapter 1), which stresses that human life and sociality have always been (and are increasingly) entangled with the chemical elements and geophysical energies of the planet.

Marxist historian Andreas Malm (2016b) cautions against the *teleology* which sometimes crops up in arguments like these. Teleology is a mode of narrating history which sees past events and processes as leading to an inevitable or pre-destined endpoint. Historians might set out to explain present conditions (such as the Anthropocene, or modern society's dependence on fossil fuels) by looking for historical precursors and precedents, and seeing these as connected to present conditions through shared essences and transhistorical chains of inevitable causation. But such teleological reasoning – which is as common in Marxist versions of history as any other (Hetherington 1997) – overlooks how history in general is characterised by change, dislocation, and discontinuity. There is no smooth path between early humans' mastery of fire and the fossil economy, and nor is the latter the outcome of the same human traits which shaped the former. Malm emphasises instead how we need to treat these two historical phenomena on their own terms, exploring the social structures and value systems which gave rise to each of them in certain places and times; they need to be analytically situated, both historically and geographically.

Malm's description of local resistance or indifference to coal mining bears this out. As he summarises it, 'far from all humans who knew how to light and slake a fire developed a fossil economy, even if they had immediate access to coal and knew it could be used for fuel... The fossil economy had to be foisted upon them' (Malm 2016b, 234). Observing that historically 'no humans have yet engaged in systematic extraction' of fossil fuels 'to satisfy subsistence needs', and that the expansion of the fossil economy over the course of the nineteenth century depended on some people having command over the labour of others, Malm concludes that 'fossil fuels are by their very definition a condensation of unequal social relations' (Malm 2016b, 235). By analysing the historical geographies of coal use within the context of nineteenth-century politics and society, and their transformations by industrialism and imperialism, we get a very different sense of the meaning of the move to fossil fuels. Rather than the continuation down a 'long path' which connects early humans and their use of fire to the atmosphere-altering combustion practices of recent centuries (Albert 2015), we can see instead important discontinuities in human uses of fire (Bowman 2015), with changes and transitions shaped by more proximate causes in their social, economic, and political contexts. A significant part of that context consists of the manifold histories and geographies of resistance to industrialisation, the fossil economy, and the 'deterioration of the Earth' (Bonneuil and Fressoz 2016).

Geographies of knowledge and resistance

While the *horizontal* expansion of colonialism and capitalism – and the associated movements of material goods, wealth, ideas and knowledge – is often taken as the hallmark of modernity (Sloterdijk 2013), Clark and Szerszynski (2020, 103–6) identify in this period the onset of a *vertical* modernity, which was rather less sure of itself than we might assume. As 'Europe' expanded horizontally, it also went deeper into the Earth, most notably of course to extract resources such as coal from subterranean strata of rock. As people dug for coal and other profitable minerals, they also encountered new kinds of fossils – fossils which told new stories about the history of the Earth, and about the potential for extinction. The eighteenth and nineteenth centuries saw fiery debates about the true age of the Earth, and arguments raged about whether Earth history was a story of slow, gradual change, or of periodic spasms of catastrophic upheaval in the lithosphere and biosphere (Rudwick 2010). For Clark and Szerszynski (2020), 'modernity' has therefore always been marked by an unease, and a sense of (or perhaps fear of) frailty and impermanence. During the time of the industrial revolution, a troubling new question began to be asked: if the Earth has the potential to throw whole species into extinction, might humans be next? While earlier, diverse mythological and religious ideas about apocalypse usually saw the end of humanity as coinciding with the end of the universe in a moment of profound spiritual significance, early nineteenth-century thinkers began to grapple with the prospect of humans disappearing while the universe would just shrug its shoulders and carry on (Moynihan 2020).

But if the Earth and the wider cosmos might be inherently blasé about human extinction, many people were far from blasé about this new fossil economy. New historical research is showing how the fossil economy was put together not in spite of struggle and resistance, but through it. It was as much an outcome of shifting, often antagonistic relations between capital, labour and 'nature' as it was an outcome of technological progress: mechanisation helped make workers interchangeable, coal could replace capricious waterpower, expensive steam-powered machines could replace truculent workers. Attending to the spaces and practices of resistance therefore helps guard against teleological conceptions of the industrial Anthropocene (Box 4.2) – the idea that it was all inevitable – and, by reminding us of alternatives and paths not followed, opens up new ways of thinking about possible Anthropocene futures which need not necessarily lead to human extinction.

This is perhaps the major historiographical contribution of Bonneuil and Fressoz's *The Shock of the Anthropocene*. Throughout their text and in dedicated chapters, they emphasise the resistance faced by the spreading world-ecology of mercantile and then industrial capitalism, and the profound knowledge and awareness that historical actors had of the environmental damage being caused. For example, they seek to rehabilitate the 'Luddites', an early nineteenth-century movement of machine-breakers whose name to this day is used pejoratively as a label for anyone deemed to be 'anti-technology', or 'anti-progress' (Mueller 2021). Machine-breaking was a common form of labour conflict around the turn of the nineteenth century across western Europe, a movement of urban artisans and rural labours who:

> expressed their refusal to see themselves dispossessed of their skills, their livelihood and a way of life that combined agriculture with manufacture. They rejected poor-quality industrial products and championed the idea of a fair price for their labour against the machines that were the cause of imbalance and inequality.
>
> (Bonneuil and Fressoz 2016, 259)

While Barbara Hahn presents the Luddites – wool croppers who engaged in a campaign of machine-breaking across the English north and midlands from 1811 to 1817 – as something of an 'unreflective "rebellion of the belly"' (Gregory 1990, 359), whose campaign had 'fundamental causes in hunger and economic misery' (Hahn 2020, 135), Bonneuil and Fressoz see in the Luddite movement a fundamental critique of the emerging industrial economy, targeting:

> a fundamental bifurcation that set-in with the entry into the Anthropocene: the profitability of costly machines led capitalists to prioritize the quantity of products over their quality or durability…This resistance was never against technology as such, but against a particular technology and its ability to crush others, and we need to unfold the spectrum of alternatives that existed at each moment: canals instead of railways; improved oil lamps instead of gas lighting; flexible and quality production instead of mass production.
>
> (Bonneuil and Fressoz 2016, 260–61)

The Luddite critique found support in the middle classes too, who worried about the misery of machine work and the 'disreputability' of its products. The Romantic poet Lord Byron sprang to their defence, perhaps seeing in the Luddites the noble hero standing against the hegemonic norms of society, which inflected much of his writing and self-image (Merchant 2023). Other Romantic writers penned hymns to the beauty of nature, and critiqued both the direct threats posed to it by the march of industrial technology, and the alienation of people from the natural world caused by urbanisation and a life spent increasingly indoors and in the shadows of what William Blake famously described in 1810 as those 'dark Satanic mills'.

Urban and rural folk got organised to push back against industrial pollution. Fressoz (2007) found that around the French city of Marseilles in the 1820s, 10% of all the cases heard in the courts were brought against factory owners for the pollution caused by the chemical industries. In 1842, the Manchester Association for the Prevention of Smoke was formed in recognition of the damage being done both to the city and its inhabitants by the 500 or so chimneys belching coal smoke into the atmosphere (see Box 4.1). Other societies were soon set up in its image, as well as a Parliamentary committee – 'the culmination of a long and arduous Parliamentary struggle to establish air pollution as a legitimate area of governmental responsibility' (Whitehead 2011, 9; Figure 4.7). Such moves had some successes – chimney heights were raised to try and dissipate the smoke over wider areas, or condensers were installed to minimise things like hydrochloric fumes. But industrial interests were arguably too strong for the problem of air pollution to be more concertedly tackled before the twentieth century, even though the science of air pollution developed rapidly over the course of the nineteenth. This nonlinear relationship between science's growing ability to know and understand environmental problems, and the political will to act on that knowledge, is something we'll return to in later chapters.

By the mid-nineteenth century, labour movements and capitalists were largely united in their embrace of machinery and 'progress' – even Marx himself denounced anti-mechanisation stances as archaic, and paved the way for productivist ideologies which would animate revolutionary left-wing political projects in the early twentieth century. Nonetheless, towards the end of the nineteenth century, 'the critique of anthropocenic activity' re-emerged,

Figure 4.7 A British cartoon from 1880 details the effects of coal burning – bronchitis, asthma, and pneumonia are mentioned under the demon's claws, while a goblet of 'carbonic acid' wafts gas over the city. (Source: https://victorianweb.org/periodicals/punch/publichealth/5.html)

organised, Bonneuil and Fressoz (2016, 268) suggest, in three distinct but related discourses:

a *Conservationism* denounced 'the merely extractive logic' of the industrial world-ecology as 'unsustainable'. Relevant here were geographical arguments, advanced by the likes of Halford Mackinder, about 'global closure' (Kearns 1993): the period of European expansion was coming to a close, with all four corners of the Earth colonised (literally and metaphorically), and resource frontiers appearing increasingly finite. The world, argues philosopher Peter Sloterdijk (2013), began to be seen – from the West at least – as an enclosed 'interior', its flows of matter, energy, and money requiring careful, scientific management. Scientific theories about atmospheric radiation, the greenhouse effect, and the implications of pumping millions of tonnes of 'carbonic acid' (carbon dioxide) into the atmosphere featured in such discourses, following discoveries in the middle of the century of the role of CO_2 in trapping heat in the atmosphere, and later calculations about the possible effects on the global climate of increasing atmospheric CO_2 concentrations (Hulme 2021).
b *Preservationism* picked-up on earlier Romantic ideals of protecting nature for its inherent aesthetic, recreational, or spiritual worth, and of protecting it from utilitarian economic exploitation. Movements for the establishment of national parks, such as the Sierra Club in the United States, mobilised such ideas, while naturalists in Europe railed against the destruction of nature both in imperial metropoles and on colonial peripheries (see Chapter 7). But while such stances were radical in their anti-utilitarianism, the establishment of things like national parks 'also participated in the expulsion of indigenous territories' in the name of protecting a mythical, unpeopled natural world, and paved the way for the 'touristic consumption of nature by world elites' (Bonneuil and Fressoz 2016, 271).
c *Back-to-nature socialism* drew on slightly different currents of Romantic thought, particularly in England: 'the desire to preserve communitarian social relations in the face of individualism, to protect the countryside against the aggression of the modern world, and to maintain artisan and artistic skills in the face of industrial standardization' (Bonneuil and Fressoz 2016, 271). These ideals were married to a rejection of capitalism and an embrace of a revolutionary 'politics of beauty and conviviality in harmony with nature'. William Morris and John Ruskin in England; Ludwig Klages and Gustav Landauer in Germany; Henri Beylie and Henri Zisly in France; thinkers who in different ways railed against the ideologies of 'progress' and industrialism which united both capitalist and Marxist politics, and sought to enact new utopian communities where egalitarian social relations would be shaped around a new conviviality with the nonhuman world.

These latter ideas evolved in dialogue with emerging anticolonial movements; they were influential for the young Mahatma Gandhi, for example, who developed a critique of industrial civilisation and British imperialism, which shaped the movement for Indian independence, with its own vision of a postcolonial Indian economy unshackled from the unequal exchanges of the industrial world-ecology.

Additional critiques of this world-ecology emanated from across the world. In East Asia, for example, much criticism and resistance was motivated by the 'gunboat diplomacy' which forcefully opened up Chinese and Japanese markets to the steamboat-powered trade of British merchants, and particularly to their cargoes of Anglo-Indian opium. For the historian Mark W. Driscoll, the First Opium War (1839–42) – the violent struggle to open up Chinese markets for the drug, in exchange for things like tea and indentured labourers – 'was a crucial event in the shift from the Holocene to the Anthropocene', as it represented the moment when British (and soon thereafter American) power ultimately 'achieved planetary pre-eminence by military subjugation of the prosperous polities of China and Japan' (Driscoll 2021, 3). Driscoll recounts how resistance to this military and commercial subjugation was couched in a language of climatological and ecological balance. His book's title, *The Whites are Enemies of Heaven* – paraphrases an 1839 letter from Qing Commissioner Lin Zexu to Britain's Queen Victoria, and which would echo through later denunciations of the myopic 'pursuit of profit', the 'destruction of local practices of spirituality', and the 'decimation of local ecologies' (Driscoll 2021, 23). Driscoll tracks these ideas through various rebellions which sprang up in East Asia over the second half of the nineteenth century, but also sees them as connected to more recent struggles by Indigenous groups worldwide against destructive resource extraction and ecological degradation; struggles which frequently start from very different ontological ideas about, for example, humanity's place within (rather than apart from or above) local ecologies and global natures. The long histories of such struggles 'offer compelling stories for earthly survival today' (Driscoll 2021, x).

Conclusion

In this chapter, we've reviewed work which has answered the 'why-of-where' question of the 'Industrial Revolution' by focusing on geographical transformations taking place at a range of spatial scales, from the factory and the home to the emerging world-ecology of industrial capitalism. Together, this work by historians and historical geographers is using the prompt of the Anthropocene proposal to re-evaluate the significance of what happened in northern England from the late eighteenth century onwards; this was not just a significant moment in the emergence of capitalism or of a global 'modernity' (Hetherington 1997), but a crucial 'detonator' of the Anthropocene as well (Tsing et al. 2020). It has prompted re-appraisals of the history of Great Britain and its relationship to the rest of the world (Otter et al. 2018), and has positioned both empire and labour relations as being crucial to understanding the local and then global shift to widespread fossil fuel use (Malm 2016a).

Much of this analysis subscribes to a variant of the 'eco-Marxist' narrative of the Anthropocene, which proposes that the chief protagonist in the story of the Anthropocene is not humanity as a whole (the *anthropos*), but the 'super-human' actor, capital (see Chapter 1). As such, as discussed in Chapter 3, the *Capitalocene* has emerged as an alternative label for this period of geohistorical transformation, which aims to more precisely identify the drivers and culprits of the environmental changes captured by the term *Anthropocene* (Moore 2022). Advocates of the Capitalocene do not necessarily agree on a precisely identified onset point in the way that geologists might wish them to – Andreas Malm, for example, points to the ascendancy of fossil capital from around 1830 onwards, while Jason Moore points to the emergence of a new capitalist world-ecology around the sixteenth century. But in naming this age after capitalism rather than the *anthropos*, blame is laid squarely at the door of certain institutions, forms of social organisation, and even individuals, with implications for how societies might think about historical responsibilities in the present:

> The conclusion that forces itself upon us, disturbing as it may be, is that our ancestors destroyed environments in full awareness of what they were doing.
>
> (Bonneuil and Fressoz 2016)

Recent research on the history of environmental knowledge and environmental concern therefore invites us to ask new questions about responsibility:

> The historical problem…is not the emergence of an 'environmental awareness', but rather, the reverse: to understand the schizophrenic [sic] nature of modernity, which continued to view humans as the products of their environment at the same time as it let them damage and destroy it
>
> (Bonneuil and Fressoz 2016, 196)

Bonneuil and Fressoz pull no punches about the wrong choices and historical responsibilities of 'our ancestors', at least some of them in Western Europe. Many would nonetheless argue that it is unfair to:

a judge historical actors by the moral standards of the present;
b judge them for doing things whose consequences they couldn't have understood (like pumping CO_2 into the atmosphere); or
c hold present day individuals, institutions or countries responsible for the actions of their forebears, over which they had no control.

But in historicising the shift to a fossil economy, the work of historians such as Malm, Hahn, Bonneuil, and Fressoz troubles simple defences of points (a) and (b). On point (a), we can now confidently say that the shift to a fossil economy was contested on moral and political grounds by numerous actors at the time, and thus, it is possible to judge historical actors according to the moral and political choices that were clearly open to them (Retz 2022). On point (b), whilst it is true that the vast majority of fossil fuel boosters in the nineteenth century would have known little of the potential for global anthropogenic climate change (despite the coeval emergence of the science of the greenhouse effect – see Hulme 2021) they did know, and were deep in political and legal struggles over, more immediately visible forms of pollution and environmental damage. And on point (c), one need only look at how recent international climate policy arguments have been structured around disagreements over how past responsibilities for emitting greenhouse gas emissions should determine countries' relative contributions to solving the problems of the present and the future (Bulkeley and Newell 2023): 'rarely have issues of the past proved so divisive for so long in so highly profiled international negotiations' (Malm 2016b, 219).

For Andreas Malm (2016b, 216), this means that research on the Anthropocene needs to be newly 'inflected by historical consciousness'. Understanding the present and future of the Anthropocene requires not just quantitative measures of things like economic growth and greenhouse gas emissions, since such numbers

> do not convey the dynamics of the history in question. No numbers can capture its texture, lay bare processes of causation, specify who did what for what reasons and in the service of whom to set the world on the path to this blaze: only historical narrative and analysis could accomplish that.
> (Malm 2016b, 221)

We would add that such historical analysis also requires the geographical sensibility advocated by Erle Ellis (2017), Jason Moore (Chapter 3), and the 'thick geographies' we described in Chapter 2. This means attending to the specifics of place (such as the particular struggles of Lancashire mule spinners, or the geographies of water supply in northern England), but also to the shifting global geographies of resource frontiers, colonisation, trade, and labour relations. But it also requires wariness of universalising histories, which take changes in one part of the world to be representative of those in others (Wu 2019) – a point we will return to in the next chapter. So on that note, let us venture into the twentieth century, to examine how the 'fossil economy' expanded from coal to oil, how a new 'consumer capitalism' was built out of the ruins of two world wars, and why a 'Great Acceleration' in everything from nitrogen transport to the spread of McDonald's restaurants marks, for many observers, the moment when the Anthropocene really got going.

Study questions

1 In what ways are fossil fuels 'a condensation of unequal social relations' (Malm 2016b, 235)?
2 Evaluate the arguments that the Anthropocene would be better known as the Plantationocene or the Capitalocene.
3 What lessons might nineteenth century environmentalisms hold for those seeking to combat the drivers of the Anthropocene in the present?

Notes

1 Williams' book was originally published in the US in 1944, but UK publishers refused to touch it for twenty years as they took exception to Williams' argument that the abolition of the slave trade was based on British economic self-interest rather than humanitarian impulses.
2 We'll return to cultural preoccupations with ruins in Chapter 9.

References

Albert, R. M. 2015. "Anthropocene and Early Human Behavior." *The Holocene* 25 (10): 1542–52.
Allen, R. C. 2011. "Why the Industrial Revolution Was British: Commerce, Induced Invention, and the Scientific Revolution1." *The Economic History Review* 64 (2): 357–84.
Angus, I. 2016. *Facing the Anthropocene: Fossil Capitalism and the Crisis of the Earth System*. New York: NYU Press.
Bacon, T. 1837. *First Impressions and Studies from Nature in Hindostan*. London: W.H. Allen.
Barak, O. 2020. *Powering Empire: How Coal Made the Middle East and Sparked Global Carbonization*. Oakland: University of California Press.
Berg, M., and P. Hudson. 2023. *Slavery, Capitalism and the Industrial Revolution*. Medford: Polity.
Blunt, A., and R. Dowling. 2022. *Home*. Abingdon: Routledge.
Bonneuil, C., and J. B. Fressoz. 2016. *The Shock of the Anthropocene*. London: Verso.
Borgström, G. 1965. *The Hungry Planet: The Modern World at the Edge of Famine*. New York: Macmillan.
Bowman, D. M. J. S. 2015. "What Is the Relevance of Pyrogeography to the Anthropocene?" *The Anthropocene Review* 2 (1): 73–76.
Bulkeley, H., and P. Newell. 2023. *Governing Climate Change*. London: Taylor & Francis.
Burnard, T., and G. Riello. 2020. "Slavery and the New History of Capitalism." *Journal of Global History* 15 (2): 225–44.
Chatterjee, E. 2019. "The Asian Anthropocene: Electricity and Fossil Developmentalism." *Journal of Asian Studies*, ahead of print.
Clark, N. 2012. "Rock, Life, Fire: Speculative Geophysics and the Anthropocene." *Oxford Literary Review* 34 (2): 259–76.

Clark, N., and B. Szerszynski. 2020. *Planetary Social Thought: The Anthropocene Challenge to the Social Sciences*. Cambridge: Polity.

Clark, N., and K. Yusoff. 2014. "Combustion and Society: A Fire-Centred History of Energy Use." *Theory, Culture & Society* 31 (5): 203–26.

Clifford, J., and S. Castonguay. 2022. "British Ghost Acres and Environmental Changes in the Laurentian Forest During the Nineteenth Century." *Journal of Historical Geography* 78 (October): 126–38.

Cook, Chris. 2005. *The Routledge Companion to Britain in the Nineteenth Century, 1815-1914*. Abingdon: Routledge.

Coen, D. R., and F. A. Jonsson. 2022. "Between History and Earth System Science." *Isis* 113 (2): 407–16.

Cronon, W. 1991. *Nature's Metropolis: Chicago and the Great West*. New York: W.W. Norton.

Crutzen, P. J. 2002. "Geology of Mankind." *Nature* 415 (January): 23.

Davies, J. 2016. *The Birth of the Anthropocene*. Oakland: University of California Press.

Davis, R., M. Hatch, S. Hoare, et al. 2025. "Earliest Evidence of Making Fire." *Nature*, December 10, 1–7. https://doi.org/10.1038/s41586-025-09855-6

Driscoll, M. W. 2021. *The Whites Are Enemies of Heaven: Climate Caucasianism and Asian Ecological Protection*. Durham: Duke University Press.

Ellis, E. C. 2017. "Physical Geography in the Anthropocene." *Progress in Physical Geography: Earth and Environment* 41 (5): 525–32.

Ellison, T. 1886. *The Cotton Trade of Great Britain: Including a History of the Liverpool Cotton Market and of the Liverpool Cotton Brokers*. Liverpool: Association. E. Wilson.

Fressoz, J. B. 2007. "Beck Back in the 19th Century: Towards a Genealogy of Risk Society." *History and Technology* 23 (4): 333–50.

Gandy, M. 2018. "Cities in Deep Time." *City* 22 (1): 96–105.

Glennie, P., and N. Thrift. 2009. *Shaping the Day: A History of Timekeeping in England and Wales 1300–1800*. Oxford: Oxford University Press.

Glikson, A. 2013. "Fire and Human Evolution: The Deep-Time Blueprints of the Anthropocene." *Anthropocene* 3 (November): 89–92.

Goudie, A. 2020. "The Human Impact in Geomorphology – 50 Years of Change." *Geomorphology* 366 (October): 106601.

Goudie, A. S., and H. A. Viles. 2016. *Geomorphology in the Anthropocene*. Cambridge: Cambridge University Press.

Gregory, D. 1990. '"A New and Differing Face in Many Places": Three Geographies of Industrialization'. In *Historical Geography of England and Wales*, 351–99. London: Elsevier.

Gregory, D. 1995. "Imaginative Geographies." *Progress in Human Geography* 19 (4): 447–85.

Griffin, E. 2013. *Liberty's Dawn: A People's History of the Industrial Revolution*. New Haven: Yale University Press.

Groot, J. 2006. 'Metropolitan Desires and Colonial Connections: Reflections on Consumption and Empire'. In *At Home With the Empire: Metropolitan Culture and the Imperial World*, edited by Catherine Hall and Sonya O. Rose, 166–90. Cambridge: Cambridge University Press.

Hahn, B. 2020. *Technology in the Industrial Revolution*. Cambridge: Cambridge University Press.

Hall, N. 2001. "The Emergence of the Liverpool Raw Cotton Market, 1800–1850." *Northern History* 38 (1): 65–81.

Hamilton, C. 2015. "Getting the Anthropocene so Wrong." *The Anthropocene Review* 2 (2): 102–7.

Harvey, D. 1996. *Justice, Nature and the Geography of Difference*. Oxford: Blackwell.

Hassam, A. 2017. 'Portable Iron Structures and Uncertain Colonial Spaces at the Sydenham Crystal Palace'. In *Imperial Cities*, edited by Felix Driver and David GIlbert, 174–93. Manchester: Manchester University Press.

Hetherington, K. 1997. *The Badlands of Modernity: Heterotopia and Social Ordering*. London: Psychology Press.

Hornborg, A. 2006. "Footprints in the Cotton Fields: The Industrial Revolution as Time–Space Appropriation and Environmental Load Displacement." *Ecological Economics* 59 (1): 74–81.

Hornborg, A. 2012. *Global Ecology and Unequal Exchange: Fetishism in a Zero-Sum World*. Abingdon: Routledge.

Horton, J., and P. Kraftl. 2014. *Cultural Geographies: An Introduction*. Abingdon: Routledge.

Hulme, M. 2021. *Climate Change*. London: Routledge.

Hund, W. D., M. Pickering, and A. Ramamurthy, eds. 2013. *Colonial Advertising & Commodity Racism*. Münster: LIT Verlag Münster.

Inikori, J. E. 2002. *Africans and the Industrial Revolution in England: A Study in International Trade and Economic Development*. Cambridge: Cambridge University Press.

Kearns, G. 1993. 'Prologue: Fin De Siècle Geopolitics: Mackinder, Hobson and Theories of Global Closure'. In *Political Geography of the Twentieth Century: A Global Analysis*, edited by Peter James Taylor and Gerard Kearns, 9–30. London: Belhaven Press.

Lähde, V. 2025. 'Are You Confusedocene?' *Aeon* 6 November. https://aeon.co/essays/declared-dead-last-year-the-anthropocene-is-very-much-alive

Law, J. 1989. "Technology and Heterogeneous Engineering: The Case of Portuguese Expansion." In *The Social Construction of Technological Systems: New Directions in the Sociology and History of Technology*, edited by Wiebe E. Bijker, Thomas Parke Hughes, and Trevor J. Pinch, 111–34. Cambridge: MIT Press.

Lawton, R., and C. G. Pooley. 2021. *Britain 1740 – 1950: An Historical Geography*. London: Routledge.

Lewis, S. L., and M. A. Maslin. 2015. "Defining the Anthropocene." *Nature* 519 (7542): 171–80.

Lewis, S. L., and M. A. Maslin. 2018. *The Human Planet: How We Created the Anthropocene*. London: Penguin.

Malm, A. 2013. "The Origins of Fossil Capital: From Water to Steam in the British Cotton Industry." *Historical Materialism* 21 (1): 15–68.

Malm, A. 2014. "Fleeing the Flowing Commons: Robert Thom, Water Reservoir Schemes, and the Shift to Steam Power in Early Nineteenth-Century Britain." *Environmental History* 19 (1): 55–77.

Malm, A. 2016a. *Fossil Capital: The Rise of Steam-Power and the Roots of Global Warming*. London: Verso.

Malm, A. 2016b. "Who Lit This Fire? Approaching the History of the Fossil Economy." *Critical Historical Studies* 3 (2): 215–48.

McCarthy, F. M. G., M. J. Head, C. N. Waters, and J. Zalasiewicz. 2025. "Would Adding the Anthropocene to the Geologic Time Scale Matter?" *AGU Advances* 6 (2): e2024AV00143.

McNeill, J. R. 2013. "Envisioning an Ecological Atlantic, 1500–1850." *Nova Acta Leopoldina* 114 (390): 21–33.

Merchant, B. 2023. *Blood in the Machine: The Origins of the Rebellion Against Big Tech*. London: Hachette.

Mintz, S. W. 1985. *Sweetness and Power: The Place of Sugar in Modern History*. New York: Viking.

Montrie, C. 2018. *The Myth of Silent Spring*. Berkeley: University of California Press.

Moore, J. 2010. "This Lofty Mountain of Silver Could Conquer the Whole World": Potosí and the Political Ecology of Underdevelopment, 1545–1800." *Journal of Philosophical Economics* IV (1): 58–103.

Moore, J. W. 2022. "Anthropocene, Capitalocene & the Flight from World History: Dialectical Universalism & the Geographies of Class Power in the Capitalist World-Ecology, 1492–2022." *Nordia Geographical Publications* 51 (2): 123–46.

Moore, J. W. 2023. 'There Is No Such Thing as a Technological Accident'. In *Technological Accidents*, edited by Jake Brower and Sjoerd van Tuinen. Leiden: V2 Publishing.

Moynihan, T. 2020. *X-Risk: How Humanity Discovered Its Own Extinction*. Cambridge: MIT Press.

Mueller, G. 2021. *Breaking Things at Work: The Luddites Are Right About Why You Hate Your Job*. London: Verso.

O'Brien, P. 1982. "European Economic Development: The Contribution of the Periphery." *The Economic History Review* 35 (1): 1–18.

Otter, C., A. Bashford, J. L. Brooke, F. A. Jonsson, and J. M. Kelly. 2018. "Roundtable: The Anthropocene in British History." *Journal of British Studies* 57 (3): 568–96.

Parthasarathi, P. 2014. *Why Europe Grew Rich and Asia Did Not: Global Economic Divergence, 1600–1850*. Cambridge: Cambridge University Press.

Pawson, E. 1990. 'British Expansion Overseas c. 1730–1914'. In *Historical Geography of England and Wales*, 521–44. London: Elsevier.

Pomeranz, K. 2021. *The Great Divergence: China, Europe, and the Making of the Modern World Economy*. Princeton: Princeton University Press.

Quark, A. A. 2008. 'Toward a New Theory of Change: Socio-Natural Regimes and the Historical Development of the Textiles Commodity Chain'. *Review (*Fernand Braudel *Center)* 31 (1): 1–37.

Ramamurthy, A. 2017. *Imperial Persuaders: Images of Africa and Asia in British Advertising*. Manchester: Manchester University Press.

Retz, T. 2022. "Teaching History on the Scale of the Anthropocene: Three Ethical Challenges." *Historical Encounters* 9 (2): 7–17.

Riello, G. 2015. *Cotton: The Fabric That Made the Modern World*. Cambridge: Cambridge University Press.

Rönnbäck, K. 2018. "On the Economic Importance of the Slave Plantation Complex to the British Economy During the Eighteenth Century: A Value-Added Approach." *Journal of Global History* 13 (3): 309–27.

Rudwick, M. J. S. 2010. *Worlds Before Adam: The Reconstruction of Geohistory in the Age of Reform*. Chicago: University of Chicago Press.

Saldanha, A. 2020. "A Date With Destiny: Racial Capitalism and the Beginnings of the Anthropocene." *Environment and Planning D: Society and Space* 38 (1): 12–34.

Satia, P. 2019. *Empire of Guns: The Violent Making of the Industrial Revolution*. Stanford: Stanford University Press.

Schandl, H., and F. Krausmann. 2007. 'The Great Transformation: A Socio-Metabolic Reading of the Industrialization of the United Kingdom'. In *Socioecological Transitions and Global Change*. Cheltenham: Edward Elgar Publishing.

Shulman, P. A. 2019. *Coal and Empire*. New York: Oxford University Press.

Sloterdijk, P. 2013. *In the World Interior of Capital: Towards a Philosophical Theory of Globalization*. Cambridge: Polity.

Steffen, W., J. Crutzen, and J. R. McNeill. 2007. "The Anthropocene: Are Humans Now Overwhelming the Great Forces of Nature." *Ambio* 36 (8): 614–21.

Thompson, E. P. 2017. 'Time, Work-Discipline, and Industrial Capitalism'. In *Class*, 27–40. Hoboken: John Wiley & Sons, Ltd.

Tocqueville, A. 1835. *Journeys to England and Ireland*. London: Faber & Faber.

Tsing, A., J. Deger, A. Saxena Keleman, and F. Zhou. 2020. *Feral Atlas: The More-Than-Human Anthropocene*. Stanford: Stanford University Press.

Whitehead, M. 2011. *State, Science and the Skies: Governmentalities of the British Atmosphere*. Oxford: Wiley-Blackwell.

Williams, E. 1944. *Capitalism and Slavery*. London: Penguin UK.

Wrigley, E. A. 2013. "Energy and the English Industrial Revolution." *Philosophical Transactions of the Royal Society A: Mathematical, Physical and Engineering Sciences* 371 (1986): 20110568.

Wu, S. X. 2019. "History Matters: Contingency in the Creation of Ecologically Unequal Exchange." In *Ecologically Unequal Exchange: Environmental Injustice in Comparative and Historical Perspective*, edited by R. Scott Frey, Paul K. Gellert, and Harry F. Dahms, 221–41. Cham: Springer International Publishing.

5 Geographies of the Great Acceleration

Introduction

The world changed rapidly in the decades between the end of the World War II and the early 2000s, when Earth System Scientists started talking about the Anthropocene and the post-war 'Great Acceleration'. In those few decades, the number of people tripled, the number of cars rose from 40 million to 850 million, and urban populations rose from 700 million to 3.7 billion. Annual plastic production rose from 1 million to 300 million tons, and the production of nitrogen (mainly for fertiliser), rose from 4 million to 85 million tons (Lane 2019). In the year 2020, it was estimated that the weight of all the stuff humans produce – 'anthropogenic mass' – exceeded all the world's biomass. All the plastic in the world now weighs more than all the animals, while buildings and infrastructure outweigh all of Earth's trees and shrubs. For each person on the planet, 'anthropogenic mass equal to more than his or her bodyweight is produced every week' (Elhacham et al. 2020, 442). Humanity weighs heavily on the world, and the scales tipped most dramatically in the years immediately following the World War II.

Observations like these increasingly led Anthropocene researchers to settle on the period around the middle of the twentieth century as the most convincing onset date of the Anthropocene. Graphs like those of Figure 5.1 dramatically show the uptick in human economic activities and their environmental impacts post-1945, many of which will remain present in geological records for millennia to come. But geologists and Earth System scientists increasingly looked to the radioactive traces of nuclear bomb detonations as a possible geological marker. When bombs started being tested and deployed from 1945 onwards, they left behind a globally synchronous marker 'in high-resolution ice cores, lake and salt marsh sediments, corals, speleothems and tree-rings' (Lewis and Maslin 2015, 176). Nuclear weaponry might not be directly related to the transformations of the Earth System which define the Anthropocene, but radioactive fallout provides a handy 'proxy measure' of humanity's new abilities to manipulate matter and leave its mark in the rocks.

DOI: 10.4324/9781003162469-7

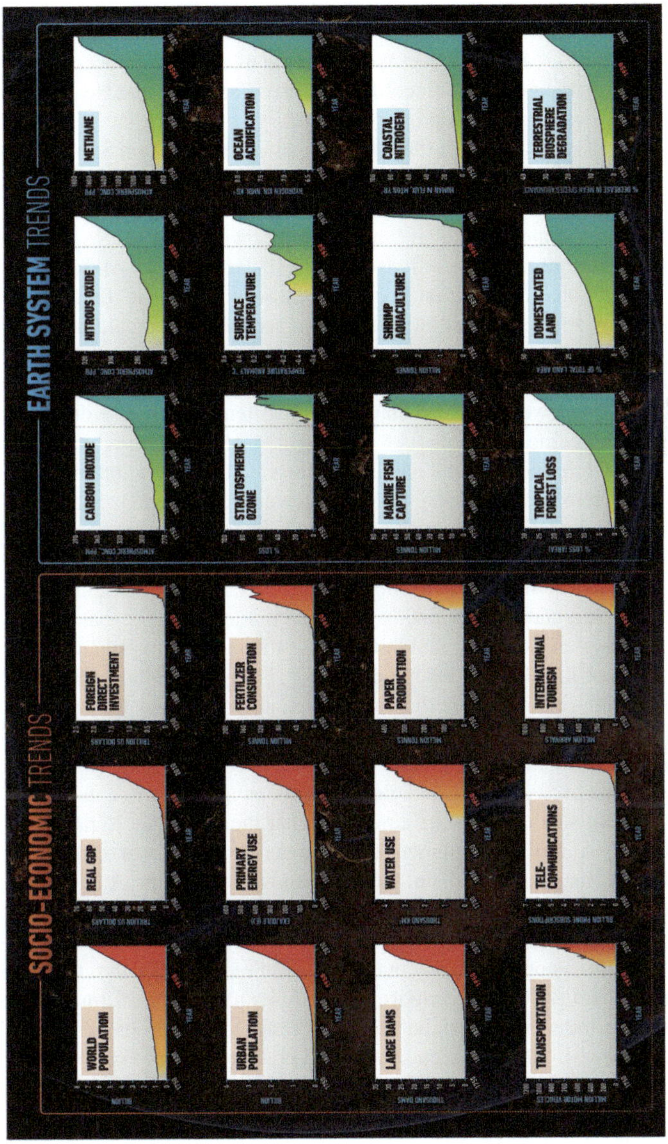

Figure 5.1 Trends in globally aggregated indicators for socio-economic development and Earth System change, 1750–2010, based on data from Steffen et al. (2015). (Source: Globaia. Design: Félix Pharand-Deschênes. Reproduced with permission.)

That mark is particularly visible in the annual sediment layers at the bottom of Crawford Lake in Ontario, Canada, where plutonium deposits sit alongside traces of increased fossil fuel combustion in the 1950s. As such, this site was proposed as the formal location for a 'Global boundary Stratotype Section and Point' (GSSP) – a point where the beginning of the Anthropocene is particularly clear in the geological record (McCarthy, Patterson, and Head 2023). However, the position of this proposed boundary so close to the present (in geological terms at least) made the International Commission on Stratigraphy (ICS) nervous about adopting it in the Geological Time Scale (GTS). To do so would be to define a unit of geological time 'with a span of less than a single human lifetime', which would sit 'uncomfortably within the GTS where the units span thousands or even millions of years' (IUGS 2024, 1). And although nuclear detonations like the first bomb test in New Mexico in 1945 can be dated precisely to the nearest hundredth of a second, the actual underlying drivers and impacts of the Great Acceleration are much more spatially and temporally variable (McCarthy, Patterson, and Head 2023).

In this chapter, we'll explore that spatial and temporal variability in the spirit of those geologists, ecologists and physical geographers who increasingly see the Anthropocene as a whole, and the Great Acceleration more specifically, as a complex, unfolding *event* (Edgeworth et al. 2023). That means examining the historical geographies of the Great Acceleration – its root causes, its situatedness in particular places and times, its unfolding across geographical scales. In the next section we explore how the idea of the 'Great Acceleration' originated from, and has evolved through, novel exchanges between Earth System scientists and historians. After that we investigate claims that the roots of the Great Acceleration lie not simply in the industrialisation of the preceding century, but in the transformations wrought by the two world wars of the first half of the twentieth century. These wars created new relationships between state, society and the environment, new energy systems, and new means of assembling and globally dispersing huge amounts of stuff. The subsequent section explores arguments that the Great Acceleration is a fundamentally American phenomenon, rooted in new attitudes to resource use and a novel consumer capitalism. Next, we move outside the Anglosphere, exploring how growth in the global North was predicated on a continuation of the global resource-gobbling we encountered in the previous chapter. Nonetheless, we also encounter arguments that it wasn't simply Western colonialism or capitalism which drove environmentally-damaging transformations in places such as India and China, but locally specific combinations of state and industrial power. In the final section we build on this insistence on the spatial heterogeneity of the Great Acceleration to explore geographies of knowledge and resistance, building on previous chapters' argument that humanity's long march into the Anthropocene has not only been wildly uneven and defined by inequalities, but has always been marked by loud calls for different directions to be taken.

Disaggregating the Great Acceleration

Like the term 'the Anthropocene' itself, the phrase 'the Great Acceleration' was first aired at an academic conference which brought together Earth System scientists, historians, geographers and others to develop an 'integrated' history of people and the environment (Hibbard, Crutzen, and Lambin 2006). In the years that followed, numerous accounts were offered of conjoined environmental and socioeconomic changes in the period following World War II. These accounts invariably foregrounded a series of graphs, originally published in 2004, depicting a rapid uptick in various indices of change: from global population to the number of McDonald's restaurants, and from global CO_2 emissions to levels of nitrogen transport to the coastal zone. The graphs tell a relentlessly striking story about an unprecedented, multifaceted and global transformation in human-environment relations (Steffen et al. 2015; Steinberg 2019).

The Great Acceleration graphs are the product of collaboration between Earth System scientists and environmental historians, and are a further example of the novel interdisciplinary collaborations occurring under the umbrella of the Anthropocene. The most prominent historical account of the Great Acceleration was offered by AWG member John McNeil and Peter Engelke (McNeill and Engelke 2014). Their book gives detailed descriptions of how indices of global energy consumption, human population growth, economic and urban growth have risen alongside indices of the environmental impacts of such processes. Yet while global correlations between human processes and environmental impacts may justify the designation of a new geological age, they don't by themselves explain the *causation* of such planet-altering transformations.

As Ted Steinberg argued in the *Geographical Review*, McNeil and Engelke were peculiarly silent on the driving forces of these changes (Steinberg 2019). The Great Acceleration, specifically its environmental costs, is posed as a mere 'unforeseen and unintended by-product of actions undertaken for routine quests for wealth, power, and contentment' (McNeill and Engelke 2014, 208–9). But just how routine were these quests?

Steinberg is at pains to point out the irony that the term 'Great Acceleration' is a play on Karl Polanyi's 1944 *The Great Transformation: The Political and Economic Origins of Our Time*, which offered an analysis of the failures of market economics in the run-up to the Great Depression and the subsequent rise of fascism. But while Polanyi's thought is relegated to a footnote in McNeil and Engelke's book – a footnote which suggests Polanyi saw 'the market' as a fairly natural part of the social order – Steinberg emphasises that Polanyi was deeply sceptical of an emerging economic orthodoxy which saw the market as a separate, autonomous entity, to which society must be reconciled by transforming social relations (and relations to the natural world) into relations of economic exchange. Polanyi saw this as a utopian project that could not come to fruition 'without annihilating the human and natural substance of society' (quoted in Steinberg 2019, 266). There was nothing 'routine' about the application of market imperatives to global society – for Polanyi, this was the root cause of the Great Depression and the rise of fascism, and a product of relatively recent transformations in political and economic thinking. These transformations, most notably the commodification of land, labour, and money, were the culmination of the material and ideational transformations described in the preceding two chapters. As such, many historians and geographers interested in identifying the patterns of historical causation *behind* moments of radical change would argue that the Great Acceleration has its roots in the industrial revolution (see Otter et al. 2018), and that the industrial revolution in turn has its roots in the earlier expansion of European empires (e.g. Moore 2017).

As such, there is less critical scholarship on the 'Great Acceleration' version of the Anthropocene as one might expect, especially compared to the flurry of interest in the histories and geographies of colonial environmental transformations and the onset of industrialisation (Lane 2019). As we'll explore below, the critical geographical and historical scholarship that does exist has offered Polanyian critiques of the intensified application of market incentives to new areas of social life, and to humanity's relationship to nonhuman natures, identifying them as the root of the dramatic environmental transformations which are taken to define the 'Great Acceleration'.

Authors of what we may call the 'standard account' of the Great Acceleration have directly responded to criticisms that their account unjustly homogenises the human drivers of planetary transformation. In a 2015 article, Will Steffen and colleagues offered revised versions of the classic graphs which break down the socio-economic trends into OECD countries, the BRICS, and 'Others' (Figure 5.2). This splitting was described as 'important in exploring equity issues in terms of the differential pressures that various groups of countries apply to the Earth System and how the distribution of these pressures among groups is changing through time' (Steffen et al. 2015, 83), and was offered as a direct response to criticisms from social science and humanities scholars of the homogenising gaze of previous constructions of the Great Acceleration (e.g. Malm and Hornborg 2014). This is an important and novel move – when else have Earth System scientists adapted their methods in response to criticism from Marxist historians? – and it is representative of a wider blurring between the Earth sciences and the social sciences as it becomes increasingly difficult and perhaps meaningless to talk about one without the other (Oreskes 2015). But we might argue that these newer graphs still obscure important geographical processes and relationships.

For instance, disaggregating fertiliser consumption in this way shows declining use in OECD countries, but rapid expansion in the BRICS and the 'others'. Perhaps this shows that environmental enlightenment is well underway in the rich world, as consumers demand more planet-friendly, organic

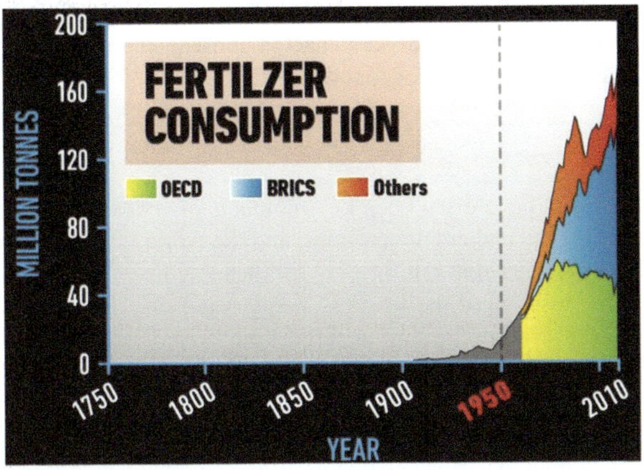

Figure 5.2 Fertilizer consumption, 1750–2010, broken down into OECD countries, the 'BRICS' (Brazil, Russia, India, China, and South Africa), and the rest of the world. Based on data from Steffen et al. (2015). (Source: Globaia. Design: Félix Pharand-Deschênes. Reproduced with permission.)

produce; the rest of the world, meanwhile, needs to 'catch-up' in its embrace of more sustainable forms of agriculture. Such a reading, however, founded upon a representation of territorially-bounded consumption practices, obscures global relationships of production and exchange. For example, it obscures the extent to which artificially fertilised crops are being grown in poorer parts of the world for export and consumption in richer countries. And that's not to mention the often-inequitable transnational politics through which products like nitrogen fertilisers have been introduced into parts of the world as part of agricultural modernisation programmes or 'Green Revolutions' – processes which often increased yields, but which also regularly disadvantaged poorer producers and favoured a marketized agri-food system (Martín 2017). Once we dig into dynamics like these, a more complex picture of the Great Acceleration emerges. To begin piecing that picture together, let's start with the Great Acceleration's early stirrings.

War by other means

While the story of the Great Acceleration can be read as one of a fundamental *break* from the period of global war and crisis of the first half of the twentieth century, new historical and geographical arguments insist instead on the continuity of these two time periods, and see the transformative processes of the Great Acceleration as fundamentally rooted in, rather than distinct from, the geographies of global warfare. Below, we explore this through the geographies of two key materials – oil and aluminium – before briefly reflecting on how the two world wars enacted consequential new relationships between state, society, and the economy.

Oil and war

Historically, wars have often been key moments of 'transition' between different technological and energy regimes, and the world wars of the twentieth century certainly bear that out. In the years leading up to World War I, coal was still king. Just 5% of the world's energy was being drawn from oil, despite huge discoveries of the stuff in parts of the United States, Russia, and the Middle East around the turn of the century. Coal still contributed some 74% of the world's energy and was the staple fuel of military powers such as Great Britain, whose Royal Navy maintained a global infrastructure of coaling stations, supplied both by British coal exports and a global network of mines, to keep its ships moving in all corners of the world (Figure 5.3).

Things started to change in 1911, when the German government reacted angrily to a French military incursion into the interior of Morocco and sent a gunboat to the region, demanding its own new slices of African territory. Britain backed the French position, and there was talk of war, but Germany backed down after negotiations with France led to the transfer of land from French Congo to German Cameroon. Nonetheless, the incident brought home to British strategists the potential trouble that a powerful German navy, equipped perhaps with an Atlantic base, might pose to the hegemony of the Royal Navy. Britain's Home Secretary at the time, Winston Churchill, proposed that the solution lay in transitioning the Navy from coal to oil, persuaded as he was by the greater efficiency and speed that the new fuel offered. The Navy agreed, and Churchill was made First Lord of the Admiralty. But the empire had no oil of its own. So, in 1914, when pan-European tensions were about to break out into war, the British government bought a 51% share in Anglo-Persian oil, along with a 20-year supply contract for the Royal Navy (Bonneuil and Fressoz 2016).

For scholars of 'sociotechnical transitions' – i.e. the processes by which societies move from being organised around one set of technologies and energy sources to another – major wars are significant moments. They represent intensified 'selection environments', where different technologies compete to help actors who are pursuing fairly singular goals – in this case, military victory. Johnstone and McLeish (2020) argue that World War I played a significant role in the move from coal- to oil-based mobility technologies. Some oft-quoted numbers illustrate the point neatly: at the war's outset, the UK military had 823 automobiles, 15 motorbikes, and 183 aircraft at its disposal; by the end of hostilities, it boasted 23,000 motorcars, 34,000 motorbikes, 63,000 trucks, and over 34,000 aircraft. We shouldn't lose sight of the fact that World War I was fought with some very old technologies too – horses,

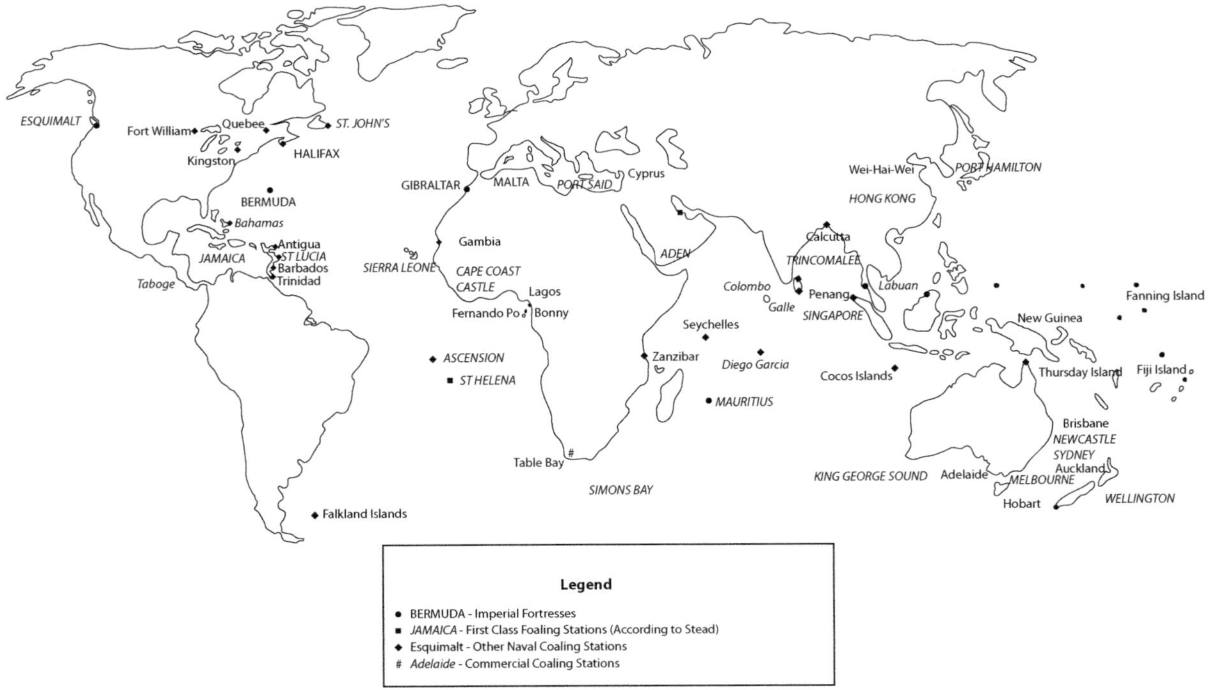

Figure 5.3 British coaling stations used by the Royal Navy, 1870–1914. (Source: Gray 2017, 96.)

trains, nineteenth-century rifles (Edgerton 2019) – and the German military remained reliant on coal and coal-derived liquid fuels. For many commentators at war's end, this difference proved decisive: 'the allied cause had floated to victory on a wave of oil', Britain's Lord Curzon reminisced soon after the Armistice (Buell 2012, 289–90). That wave had largely come from America, and while Britain began the war as a net energy exporter, it now risked a new dependence on American oil. The identification of oil as a new 'strategic resource' led to the opening of new resource frontiers and land grabs in the 'Middle East', as Britain, France, and the United States tussled for access and control over oil-rich areas of the collapsing Ottoman Empire. By 1929, British companies were producing 41% of all the oil produced outside of the United States, but supply routes through the Mediterranean would be severely hampered at the outset of World War II, and American oil would prove crucial to Allied war efforts again (Johnstone and McLeish 2020).

If World War I represented a victory of 'trucks over trains', that struggle would continue on the cultural terrain during the interwar years, particularly in North America. A 'culture war' broke out over whether road or rail was the mobility system of the future (Mom 2014), after World War I had 'popularised a product [petroleum] that had heretofore been viewed as elitist' (Johnstone and McLeish 2020, 7). Automobility became both a means of getting to work and of taking enjoyable jaunts in the countryside (Urry 2007), at the same time as it became central to military planning. The United States engaged in a comprehensive programme of highway construction, in which matters of military mobility were foremost. Nonetheless, World War II saw 'a marked increased dependence on oil compared to the First World War' (Johnstone and McLeish 2020, 8), with American forces in Europe, for example, burning through one hundred times more gasoline than they had during World War I, and the average American soldier consuming '228 times more energy than its predecessor' (Bonneuil and Fressoz 2016, 144).

The petrol-fuelled combat of World War II required the maintenance of a constant and abundant supply, which meant a rapid transformation of oil logistics (Figure 5.4). Britain became dependent on the United States for 90% of its oil supply, supplied through new pipelines in the United States such as Big Inch and Little Big Inch, which to this day connect Texan oilfields to the Atlantic coast in New Jersey. In Britain itself, hundreds of miles of underground oil pipelines were secretly constructed to connect docks and refineries to airfields along the east coast, an infrastructure that would form the basis of Britain's post-war civil aviation boom.

World War II saw oil seep into other aspects of social life too, beyond the fuel tanks of military aeroplanes and trucks (Johnstone and McLeish 2022). Concerted efforts to increase agricultural productivity saw horses replaced with tractors, and oil-derived fertilisers and pesticides proliferated. The production of plastics in the United States almost tripled between 1940 and 1945, and artificial fabrics became clothing staples. By war's end, oil did 'not just

Figure 5.4 A World War II propaganda poster produced by the U.S. Petroleum Industry War Council. (Source: National Archives at College Park, MD. Public domain, via Wikimedia Commons.)

fuel Americans' vehicles', but had 'changed their diet, their clothes, their neighbourhoods, their jobs, their fun – in fact, everything about U.S. society' (Black 2012, quoted in Johnstone and McLeish 2020, 9). Johnstone and McLeish are nonetheless at pains to stress the historical contingency of the transition to oil. Their socio-technical perspective foregrounds the political choices involved in moves from one 'regime' to another, and the 'historically contingent factors' – including things like wars – 'whereby certain energy trajectories become dominant and alternatives are 'closed down'' (Johnstone and McLeish 2020, 2). What is so significant about the two world wars in this view is not the oil that was directly consumed by them (or the resulting greenhouse gas emissions), but the socio-technical trajectories they set the world upon. To get a clearer view of this, let's turn to another material that was central to military strategy and to subsequent patterns of resource use and consumption – aluminium.

Baked electricity: repurposing the war machine

The recognition of the growing military and geopolitical significance of aerial warfare was recognised in World War I, and by the eve of World War II, strategists conceived the coming conflagration as one whose outcome would be ultimately decided by aerial supremacy (Adey 2010). That meant lots of planes, and lots of planes meant lots of aluminium – a uniquely lightweight, flexible and durable metal, one of the world's most common elements but whose ore, bauxite, is only available at industrial strength in a few locations. As British aircraft manufacturers lost access to southern European bauxite with the outbreak of World War II, the British government turned to bauxite reserves in British Guiana and to expanding production facilities in Canada. The Canadian aluminium producer Alcan benefited from imperial preferences for commonwealth companies in British Guiana, where it opened new bauxite mines along the Demerara River. Geographer Matthew Evenden has shown how this development of a new bauxite mining frontier 'reads like a caricature of dependency theory', with a 'light tax burden…cheap wage labour, readily available in the context of a declining plantation economy…[a] company town divided along racial lines' (Evenden 2011, 73–75). Expanding the mines devastated tropical forests and left legacy landscapes 'best described as lunar: pocked, mineralized surfaces, devoid of topsoil, fora, or fauna' which, even with the best rehabilitation programmes (which certainly didn't exist in colonial British Guiana) would take decades to recover (Evenden 2011, 83).

From British Guiana, bauxite was gathered together at Trinidad, another British colony, and then shipped north towards Canada, where the British government had financed the expansion of aluminium production facilities in Quebec. The enormous amounts of energy required to extract aluminium from bauxite had led to it being christened 'baked electricity', and the smelting plants were served by a new regional power grid stretching across Quebec and Ontario and linking together numerous dams and hydroelectric plants. In ways not dissimilar to the nineteenth century expansion of the Canadian timber frontier (see Chapter 3), this new power system fundamentally transformed riverine ecologies and intruded on Indigenous lands and livelihoods, while the smelting process released known carcinogens into the air and water, which have since been linked to heightened cancer rates in the region in both humans and in St Lawrence beluga whales.

In a few short years, the ecologies and economies of British Guiana and Canada were 'practically conjoined', through a 'superstructure of corporate organization, capital investment, and complementary political jurisdictions within the British Commonwealth' (Evenden 2011, 80). Bauxite extraction in British Guiana increased fourfold over the course of the war, while the local government saw a 'rent' on the material of only 1% of the aluminium market value. Over the same period, Canada provided 90% of British and Commonwealth aluminium, and global production tripled. Much of it went into aeroplanes and ships, to be mangled in the course of battle, to be repurposed afterwards, or else discarded in one of the many military waste dumps which would spring up after the war; just one instance of the 'unprecedented capacity of the Second World War to gather and scatter materials with untold human and environmental consequences' (Evenden 2011, 86; see also Robertson et al. 2020).

Evenden's account of the historical geographies of the aluminium 'commodity chain' has much in common with Jason Moore's analysis of commodity frontiers. The bauxite frontier in British Guiana, as well as in other colonial settings like Suriname and Jamaica, was defined by the relatively easy appropriation of land, of low-cost labour, and of 'Cheap Natures' in the form of the commodity itself and the degraded ecologies for which little to no compensation was payable (see Chapter 3). It is also a reminder, following Clifford and Castonguay (2022), that it is not enough to simply quantify the colonial and overseas resources flowing into 'core' countries like Great Britain; historical geographies of the Anthropocene need also attend to the local transformations of societies and natures at restless commodity frontiers and at other points along increasingly global commodity chains, with their interconnected sites of extraction, processing, manufacturing, distribution, consumption, and disposal. This is an approach which has animated much recent geographical work on the social and environmental transformations wrought by present-day as well as historical commodity chains (Hulme 2015; Parsons, Safra de Campos, and Moncaster 2022).

But what to do with all this aluminium – or more pertinetly, this aluminium production capacity – once the war ended? As Evenden puts it, 'domestic consumption rose to absorb the new surplus', helped along the way by marketing campaigns which closely aligned aluminium with modernity. Companies like Alcoa publicised the material 'as a reliable and desirable domestic material useful in a host of consumer goods', from saucepans to motor scooters, some military aircraft – and aircraft factories – were given over to the booming civil aviation market, and the UK government 'employed aluminium surplus as housing material, drawing tightly the associations among social reform, housing, and modern materials' in its efforts to remake urban geographies which had been transformed by bombs dropped from aluminium planes (Evenden 2011, 85). Aluminium was but one element of a wider repurposing of wartime industrial production, which in the United States had tripled during the war years, spurred by the sudden influx of 70% of all government spending. Chemical production tripled, shipbuilding rose by an order of magnitude, and aircraft production was multiplied fifteen times (Bonneuil and Fressoz 2016). From this industrial surplus, historians have argued, emerged a radically new 'consumer capitalism' which would drive the Great Acceleration.

The state, economy, and society

This 'governmental coordination of scientific research and industrial development' (Holmes, Bolen, and Kirkbride 2021, 184) continued into the postwar and Cold War period in the US, and can be detected in projects such as those of the Tennessee Valley Authority to remake environments to generate vast amounts of hydroelectricity (on an even grander scale than Quebec's aluminium complex), and in the vast infrastructure of nuclear weapons and energy production which came to define Cold War technoindustrialism, and whose radioactive traces will signal the onset of the Great Acceleration in the rocks for millennia to come (Head, Steffen, and Fagerlind 2021). Holmes et al. dub this marriage of the state, science, and industry 'the Anthropocene mode of production', and argue that it helped kick-start the Great Acceleration, first in the United States, and then elsewhere in the world as it was exported, through diplomacy and 'development' programmes, as a model for other countries to follow. This is an argument that challenges conventional Marxist analyses of the internal logics of capitalism, forcing analysts to pay close attention to the role of the state in processes of capital accumulation and environment-making (Parenti 2015). This is a relationship that is often particularly strong in times of war (e.g. Robertson et al. 2020), but is a factor that is often backgrounded in accounts like Jason Moore's.

World War II saw other transformations in the relationship between state, society, and economy too, at least in some of the main belligerent countries. The historian Thomas Hippler argues that with the onset of aerial bombing in the European theatre of war, and the enrolment of entire national populations in the war machine – as fighting and labouring bodies, and as potential targets of attack –the state had to take a much more direct role in people's lives. The state had to keep people alive and happy so they could adequately contribute to the war effort, as well as to resolve or neuter the social tensions caused by the burdens of warfare falling most heavily on the working class. In countries like Great Britain and Germany, populations had to be newly 'nationalized', and 'a national social state had to be created in order to create this nationalized 'people', ready not only to endure famine and bombs, but also to fight against other peoples, similarly nationalized'. As Hippler puts it, bombing was 'the hell of a world whose paradise is social security' (Hippler 2017, 117). That is, the state had to strike a new bargain with its nationalized people. Social security, social housing, free healthcare – the things now recognised as products of the post-war 'welfare state' – were the compensation for the participation of 'the people' in the conduct of global war.

Historian and influential Anthropocene analyst Dipesh Chakrabarty (2014) has argued that this emergence of the national population as a new kind of biopolitical object to be managed is directly connected to the environmental transformations that came next. Ballooning populations in many parts of the world, helped by medical breakthroughs and new attitudes to 'public health' which have their origins in war, led to the increased levels of production and consumption that triggered the Great Acceleration. These population explosions weren't driven by 'unbridled capitalism', but by new relationships between the state and society. Population has since become a controversial topic of environmental debate. For many, the 'population problem' now is not so much how to keep it healthy and productive, but how to manage populations at national or global scales so that their environmental impacts can be reined in (Chakrabarty 2014, 11). Others argue that focusing on human population as an environmental problem simply shifts blame to the more populous (and often poorer) regions of the world, and away from those with greater historical responsibility for environmental change (e.g. Egerö 2013).

An American Anthropocene?

While the nuclear explosions and radioactive fallout which geologists proposed as the marker of the Anthropocene's onset may have been little more than a useful 'signifier' in the rocks, having 'nothing directly to do with the Anthropocene' and the human transformation of the Earth System, they nonetheless tell an important related story of 'the dawn of the era of global economic domination by the United States of America' (Hamilton 2015, 105). This new dawn was a product both of material transformations and of transformational ideas. In particular, new ideas about economies and resources helped kick-start new processes of accumulation. Attending to these conjoined histories of idea-making, power-creating, and Earth-moving (Moore 2015) can give a more refined sense of the historical geography of the Great Acceleration. It was a set of transformations which began in particular places; a situated product of particular minds, materials, and markets.

Resources for freedom

World War II's orgy of raw materials consumption meant that the early post-war years in the United States and elsewhere were marked by concerns about resource scarcity among scientists and policymakers. Popular books like Fairfield Osborn's *Our Plundered Planet* and William Vogt's *Road to Survival*, both published in 1948, offered pessimistic readings of the exhaustion of global resources and its troubling social and economic implications. They formed part of a new imaginative geography, summoned into being by geological prospectors and economic policymakers as well as popular authors, which intensified the late-nineteenth century discourse of 'global closure' and of finite geographical and resource frontiers. This imaginative geography 'helped reduce the vast empty spaces revealed by Mercator's maps into a more tightly interconnected, densely packed and closer planet' (Lane 2019, 13).

Against the background of these neo-Malthusian concerns about resource scarcity (Linnér 2023), the influential U.S. President's Materials Policy Commission (PMPC) – known colloquially as the Paley Commission after its chairperson – was launched to investigate how the United States could navigate this new world of scarcity. The Commission's 1952 report, titled *Resources for Freedom: Foundations for Growth and Security*, argued that the United States need not worry. It offered a new way of accounting for resource scarcity or abundance that was based not on scientific analyses of stocks of available resources (like geological estimates of national oil reserves), but on economic analyses of the costs of extraction and the prices of end products. Resource depletion was therefore expressed not through the absolute depletion of physical stocks, but through rising costs determined by the 'relative depletion of stocks with respect to each other'. This was a conceptual shift from *absolute* to *relative* scarcity. Material resources were 'publicly reconceptualised…from the geological metaphor of an absolutely exhaustible reservoir, to a business accounting metaphor of replenishable working inventory' (Lane 2019, 16). And crucially, that 'inventory' of economically viable resources could be taken to include resources located anywhere, not just within national or imperial borders. While neo-Malthusians advocated resource conservation and national self-sufficiency, the Paley Commission advocated that the United States expand the extraction of overseas resources. Securing 'resources for freedom' meant that 'we must find or create that environment outside our shores' (*Fortune* magazine, 1952, quoted in Lane 2019, 16).

Growing that resource base was seen as key to growing the economy. This was the period when *growth* became the central goal of economic policymaking: up until the 1940s, 'growth' had hardly featured in economic discourse; in the early 1950s, the 'growth paradigm' became policy orthodoxy across the non-communist world, while 'development' was articulated as the new goal for emerging postcolonial countries as formal European empires began to retreat. New economic accounting systems meant that, for the first time, the size and growth rates of national economies could be directly compared, spurring a dash in the United States to outgrow the USSR, and enabling the articulation of growth targets as explicit policy goals at national and super-national scales (Mitchell 2014).

Much of that growth, the Paley report influentially argued, could be achieved by allowing private enterprises 'unfettered' access to the raw materials of the whole world, an idea which had already been articulated at the highest levels of U.S. policymaking at the close of World War II. In May

1945, U.S. Secretary of the Interior Harold Ickes argued, in a letter to President Roosevelt:

> It is essential...that we fulfil the Atlantic Charter declaration of providing access, on equal terms, by all [Western] nations to the raw materials of the world
>
> (quoted in Bonneuil and Fressoz 2016, 246)

Western governments, particularly the US, directly supported the 'unfettered access' of private enterprise to the 'raw materials of the world', and foreign investment in the extractive industries, most notably for oil and minerals in the global South, ballooned. It is for these reasons, both material and ideational, that Lane (2019) dubs the Great Acceleration the 'American Anthropocene' – situating the historical changes in a particular place, and tracing their unfolding through the relationships of that place to the rest of the world.

The rise of consumer capitalism

Humans have always 'consumed' the necessities of life – food, water, fuel, and clothing to keep warm – and have worked in order to obtain them. But it was in eighteenth century England that widespread and concerted attempts began to be made to dislodge a culture of thrift and to instead encourage elaborate spending on industrially manufactured goods, in part by appealing to people's status anxieties and by fostering envy of those who had more (Gregory 1990; McKendrick, Brewer, and Plumb 1982). Nonetheless, the consumption of the nineteenth century's sumptuous clothing, furniture, and pottery remained largely a preserve of the upper-middle classes (Vries 2008). In the second half of that century, industrialisation took off in North America, and by the 1920s, concern abounded among business leaders and economists about the spectre of overproduction – the United States was producing too much stuff, and people needed to be persuaded to buy it. 'Goods are plentiful', the economist John Kennett Galbraith remarked in 1958, so 'demand for them must be elaborately contrived' (quoted in Higgs 2016, 74). It's in this period that we can see the consequent rise of the advertising industry, department stores, mail-order catalogues, and financial arrangements which allowed people to 'buy now, pay later' (Leach 2011). Widespread electrification enabled the democratisation of energy-intensive consumption, powering devices like home radios, which in turn gave advertisers unprecedented access to the public mind, helping to 'democratize desire' and spread consumerist urges more widely (Higgs 2016).

As the pioneer of Public Relations (PR), Edward Bernays put it in his 1928 book *Propaganda*:

> Mass production is profitable only if its rhythm can be maintained—that is if it can continue to sell its product in steady or increasing quantity.... Today

> supply must actively seek to create its corresponding demand … [and] cannot afford to wait until the public asks for its product; it must maintain constant touch, through advertising and propaganda … to assure itself the continuous demand which alone will make its costly plant profitable.
>
> (quoted in Higgs 2016, 71)

Bernays was a nephew of Sigmund Freud, and reasoned that the work of advertising and PR was to excavate and titillate the repressed desires which, Freud famously argued, shape the mental life of citizens of modernity. Bernays and his fellow propagandists likewise saw their role as helping industry respond to the new wave of suffrage spreading through the industrialised world. As more people earned a democratic claim on the conduct of government, it was the job of PR to ensure that public sentiment remained sympathetic to the interests of corporations and capital, lest they fall prey to more revolutionary, anti-capitalist ideas. They were a self-proclaimed 'invisible government' pulling the strings of the public mind (Ewen 1996).

The financial crash and Great Depression from 1929 onwards put the brakes on this expansion of the 'consumer society', but Bernays' thought would find new applications in the post-war period. The massive increase in industrial production during the war years threatened a new crisis of overproduction, and consumer desire and demand needed to be hoisted to new levels. This time around, in the United States at least, relatively high-paying, secure, and unionised jobs in manufacturing industries gave workers new levels of spending power. Henry Ford offered double the wages of his car-making competitors both to retain his workforce and to ensure they could afford the cars they were making. Advertisers were quick to try and direct this new spending power towards new consumer goods. The Paley report's promise of infinite growth was accompanied by the promise of ever more luxurious standards of consumption, and increasingly the site of that consumption was envisaged as the suburbs (Box 5.1).

Cheap petroleum helped car use explode, and places like roadside drive-thru restaurants and drive-in cinemas meant that the consumption of both food and culture could take place within a car (Black 2012). Afterwards, one could repair to one's suburban home, which was envisaged as fully electrified and replete with the latest refrigerators, self-cleaning ovens, and television sets. Bonneuil and Fressoz (2016) identify what they call a new form of 'disciplinary hedonism' in efforts to get these new products into the American home. Innovations in consumer credit and credit ratings, trademark laws, and advertising – most notably by this point on television – helped pull the strings of consumer desire and transform cultural values. The television was perhaps 'the most powerful weapon of the dominant producers', enabling the 'intensive indoctrination' of whole families in their living rooms (Lebow 1955, 7). Meanwhile, practices like 'repairing, economizing and saving were presented as outdated habits harmful to the national economy, while repeated and ostentatious consumption, fashion and obsolescence became respectable objectives' (Bonneuil and Fressoz 2016, 157).

Box 5.1 Situating the Anthropocene: the shopping mall

Geographer David Harvey situates the emergence of a new bourgeois consumer culture in mid-nineteenth century Paris as a direct consequence of capital over-accumulation (Harvey 2018). The German philosopher Walter Benjamin famously spent years studying the spatial expression of this new culture in his unfinished 'Arcades' project. As a guidebook to Paris described them:

> These arcades, a recent invention of industrial luxury, are glass-roofed, marble-panelled corridors extending through whole blocks of buildings, whose owners have joined together for such enterprises. Lining both sides of the corridors, which get their light from above, are the most elegant shops, so that the arcade is a city, a world in miniature, in which customers will find everything they need.
>
> (quoted in Benjamin 1999, 31)

In post-war North America, these 'worlds in miniature' would migrate to the suburbs, following the housing developers' pursuit of cheap land and, of course, the incoming suburban population's newfound spending power. 'Developers methodically surveyed, divided, and appropriated suburban cornfields and orange groves to create a new landscape of consumption' (Crawford 1992, 7), replicating standardised designs across suburban locations with promises of spectacular returns on investment (Kunstler 1994; Woodward 1998). Conventionally, two department stores would be the 'anchors' at either end of the mall, with a series of smaller shops in between to tempt shoppers wandering between the two poles. But as well as shopping destinations, mall designers also report having a strong sense of civic duty to produce 'new town squares' (Woodward 1998), manipulating light and space to simulate the bustle of downtown (Figure 5.5). Nonetheless, critics have argued that this offered little more than a 'fantasy urbanism' which purported to strip away "the city's negative aspects: weather, traffic, and poor people" (Crawford 1992, 22).

Crucially, the suburban shopping mall offered consumption as a recreational leisure activity – shopping as something to enjoy and take one's time over, rather than a chore to be completed before other aspects of life could be enjoyed (Miles 2010). More recent malls like the giant West Edmonton Mall in Canada took this even further, offering the chance to voyage around the world in sight, touch, and smell, in a refrain of earlier colonial conceits of gathering together the riches and resources of the world under one roof. At West Edmonton, one can even go aboard a replica of the Santa Maria, the ship on which Columbus discovered the 'New World' and perhaps set humanity on the path towards the Anthropocene, and towards the kind of world-ecology that made something like the West Edmonton Mall possible (see Chapter 3).

This 'Disneyfication' of consumption spaces blurred the line between consumption and experience, most notably through the engineering of 'atmospheres' – the manipulation of things like sound and smell (ever noticed the fresh cookie smell coming through the air vents?) – which encourages people to part with their cash (Horton and Kraftl 2014). But while shopping mall architects strove for a simulation of public space, social life within mall spaces was tightly controlled, with anything from sitting on the floor to political leafletting likely to get you moved along by a security guard (Miles 2010). Mall space was tightly disciplined, and as such became a site of social exclusion. Some have argued that in consumer societies, the most important social distinction is not between the capital-owning bourgeoisie and the labouring proletariat, but between the creditworthy and the 'flawed consumers' who are unable to access a life of conspicuous consumption. Advertising and consumption spaces were geared around the 'seduction' of the former, while the latter were actively excluded from credit arrangements and, through the actions of CCTV, security guards, and consumer profiling, from consumption spaces such as shopping malls (Clarke 2004).

While we shouldn't lose sight of other spaces where mundane acts of consumption occur – like supermarkets and corner shops, and more recently of course the internet – shopping malls undoubtedly capture some important elements of geographies of post-war consumer capitalism: the production of suburban space, the spatial design of 'disciplinary hedonism', and the transformation of consumption into an experiential leisure activity.

It is in this period that 'planned obsolescence' (London 1932) began to fully creep into consumer-driven economies – i.e. the production of commodities designed not for durability but for rapid exhaustion, thus requiring their frequent replacement (Slade 2009). Retail economist and marketing consultant Victor Lebow summed up the new status of consumption in 1955:

> Our enormously productive economy demands that we make consumption our way of life, that we convert the buying and use of goods into rituals, that we seek our spiritual satisfactions, our ego satisfactions, in consumption. The measure of social status, of social acceptance, of prestige, is now to be found in our consumption patterns.
>
> (Lebow 1955, 7)

Figure 5.5 Newspaper advert for the opening of a new mall in Birmingham, Alabama in 1960. (Source: *The Birmingham News*, via Wikimedia Commons.)

The new measures of the consumer's social status – 'what he [sic] wears, drives, eats' etc., needed to be offered with 'a special urgency. We need things consumed, burned up, worn out, replaced, and discarded at an ever-increasing pace', Lebow argued (1955). But planned obsolescence wasn't just built into the material fabric of consumer commodities, but into the consumer's mind too; the consumer needed to *feel* the obsolescence of a product, and its need for replacement, even if the product was still perfectly capable of fulfilling its purposes (Higgs 2016).

Bonneuil and Fressoz's notion of 'disciplinary hedonism' calls to mind philosopher Michel Foucault's observation about how discipline, in modern societies, often works not through the application of rules, and punishment and violence when they're broken, but through the creation of spaces in which people effectively discipline themselves into behaving in certain ways, without feeling that they are being directly conditioned by powerful others (Foucault 1979). But within those spaces we can also see counter-hegemonic practices emerging (Box 5.2); practices which perhaps point the way towards more sustainable futures.

Box 5.2 Resisting consumerism?

Francine: Why did they come here?
Stephen: Some kind of instinct, memory of what they used to do. This was an important place in their lives.

Stephen's answer to Francine's question in George A. Romero's 1978 horror flick *Dawn of the Dead* (Figure 5.6) comes as our heroes watch a gaggle of zombies staggering around a typical suburban shopping mall. Critics disagree over whether Romero's script offers a critique of consumerism, whereby the zombies are metaphors for those of us duped and seduced by the disciplinary hedonism of consumer capitalism, or whether the zombies are instead a metaphor for capitalism itself, and its mindless feasting on the labouring bodies of the proletariat and on the very flesh of the Earth.[1] These two interpretations invite questions of agency within consumer capitalism: is the 'disciplinary hedonism' of post-war capitalism so powerful that people can be essentially turned into mindless zombies? This is certainly what people like Edward Bernays thought – people were largely incapable of logical thought, and could therefore be manipulated into thinking and doing (and buying) almost anything. Geographers, though, aren't so sure, and have been keen to restore some agency to those who have often been seen by the powerful as simple dupes.

The historical and cultural geographies of consumption have become a lively research area. Cultural geographers emphasise consumers' own agency, and ask *what consumers actually do, why, and how is consumption made meaningful*? While a 'consumer society' might be defined as one where meaning and identity are primarily derived from acts of consumption rather than production, geographers have sought to complicate the consumption-meaning relationship (Jayne 2006).

Figure 5.6 Poster for Romero's Dawn of the Dead. (Source: Sufefa, CC BY-SA 4.0 https://creativecommons.org/licenses/by-sa/4.0, via Wikimedia Commons.)

Work in this vein has stressed, in the words of cultural theorist Paul Willis, how consumption is central to the 'vibrant symbolic life and symbolic creativity in everyday life, activity and expression… through which individuals and groups seek creatively to establish their presence, identity and meaning' (quoted in Horton and Kraftl 2014, 67).

Rather than mindlessly consuming whatever advertisers or architects have prompted them to consume, this reading suggests that consumers, through the choice of material objects they collect and display on or about their person, are engaged in a profoundly creative act of establishing and communicating identities. Often, these identities are counter-hegemonic and defined in opposition to dominant cultural values. The appropriation and redeployment of the humble safety pin as an item of bodily adornment symbolising belonging to the punk movement is but one example of how acts of consumption can function as small acts of resistance to dominant cultural norms.

Resistance in the name of sustainability is another example. From consumers spending a little more for organic or carbon-neutral cotton t-shirts, through to hundreds of protestors lying down on the floor of a shopping mall to protest against unsustainable production and consumption practices, spaces of consumption became, over the course of the Great Acceleration period, spaces of resistance too.

Whether encouraging consumers to make more 'ethical' or 'green' choices in their consumption practices is enough to make any real difference to consumer capitalism's environmental impacts is a lively debate – 'sustainable consumption' might instead mean entirely delinking one's consumption practices from the material flows and cultural values of global capitalism (Middlemiss 2018; Whitehead 2014), and fostering the kind of alternative economic systems and spaces which we'll encounter in Chapter 10. As the role of malls and department stores has started to wane in an era of online shopping, some such spaces are being repurposed for more community-led and sustainability-oriented uses and might even point to possibilities for post-capitalist urban futures (see Miller 2026).

Beyond the Anglozone

While arguments about the Great Acceleration being a process driven in large part by political and economic changes in the United States are compelling, it is insufficient to view the Great Acceleration only from a vantage point in North America. The Great Acceleration, like the Industrial Revolution and colonial transformations which preceded it, was a global process; an epiphenomenon of changing patterns of trade, production, and consumption, and, as postcolonial scholarship shows, it was a set of processes which were experienced very differently in different parts of the world. For example, emerging historical scholarship is highlighting how most Anglophone accounts of the Great Acceleration overlook the economic logics, colonial practices, and attitudes to 'nature' which defined the 'socialist Anthropocene' of the USSR and allied countries (Fowkes and Fowkes 2025). To begin building a fuller geography of the Great Acceleration, we need to attend to this wider, global story.

National electricity: the Great Acceleration beyond the Anglozone

Holmes, Bolen, and Kirkbride's (2021) identification of a distinctive 'Anthropocene mode of production' defined by state coordination of scientific research and industrial development was far from just a Western phenomenon. We might consider here the positioning of nature as a 'class enemy' of the Soviet state, to be subdued by state-directed technoscience and industrial development (see Josephson 2010; Olšáková 2016). Relatedly, Elizabeth Chatterjee (2020) argues that a defining feature of the Great Acceleration is the emergence of a comparable mode of economic production in the major economies of South and East Asia. Chatterjee argues that Marxist-informed work on the Capitalocene, which identifies the origins of the Anthropocene in the commodification of land and labour during early European colonial expansion (e.g. Moore 2017, Chapter 3), or in the projection of British steam power onto the rest of the world (e.g. Malm 2016, Chapter 4), problematically assumes that a Western world-ecology, or fossil capitalism, was simply projected from the North Atlantic region onto a largely passive world (see also Kumar 2022). While such narratives challenge the universalism of standard accounts of the Anthropocene, Chatterjee suggests they do so by replicating an 'old, Eurocentric master narrative of modernity' (Chatterjee 2020, 6), which largely erases Asian experiences of colonialism, industrialisation, and modernisation (see also Curley and Smith 2024).

Chatterjee challenges any simple equation of colonialism and fossil capitalism by focusing on the geographical spread of grid electricity – 'the key macro-technology of the twentieth century', a 'crucial component of the fossil economy's spread', and a key driver of fossil fuel consumption during the Great Acceleration (Chatterjee 2020, 4). In British India, for example, she explores how electrification efforts were, rather predictably, skewed towards the interests of colonisers and capital. While electrification favoured things like factories and warehouses in areas under direct British rule, electrification was more widespread and reached into more private homes in areas where British control was looser. In China and Japan, she documents how nationalist, anti-imperial projects were often pursued through efforts to further exploit things like coal reserves. In mid-twentieth century Brazil, the development of a national oil industry was similarly an effort to shake off the shackles of colonial resource exploitation and cultivate a new kind of national economic freedom (Acker 2020). In these historical examples, the development of fossil capitalism was explained and justified in overtly *anti*-colonial terms.

By the time of India's independence in 1947, it had the electricity generation capacity of a medium-sized coal plant of today, and just 0.4% of its 2018 capacity. Following decolonisation, electrification – through both coal and hydropower – was seen as central to state-led development and modernisation, and an electricity supply was increasingly seen as a fundamental right of citizens. As such, Chatterjee contends, electricity was decommodified, its production and distribution taken away from the market and placed in state hands, and its distribution to private citizens and households often prioritised over its distribution to industrial users:

> Postcolonial efforts at accelerated late development were not driven by an elite of capitalist producers, nor did they neatly entail the commodification of energy. Instead, the state spearheaded electrification, now conceived of as a moral project essential for the flourishing and legitimacy of the new nation.
>
> (Chatterjee 2020, 11–12)

In India, this saw the developmentalism of Jawaharlal Nehru, India's first Prime Minister upon independence, triumph over Gandhian doctrines of simplicity and *swadeshi*, or self-sufficiency. This developmentalism prioritised technology-led economic growth and was expressed in huge dam projects, widespread electrification, and efforts to 'modernise' agriculture. It was a process which was repeated and adapted across a range of different socioeconomic and political contexts too, 'from democratic India's mixed economy to East Asian developmental states and authoritarian petro-states… across much of southern and eastern Asia a shared postcolonial project of fossil developmentalism became visible' (Chatterjee 2020, 12). Chatterjee defines 'fossil developmentalism' in contrast to 'fossil capitalism', to emphasise the state-led acceleration of fossil fuel use and electricity consumption which took place in a range of societies distinct from those of the capitalist 'Anglozone', in a process which cut across the logics of capitalist commodification (see also Chakrabarty 2014).

These kinds of arguments echo scholarship in historical geography on 'modernity', which has argued against the idea that a singular European modernity (as defined by things like new technologies and sciences, forms of political organisation, or forms of artistic and architectural expression) simply diffused to other parts of the world. Historical geographers have instead emphasised the 'heterogeneous panoply of modernities' existing and experienced across 'landscapes, networks and frontiers' and across regional, national and global scales (Ogborn 2001, 121). We might make similar arguments about the Great Acceleration or, perhaps more usefully, its proposed drivers – colonialism, fossil capitalism, fossil developmentalism, or the 'Anthropocene mode of production' (Holmes, Bolen, and Kirkbride 2021). All these drivers and their consequences have played out and been experienced differently in different parts of the globe.

For example, Regulation Theory (RT) argues that capitalism doesn't unfold according to a series of universal laws, but is shaped in important ways by *local* (i.e. national or transnational) cultures and institutions. It is a theory 'of non-linear capitalist development, which analyses the history of capitalism not as a linear unfolding of underlying economic drivers but as a sequence of discrete growth models stabilized by political and cultural institutions' (Görg, Plank, and Wiedenhofer 2020, 44). It draws attention to how capitalism isn't just the same everywhere; rather there exist many more or less compatible capitalisms, all being regulated in different ways by states, and thus displaying different characteristics and geographical outcomes. We might therefore point to the distinctive economic growth of Japan over the second half of the twentieth century, and how new relationships between state and industry (through things like joint ventures and long-term contracts) underpinned huge infrastructural investments and gave Japanese firms new levels of access to global resource frontiers, particularly for steel production (Bunker and Ciccantell 2007). Görg, Plank, and Wiedenhofer (2020) cite the related example of recent economic growth in China – at first glance, China's recent economic trajectory might look very similar to Japan's between the 1960s and '80s, or even more so to Great Britain's in the nineteenth century: there is a high proportion of coal in the energy supply; per capita energy and material use is rising; there is a rapid uptick in the use of mineral resources being consumed by heavy industries (e.g. Figure 5.7). However, there are important differences too. The rate of growth of material use in China is unprecedented – if we look at all the materials extracted by all countries during the twentieth century, China has extracted one-third of that amount just since the year 2000 (Görg, Plank, and Wiedenhofer 2020, 50), while using as much concrete in a three-year period as the United States did in the entire preceding century (Smil 2013). Much of that growth in material use and mineral extraction has been driven by rapid urbanisation – what the geographer You-tien Hsing tellingly calls China's 'great urban transformation' (You-tien 2010). This urbanisation process is itself a product of a distinctive political economy.

McNally (2017) argues that contemporary Chinese capitalism can be characterised by a unique dialectic of top-down state-centric governance and bottom-up forms of local entrepreneurship based on market competition. This 'bottom-up' part of the equation extends to local government, where

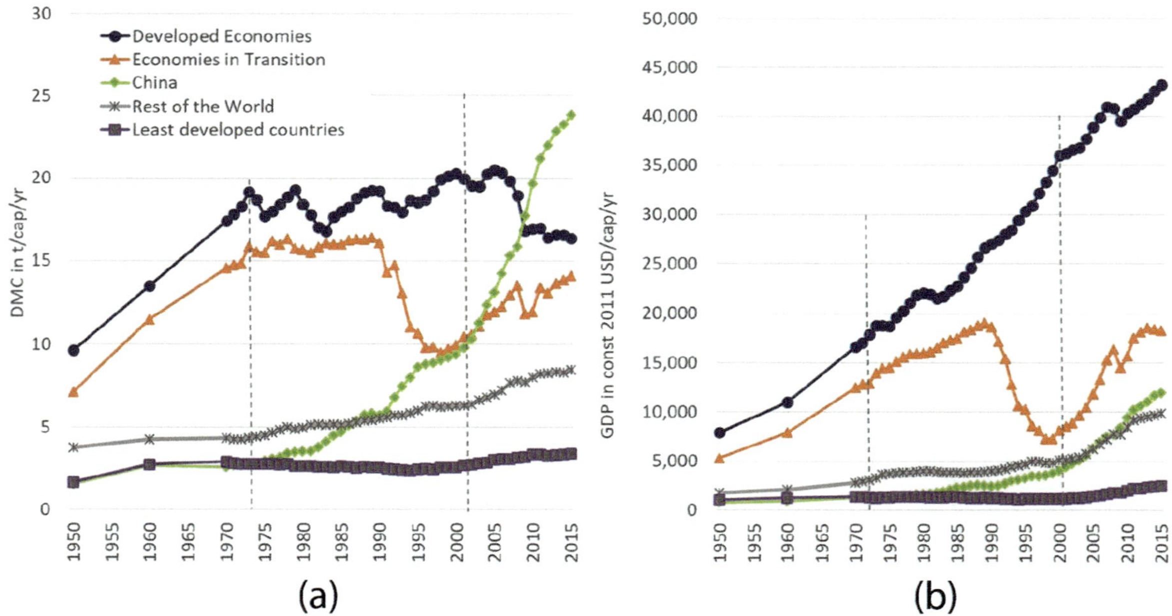

Figure 5.7 (a) Development of domestic material consumption (DMC) and (b) gross domestic product (GDP) per capita of global population for country groups. The vertical dashed lines delineate three distinct phases of the Great Acceleration: an initial postwar phase of rapid growth, a more crisis-ridden phase from the 1970s onwards, and a third phase of rapid growth and consumption. (Reproduced from Görg, Plank, and Wiedenhofer (2020, 48).)

economic decision-making is relatively decentralized and governing elites are incentivized to engage in inter-jurisdictional competition (McNally 2019). Urban mega-projects have become one such tool in the competition for local economic growth; revenue from land sales furnished local governments with the required capital, sparking frenetic levels of development, speculation, and capital (and debt) accumulation (You-tien 2010). As urbanisation and industrial growth subsequently began to slow, China's 'One Belt, One Road' strategy of massive infrastructural investment across Asia, Europe and Africa emerged as a 'spatial fix' (Harvey 1981) to a nascent crisis of domestic overproduction and declining returns on capital (Zhang 2017), echoing earlier moments of economic and infrastructural expansion which helped drive the world into the Anthropocene (see Chapters 3 and 4).

Nonetheless, these conjoined urban and infrastructural transformations (see Williams, Robinson, and Bouzarovski 2020) are not being driven by the same logics as British industrialisation and urbanisation in the nineteenth century, nor of the distinctive North American patterns of urbanisation of the twentieth century. As regulatory theory suggests, there exist many capitalisms, each driven and shaped by local institutional formations and cultures – such as the relative power of central and local governments and the incentive structures which direct decision-making.

Again, we can see here the evolution of the Anthropocene as a spatially heterogeneous condition (Head 2016), with the changing and varied role of the state playing a key role in shaping many different Great Accelerations. But that heterogeneity was – and continues to be (Liu and Bennett 2022) – shaped by changing geographies of environmental knowledge and resistance, to which we now turn.

Geographies of knowledge and resistance

Accounts of the Great Acceleration commonly feature a hopeful observation that whilst all this environmental transformation and devastation has been going on, people's awareness or consciousness of environmental change, and of the need to do something about it, has spread around the globe with the same kind of rapidity as the McDonald's restaurants or nitrogen run-off of the Great Acceleration graphics (e.g. Steffen, Crutzen, and McNeill 2007). No one has yet managed to graphically present this apparent transformation in the geographies of environmental knowledge in the same mode – how would we measure such a thing, and against what baselines? But the story of how collectively-held knowledge of environmental change has developed and circulated is central to the story of the Anthropocene, a geological age whose uniqueness rests in part on the fact that its chief protagonist, 'humanity', is 'in one way or another, becoming a self-conscious, active agent in the operation of its own life support system' (Steffen, Crutzen, and McNeill 2007, 619).

'Useless housewife data': the struggles of activist science

Thinking geographically about the rise and spread of new forms of environmental knowledge helps us refine the idea that global societies have only recently been enlightened by the efforts of the Anthropocenologists. Around the world, both scientists and citizens have spent much of the Great Acceleration piecing together new pictures of environmental change, often in direct response to some of the material and ideational transformations narrated above.

Let's start in New Delhi, where, in 1980, the Centre for Science and Environment (CSE) was founded by environmental activist Anil Agarwal. Two years later, CSE published its first *Citizen's Report on the State of the Indian Environment*. This was the first time that a resolutely *national* vision of environmental change had been articulated in post-independence India. But this wasn't a product of governmental data collection and environmental monitoring; instead, successive CSE 'State of the Indian Environment' reports have drawn on a vast network of non-governmental actors – activists, citizen scientists, journalists, agriculturalists and others – feeding information into CSE, which quickly became – and remains – a national 'centre of calculation' (Latour 1987) for both place-based environmental data and narrative accounts of social and environmental injustice (see Chapter 6). The reports foregrounded local struggles and political, normative concerns in the presentation of information, and emphasised the primacy of equitable, sustainable development and poverty alleviation in a way reminiscent of the arguments made by Prime Minister Indira Gandhi at the UN Conference on the Human Environment in Stockholm, 1972. But while on the international stage, CSE has often closely supported Indian government representatives in processes like climate change negotiations, on the national scale, CSE reports have also contested and challenged state knowledges and commitments. Contesting government data and challenging regulatory insufficiency have been animating themes of CSE's national report cards, and together the documents have challenged the legacies of Nehruvian developmentalism (expressed in things like huge dam projects and agricultural modernisation) in favour of the preservation of traditional land management techniques and the protection of air and water in both rural and urban environments.

CSE's geographies of knowledge-production – place-based, drawing on a network of 'non-official' scientists and informants, and working across geographical scales – is mirrored in other instances of environmental controversy during the Great Acceleration. Geographers Patrick Bresnihan and Naomi Milner have documented the diverse social movements that helped drive environmental politics during this period, and note that, especially in the global South and formerly colonised countries, such movements might not fit conventional (Western) ideas of 'environmentalism'. Movements to protect sustainable livelihoods, to practice care for diverse ecologies, and to encourage the 'cultivation of ways of life where humans and nonhumans thrive together' have rarely mobilised under the banner of protecting a 'global environment' (Bresnihan and Millner 2023, 3–4). Rather, they have been core parts of local, national, and transnational struggles over things like land rights and labour conditions, within social movements which don't make distinctions between 'the environment' on one side and 'social systems' on the other. As many Anthropocene commentators have been at pains to point out, dealing with environmental crises cannot be separated from dealing with social and economic injustices – a point we explore in more depth in the following chapter.

In the United States, Rachel Carson's influential 1962 work *Silent Spring*, on the effects of agricultural chemicals on wildlife, was another spur for environmental protection, which emanated from outside the halls of government or university science, and which mobilised scientific uncertainty to invite ordinary citizens into the process of documenting ecological change (Walker and Walsh 2012). Agricultural chemicals were another material legacy of the two world wars, which saw a new 'logic of extermination' applied to agricultural 'pests' as well as to human populations, whether in the trenches of World War I or the gas chambers of Nazi Germany (Bonneuil and Fressoz 2016, 134). A postwar intensification of efforts to purify agricultural environments of unwanted organisms led to considerable pushback from observers worried about the wider ecological side effects of these technologies of extermination. Nonetheless, the remarkable impact of *Silent Spring*, which documented the devastating ecological impacts of agrochemicals like DDT, has tended to suggest that 'modern environmentalism', as a mass social movement, was born in the American suburbs, when the white middle classes began to feel the effects of the industrial world-ecology in their own backyards.

This was to some extent the case in upstate New York in the 1960s, where evidence was starting to emerge of the toxic effects of industrial chemical wastes which had been dumped in a disused canal, before the land was given over to the town of Niagara Falls for the construction of a school. The ensuing 'Love Canal' controversy pitted ordinary citizens against the state and the chemical company. A group of suburban mothers, mobilised by increasingly evident threats to their children's health, conducted their own epidemiological surveys to map the different effects of the dumped chemicals. In return, state and private actors together mobilised sexist and racist tropes to dismiss this activist science as, for example, 'useless housewife data' (O'Brien 2022) – a telling example of what sociologists call the 'boundary work' by which some knowledge claims are strategically dismissed as non-scientific when they pose a challenge to established interests (Gieryn 1999).[2]

In many such controversies, this 'non-scientific' knowledge eventually gets vindicated, but not until after long struggles against condescension and suppression. But the Love Canal controversy also saw members of the African American community articulating the battle against the toxic legacies of industrial chemicals as integral to the wider struggle for civil rights (Blum 2008). This echoes a longer history, only recently excavated, of working-class and Black 'environmentalism' which can be traced back into the nineteenth century (Montrie 2018), and which helped crystallise concerns about 'environmental justice' (see Chapter 6)

All of this is not to say that states and private companies were ignorant of environmental problems during the Great Acceleration; on the contrary, the history of 'official' environmental science is surprisingly long, and exhibits some intriguing geographies.

The eye of power

The environmental sciences in the era of the Great Acceleration had a curiously Janus-faced character. On the one hand, disciplines based on various forms of geophysics were closely tied to military and industrial interests; on the other hand, disciplines like Ecology often had a more 'insurgent' character, concerning themselves with describing the negative impacts of things like industrialised agriculture and agrochemistry. Professional ecologists and geographers during this period grappled with their identity as scientists, and whether it meant they should stay out of politics altogether or rather use their knowledge to support the causes of communities in places like Love Canal. Historian Stephen Bocking shows how certain places, like government laboratories and threatened ecosystems, were central to shaping disciplinary identities in Ecology. Some ecologists sought to lean on the new authority of the physical and lab-based sciences, while others threw in their lot with communities trying to protect treasured landscapes and ecosystems (Bocking 2017).

While sciences like Ecology and Physical Geography busied themselves with understanding local- and landscape-scale environmental changes wrought by humans, sciences like Meteorology and Climatology were increasingly thinking in global terms. In part, this was due to scientific and technological advances, which allowed things like the circulation of the atmosphere to be rendered in numerical equations and simulated by new computers. But this new global vision was also a product of military interests. As the peace at the end of World War II quickly gave way to the Cold War, scientists in the United States, the Soviet Union, and their respective allies quickly became concerned with monitoring and understanding the planetary environments where hostilities might play out – the atmosphere, the poles, the oceans (Lehman 2020). Satellites and computers were enrolled into monitoring and modelling how radiation from nuclear bomb tests travelled through the atmosphere, contributing both to military strategy and to scientific understandings of how the weather worked (Edwards 2012). Efforts to measure radioactive particles in the air helped establish long-term monitoring of atmospheric CO_2

concentrations (Howe 2014). And satellite technology developed for Cold War surveillance and communication helped scientists monitor things like land use change and, later on, global temperatures (DeLoughrey 2014). Together, such technologies constituted a new 'eye of power' (Ashley 1983). While in earlier periods maps had been powerful tools of exploration and conquest, these new technologies of planetary monitoring gave powerful states new means of visualising, monitoring, understanding and perhaps even controlling the globe as a whole (Cosgrove 2001).

Indeed, one can even argue that the very idea that humans could catastrophically alter global environments came not from environmentalist movements, but from military scientists and planners who actively planned and experimented with anthropogenic environmental catastrophe as a form of warfare. The historian Jacob Hamblin (2013) shows how environmental warfare became a crucial strategic tool for Cold Warriors. British generals consulted university botanists on how to destroy enemy crops in British Malaya; American weapons in Vietnam targeted the enemy's environments (trees, crops, the weather) more often than their bodies; and schemes were devised to generate tsunamis and flood coastal cities, promising more devastation to one's enemies than bombs alone ever could.

By the 1980s, such lines of thought had converged around the potential for global anthropogenic climate change, and the U.S. Central Intelligence Agency (CIA) began to speculate about its impacts on things like Soviet harvests. Alarms had been raised about the potentially negative effects of global warming caused by anthropogenic greenhouse gas emissions throughout the 1950s, 1960s, and 1970s. These alarms were sounded by government committees and state-employed scientists, and fossil fuel companies themselves conducted sophisticated modelling studies in the 1970s, which, it turns out, quite accurately predicted the future course of global climate change. But while Rachel Carson used the inevitable existence of scientific uncertainty to invite others into the process of making environmental knowledge, uncertainties about future climate change were quickly mobilised by actors like the American Petroleum Institute in the 1980s to cast doubt on the entirety of the emerging science, or to suggest that regulatory action should wait until more certainty was achieved (Franta 2021).

Global climate modelling was therefore not a science that rose to prominence independently of its geopolitical context. Throughout the 1980s, American scientists, along with like-minded colleagues in the UK and Europe, developed computational climate models as a new way of peering into the future. Behind the 'Iron Curtain' in the Soviet Union, different ways of thinking about and predicting global climate had prominence. While the scientists of the North Atlantic world favoured theoretically-informed computer models as a means of prediction, Soviet scientists favoured the use of palaeoclimatological analogues as a guide to future conditions. It was only by understanding past climatic changes and their impacts on people, the Soviet scientists argued, that reliable knowledge could be generated about what the future might look like (Oldfield 2021).

While collaboration between the United States, European, and Soviet climate scientists had been encouraged from the early 1970s, these disputes about the right way to know the future became a key part of 1980s Cold War geopolitics. Stefan Schäfer (2021) argues that U.S. scientists and officials, including the CIA, became increasingly sceptical about Soviet ways of understanding and predicting environmental change not just (if at all) because they were considered wrong, or scientifically inferior, but precisely because they had geopolitical power. Throughout the 1980s, Soviet influence in UNESCO, the United Nations Educational, Scientific and Cultural Organisation, had been growing, and it was feared that Soviet ways of understanding and shaping human-environment relations were gaining influence in parts of the developing world, particularly among newly independent postcolonial states. Many of these new states leaned ideologically towards collectivism and away from the liberal capitalism of the West, and such views were gaining prominence in UNESCO. In 1984, a disgruntled United States broke away from the organisation, and its scientific energies were turned towards the new International Geosphere-Biosphere Program (IGBP), a US-dominated infrastructure to support a new Earth System science, and the institution from which would later emerge so many of the foundational arguments about the Anthropocene and the Great Acceleration (see Chapter 1).

When the Intergovernmental Panel on Climate Change (IPCC) published its first report on global climate change in 1990, it appeared to usher in a new period of global environmental consciousness and, perhaps, of global cooperation in the protection of the natural world. The Berlin Wall had fallen, and the capitalist West seemed to have won the great ideological battle of the post-war world. This was the supposed 'end of history' – the end of the great struggle to define the ideal form of nation-state governance and international relations. Liberal capitalism had outmanoeuvred communism, and a new global political consensus would, it was hoped by many in the West, now move beyond bipolar struggles of the past and concern itself with *global* projects like securing universal human rights, protecting the environment, and stewarding humanity through the Anthropocene (Miller 2004).

The Western scientists 'won' too – global climate observation and modelling became the remit of Working Group I of the IPCC, the 'physical science basis' of all talk about climate change, while Soviet climate sciences – along with many geographical and ecological approaches to understanding environmental risk – were left to IPCC Working Group II, concerned with the impacts of global climate change on peoples and environments. But this triumph of 'Western' science wasn't a product of the *end* of the Cold War and the cooling of geopolitical tensions in the late 1980s; rather, it was very much a product of *heightening* geopolitical tensions during that decade, and of deeply political struggles over the best way to understand and study the relationships between humanity and the global environment.

As shown in Chapters 3 and 4, there are long histories of 'environmental reflexivity': of societies in different times and places being acutely aware of the environmental costs of new forms of production and consumption, and of political struggles to contest and even suppress knowledge that might get in the way of certain powerful actors' abilities to pursue their own interests. This brief sketch of the geographies of environmental knowledge during the formative years of the Great Acceleration shows that far from a straightforward, linear process of gradual environmental enlightenment, this period was marked by different kinds of knowledge being produced in different kinds of places – from subsistence farming communities and suburban homes to military satellites and global organisations – for many different reasons. Scientific practices of various kinds furthered the co-evolution of environmental knowledge and environmental concern at different spatial scales, from Indian farms to the global atmosphere, and political struggles over who counted as an authoritative knower were never far away. Environmental knowledge is a conflictual space, and arguably always 'political' (Liu and Bennett 2022) – the examples narrated above are all instances of ontological politics, where claims were being made that new entities (ecosystems, climates, marginalised bodies) should be taken into account in political decision-making (Latour 2012). They were exercises in re-drawing the boundaries of political space, and were central to the emergence of a new kind of multi-scalar Anthropocene politics.

Conclusion

This chapter has introduced the notion of a post-war Great Acceleration, and arguments that this marks the most convincing moment when the Anthropocene, as a slice of geological time, began. But while some histories of the Great Acceleration tend to naturalise its occurrence, new work in Geography and Environmental History has shown that the societal origins of the Great Acceleration can be found in the wars of the first half of the twentieth century, in new ideas about the nature of economic growth, new geopolitical relationships, and in struggles over emerging environmental catastrophes.

Together, these accounts of the causes and paths of the Great Acceleration largely bear out Marxist analyses of the Anthropocene, which see the Great Acceleration as the intensification of processes of unequal exchange that began in the preceding centuries. Nonetheless, recent research offers two major caveats to the Marxist story of a destructive capitalist system. The first concerns the significant role of the state in the social and environmental transformations of the second half of the twentieth. The second is related and concerns the variety of state-led developmental projects in places like India and China that have shaped the trajectory of the Great Acceleration, particularly in its later phases. A fuller geographical account of the Great Acceleration needs to balance attention to its significant roots in places such as American oil wells, shopping malls, and policy documents, with attention to the diverse manifestations and experiences of fossil capitalism, fossil developmentalism, and resistance movements in different parts of the world. Emerging research in Geography, History, and related fields is starting to build this picture, to complement and complicate the story of the Great Acceleration told by physical and Earth System scientists. Yet these geographies of difference, connection, and transformation are vital not only for understanding the complexity of the Great Acceleration; they are also central, as we'll explore in the next chapter, for thinking in new ways about environmental justice in the Anthropocene. The same applies to how we think about the preservation, restoration, and conservation of the non-human realm, our focus in Chapter 7.

Study questions

1 What role did the twentieth century's two world wars play in the onset of the Great Acceleration?
2 To what extent does the notion of the 'Anthropocene mode of production' (Holmes, Bolen, and Kirkbride 2021) capture the geographical complexities of what Chatterjee (2020) calls 'the Asian Anthropocene'?
3 Compare the roles played by 'official' and 'activist' sciences in challenging the drivers of the Great Acceleration.

Notes

1 On zombie narratives in general as a form of environmental critique, see Lauro (2011).
2 Beverley Paigen, a state-employed scientist who supported the activists and who had also conducted pioneering work on the links between cigarette smoking and cancer, would later wear a badge to meetings announcing herself as 'expert on useless housewife data' (O'Brien 2022)

References

Acker, A. 2020. "A Different Story in the Anthropocene: Brazil's Post-Colonial Quest for Oil (1930–1975)." *Past & Present* 249 (1): 167–211.
Adey, P. 2010. *Aerial Life: Spaces, Mobilities, Affects*. Oxford: Wiley-Blackwell.
Ashley, R. K. 1983. "The Eye of Power: The Politics of World Modeling." *International Organization* 37 (03): 495–535.
Benjamin, W. 1999. *The Arcades Project*. Cambridge: Harvard University Press.
Black, B. C. 2012. "Oil for Living: Petroleum and American Conspicuous Consumption." *Journal of American History* 99 (1): 40–50.
Blum, E. D. 2008. *Love Canal Revisited: Race, Class, and Gender in Environmental Activism*. Lawrence: University Press of Kansas.
Bocking, S. 2017. *Ecologists and Environmental Politics: A History of Contemporary Ecology*. Morgantown: West Virginia University Press.

Bonneuil, C., Jean, and F. Baptiste. 2016. *The Shock of the Anthropocene*. London: Verso.

Bresnihan, P., and N. Millner. 2023. *All We Want Is the Earth: Land, Labour and Movements Beyond Environmentalism*. Bristol: Policy Press.

Buell, F. 2012. "A Short History of Oil Cultures: Or, the Marriage of Catastrophe and Exuberance." *Journal of American Studies* 46 (2): 273–93.

Bunker, S. G., and P. S. Ciccantell. 2007. *East Asia and the Global Economy: Japan's Ascent, With Implications for China's Future*. Baltimore: JHU Press.

Chakrabarty, D. 2014. "Climate and Capital: On Conjoined Histories." *Critical Inquiry* 41 (1): 1–23.

Chatterjee, E. 2020. "The Asian Anthropocene: Electricity and Fossil Developmentalism." *The Journal of Asian Studies* 79 (1): 3–24.

Clarke, D. B. 2004. *Consumer Society and the Post-Modern City*. Abingdon: Routledge.

Clifford, J., and S. Castonguay. 2022. "British Ghost Acres and Environmental Changes in the Laurentian Forest During the Nineteenth Century." *Journal of Historical Geography* 78 (October): 126–38.

Cosgrove, D. 2001. *Apollo's Eye: A Cartographic Genealogy of the Earth in the Western Imagination*. Vol. 0. London: John Hopkins University Press.

Crawford, M. 1992. 'The World in a Shopping Mall'. In *The City Cultures Reader*, edited by Malcolm Miles and Tim Hall. Abingdon: Routledge.

Curley, A., and S. Smith. 2024. "The Cene Scene: Who Gets to Theorize Global Time and How Do We Center Indigenous and Black Futurities?" *Environment and Planning E* 7 (1): 166–88.

DeLoughrey, E. 2014. "Satellite Planetarity and the Ends of the Earth." *Public Culture* 26 (2 [73]): 257–80.

Edgerton, D. 2019. *The Shock of the Old: Technology and Global History Since 1900*. London: Profile.

Edgeworth, M., P. Gibbard, M. Walker, D. Merritts, S. Finney, and M. Maslin. 2023. "The Stratigraphic Basis of the Anthropocene Event." *Quaternary Science Advances* 11 (July): 100088.

Edwards, P. N. 2012. "Entangled Histories: Climate Science and Nuclear Weapons Research." *Bulletin of the Atomic Scientists* 68 (4): 28–40.

Egerö, B. 2013. "Population Policy: A Valid Answer to Climate Change? Old Arguments Aired Again Before COP15." *ACME: An International Journal for Critical Geographies* 12 (1): 88–101.

Elhacham, E., L. Ben-Uri, J. Grozovski, Y. M. Bar-On, and R. Milo. 2020. "Global Human-Made Mass Exceeds All Living Biomass." *Nature* 588 (7838): 7838.

Evenden, M. 2011. "Aluminum, Commodity Chains, and the Environmental History of the Second World War." *Environmental History* 16 (1): 69–93.

Ewen, S. 1996. *PR!: A Social History of Hype*. New York: Basic Books.

Foucault, M. 1979. *Discipline & Punish: The Birth of the Prison*. London: Penguin.

Fowkes, M., and R. Fowkes. 2025. "The Socialist Anthropocene in the Visual Arts." *ARTMargins* 14 (2): 3–10.

Franta, B. 2021. "Early Oil Industry Disinformation on Global Warming." *Environmental Politics* 30 (4): 663–68.

Gieryn, T. F. 1999. *Cultural Boundaries of Science: Credibility on the Line*. Chicago: University of Chicago Press.

Görg, C., C. Plank, D. Wiedenhofer, et al. 2020. "Scrutinizing the Great Acceleration: The Anthropocene and Its Analytic Challenges for Social-Ecological Transformations." *The Anthropocene Review* 7 (1): 42–61.

Gray, Steven. 2017. "Fuelling Mobility: Coal and Britain's Naval Power, c. 1870–1914." *Journal of Historical Geography* 58 (October): 92–103.

Gregory, D. 1990. '"A New and Differing Face in Many Places": Three Geographies of Industrialization'. In *Historical Geography of England and Wales*, 351–99. London: Elsevier.

Hamblin, J. D. 2013. *Arming Mother Nature: The Birth of Catastrophic Environmentalism*. Oxford: Oxford University Press.

Hamilton, C. 2015. "Getting the Anthropocene so Wrong." *The Anthropocene Review* 2 (2): 102–7.

Harvey, D. 1981. "The Spatial Fix – Hegel, Von Thunen, and Marx." *Antipode* 13 (3): 1–12.

Harvey, D. 2018. *Marx, Capital and the Madness of Economic Reason*. Oxford: Oxford University Press.

Head, L. 2016. *Hope and Grief in the Anthropocene: Re-Conceptualising Human-Nature Relations*. London: Routledge.

Head, M. J., W. Steffen, D. Fagerlind, et al. 2021. "The Great Acceleration Is Real and Provides a Quantitative Basis for the Proposed Anthropocene Series/Epoch." *Episodes Journal of International Geoscience*, ahead of print, November.

Hibbard, K., P. Crutzen, E. Lambin, et al. 2006. 'Decadal-Scale Interactions of Humans and the Environment'. In *Sustainability or Collapse? An Integrated History and Future of People on Earth*, edited by Robert Costanza, Lisa J. Graumlich, and Will Steffen. Cambridge: MIT Press.

Higgs, K. 2016. *Collision Course: Endless Growth on a Finite Planet*. Cambridge: MIT Press.

Hippler, T. 2017. *Governing from the Skies: A Global History of Aerial Bombing*. London: Verso.

Holmes, B., J. Bolen, and B. Kirkbride. 2021. "Born Secret (Cash for Kryptonite): A Field Guide to the Anthropocene Mode of Production." *The Anthropocene Review* 8 (2): 183–95.

Horton, J., and P. Kraftl. 2014. *Cultural Geographies: An Introduction*. Abingdon: Routledge.

Howe, J. P. 2014. *Behind the Curve: Science and the Politics of Global Warming*. London: University of Washington Press.

Hulme, A. 2015. *On the Commodity Trail: The Journey of a Bargain Store Product from East to West*. London: Routledge.

IUGS. 2024. "The Anthropocene." https://www.iugs.org/_files/ugd/f1fc07_40d1a7ed58de458c9f8f24de5e739663.pdf?index=true. Accessed 14 May 2026.

Jayne, M. 2006. "Cultural Geography, Consumption and the City." *Geography* 91 (1): 34–42.

Johnstone, P., and C. McLeish. 2020. "World Wars and the Age of Oil: Exploring Directionality in Deep Energy Transitions." *Energy Research & Social Science* 69 (August): 101732.

Johnstone, P., and C. McLeish. 2022. "World Wars and Sociotechnical Change in Energy, Food, and Transport: A Deep Transitions Perspective." *Technological Forecasting and Social Change* 174 (January): 121206.

Josephson, P. 2010. 'War on Nature as Part of the Cold War: The Strategic and Ideological Roots of Environmental Degradation in the Soviet Union'. In *Environmental Histories of the Cold War*, edited by Corinna R. Unger and J. R. McNeill. Cambridge: Cambridge University Press.

Kumar, A. 2022. "Energy Geographies in/of the Anthropocene: Where Now?" *Geography Compass* 16 (10): e12659.

Kunstler, J. H. 1994. *Geography Of Nowhere: The Rise And Decline of America's Man-Made Landscape*. New York: Simon and Schuster.

Lane, R. 2019. "The American Anthropocene: Economic Scarcity and Growth During the Great Acceleration." *Geoforum* 99: 11–21.

Latour, B. 1987. *Science in Action: How to Follow Scientists and Engineers Through Society*. Cambridge: Harvard University Press.

Latour, B. 2012. "Reflexive Modernity Brings Us Back to Earth." In *Futures of Modernity: Challenges for Cosmopolitical Thought and Practice*, edited by Michael Heinlein, Cordula Kropp, Judith Neumer, Angelika Poferl, and Regina Römhild, 65–75. Bielefeld: Transcript Verlag.

Lauro, S. J. 2011. 'The Eco-Zombie: Environmental Critique in Zombie Fiction'. In *Generation Zombie: Essays on the Living Dead in Modern Culture*, edited by Stephanie Boluk and Wylie Lenz, 54–66. London: McFarland.

Leach, W. R. 2011. *Land of Desire: Merchants, Power, and the Rise of a New American Culture*. New York: Knopf Doubleday Publishing Group.

Lebow, V. 1955. "Price Competition in 1955." *Journal of Retailing* 31 (1): 3–10.

Lehman, J. 2020. "Making an Anthropocene Ocean: Synoptic Geographies of the International Geophysical Year (1957–1958)." *Annals of the American Association of Geographers* 110 (3): 606–22.

Lewis, S. L., and M. A. Maslin. 2015. "Defining the Anthropocene." *Nature* 519 (7542): 171–80.

Linnér, B. O. 2023. *The Return of Malthus: Environmentalism and Post-War Population-Resource Crises*. Cambridge: White Horse Press.

Liu, X., and M. M. Bennett. 2022. "The Geopolitics of Knowledge Communities: Situating Chinese and Foreign Studies of the Green Belt and Road Initiative." *Geoforum* 128: 168–80.

London, B. 1932. *Ending the Depression Through Planned Obsolescence*. New York. https://babel.hathitrust.org/cgi/pt?id=wu.89097035273&seq=5

Malm, A., and A. Hornborg. 2014. "The Geology of Mankind? A Critique of the Anthropocene Narrative." *The Anthropocene Review* 1 (1): 62–69.

Malm, A. 2016. *Fossil Capital: The Rise of Steam-Power and the Roots of Global Warming*. London: Verso.

Martín, W. S. 2017. "Nitrogen, Science, and Environmental Change: The Politics of the Green Revolution in Chile and the Global Nitrogen Challenge." *Journal of Political Ecology* 24 (1): 1.

McCarthy, F. M. G., R. T. Patterson, M. J. Head, et al. 2023. "The Varved Succession of Crawford Lake, Milton, Ontario, Canada as a Candidate Global Boundary Stratotype Section and Point for the Anthropocene Series." *The Anthropocene Review* 10 (1): 146–76.

McKendrick, N., J. Brewer, and J. H. Plumb. 1982. *The Birth of a Consumer Society: The Commercialization of Eighteenth-Century England*. Bloomington: Indiana University Press.

McNally, C. A. 2017. 'Rebalancing the Economy, Refurbishing the State: The Political Economic Logic of Sino-Capitalism in Contemporary China'. Revue de La Régulation. Capitalisme, Institutions, Pouvoirs 21(June): 21. http://journals.openedition.org/regulation/12294

McNally, C. A. 2019. "Theorizing Sino-Capitalism: Implications for the Study of Comparative Capitalisms." *Contemporary Politics* 25 (3): 313–33.

McNeill, J. R., and P. Engelke. 2014. *The Great Acceleration: An Environmental History of the Anthropocene Since 1945*. Cambridge: Harvard University Press.

Middlemiss, L. 2018. *Sustainable Consumption: Key Issues*. London: Routledge.

Miles, S. 2010. *Spaces for Consumption*. London: SAGE.

Miller, C. A. 2004. 'Climate Science and the Making of a Global Political Order'. In *States of Knowledge: The Co-Production of Science and Social Order*, edited by Sheila Jasanoff, 46–66. London: Routledge.

Miller, J.C. (2026). Retail Ruins, Revisited. In *The Wiley Blackwell Companion to the City, Retail, and Consumption*, edited by Daniel Paiva and Pedro Guimarães, 114–134. Oxford: John Wiley & Sons Ltd. https://doi.org/10.1002/9781394278657

Mitchell, T. 2014. "Economentality: How the Future Entered Government." *Critical Inquiry* 40 (4): 479–507.

Mom, G. 2014. 'Clashes of Cultures: Road vs. Rail in the North Atlantic World During the Interwar Coordination Crisis'. In *The Organization of Transport*, edited by Massimo Moraglio and Christopher Kopper, 8–31. London: Routledge.

Montrie, C. 2018. *The Myth of Silent Spring*. Berkeley: University of California Press.

Moore, J. W. 2015. *Capitalism in the Web of Life: Ecology and the Accumulation of Capital*. London: Verso.

Moore, J. W. 2017. "The Capitalocene, Part I: On the Nature and Origins of Our Ecological Crisis." *The Journal of Peasant Studies* 44 (3): 594–630.

O'Brien, K. 2022. *Paradise Falls: The True Story of an Environmental Catastrophe*. New York: Knopf Doubleday Publishing Group.

Ogborn, M. 2001. "Book Review: The Badlands of Modernity: Heterotopia and Social Ordering." *Cultural Geographies* 8 (1): 120–21.

Oldfield, J. D. 2021. *The Soviet Union and Global Environmental Change: Modifying the Biosphere and Conceptualizing Society-Nature Interaction*. Abingdon: Routledge.

Olšáková, D. 2016. *In the Name of the Great Work: Stalin's Plan for the Transformation of Nature and Its Impact in Eastern Europe*. New York: Berghahn.

Oreskes, N. 2015. "How Earth Science Has Become a Social Science." *Historical Social Research* 40 (2): 246–70.

Otter, C., A. Bashford, J. L. Brooke, F. A. Jonsson, and J. M. Kelly. 2018. "Roundtable: The Anthropocene in British History." *Journal of British Studies* 57 (3): 568–96.

Parenti, C. 2015. "The Environment Making State: Territory, Nature, and Value." *Antipode* 47 (4): 829–48.

Parsons, L., R. Safra de Campos, A. Moncaster, et al. 2022. "Trading Disaster: Containers and Container Thinking in the Production of Climate Precarity." *Transactions of the Institute of British Geographers* 47 (4): 990–1008.

Robertson, T., R. P. Tucker, N. B. Breyfogle, P. Mansoor, eds. 2020. *Nature at War: American Environments and World War II*. Cambridge: Cambridge University Press.

Schäfer, S. 2021. 'Ending History: Empiricism, Earth System Science and Empire'. Paper presented at Science and Democracy Network Anual Meeting.

Slade, G. 2009. *Made to Break: Technology and Obsolescence in America*. Cambridge: Harvard University Press.

Smil, V. 2013. *Making the Modern World: Materials and Dematerialization*. Oxford: Wiley.

Steffen, W., W. Broadgate, L. Deutsch, O. Gaffney, and C. Ludwig. 2015. "The Trajectory of the Anthropocene: The Great Acceleration." *The Anthropocene Review* 2 (1): 81–98.

Steffen, W., J. Crutzen, and J. R. McNeill. 2007. "The Anthropocene: Are Humans Now Overwhelming the Great Forces of Nature?" *Ambio* 36 (8): 614–21.

Steinberg, T. 2019. "Can Karl Polanyi Explain the Anthropocene? The Commodification of Nature and the Great Acceleration." *Geographical Review* 109 (April): 265–70.

Urry, J. 2007. *Mobilities*. Cambridge: Polity.

Vries, J. 2008. *The Industrious Revolution: Consumer Behaviour and the Household Economy, 1650 to the Present*. Cambridge: Cambridge University Press.

Walker, K., and L. Walsh. 2012. "No One Yet Knows What the Ultimate Consequences May Be": How Rachel Carson Transformed Scientific Uncertainty into a Site for Public Participation in Silent Spring." *Journal of Business and Technical Communication* 26 (1): 3–34.

Whitehead, M. 2014. *Environmental Transformations: A Geography of the Anthropocene*. Abingdon: Routledge.

Williams, J., C. Robinson, and S. Bouzarovski. 2020. "China's Belt and Road Initiative and the Emerging Geographies of Global Urbanisation." *The Geographical Journal* 186 (1): 128–40.

Woodward, I. 1998. "The Shopping Mall, Postmodern Space and Architectural Practice: Theorising the Postmodern Spatial Turn Through the Planning Discourse of Mall Architects." *Architectural Theory Review* 3 (2): 45–56.

You-tien, H. 2010. *The Great Urban Transformation: Politics of Land and Property in China*. In *The Great Urban Transformation*. Oxford: Oxford University Press.

Zhang, Xin. 2017. "Chinese Capitalism and the Maritime Silk Road: A World-Systems Perspective." *Geopolitics* 22 (2): 310–31.

Part III

Geo-politics

6 Scaling Justice in the Anthropocene

Introduction

The previous section of the book analysed the historical-geographical transformations that have led the world into the Anthropocene. Now we need to ask: how can societies find ways to live well in the Anthropocene, to stave off its worst effects, and build better futures? The chapters in this section begin developing answers to these fundamentally political and moral-ethical questions, beginning with this chapter's focus on justice. As we noted towards the end of Chapter 5, research and activism around 'environmental justice' (EJ) became a key feature of the politics of the period now known as the Great Acceleration. As a key conceptual concern of political and environmental geographers, justice seems like a good concept with which to start this section's exploration of some of the geo-politics of life in the Anthropocene.

At the drive-in

For most of the period known as the Great Acceleration, cinephiles in the South African city of Johannesburg could take in the latest blockbuster film in one particularly (in)famous example of that icon of post-war suburbanisation: the drive-in cinema. From the comfort of their private cars, patrons of the Top Star drive-in could enjoy the fruits of Hollywood's burgeoning film industry alongside enviable views over the city skyline. But Top Star wasn't your average suburban drive-in cinema – a phenomenon which remade both the suburbs and the private automobile as spaces of cultural consumption and leisure. For a start, Top Star was for white people only. What's more, it sat atop an enormous waste dump left behind by one of the city's many gold mines which, from 1886, had driven the development of Johannesburg as an urban centre within a new global commodity chain, and which left behind a toxic legacy of acid mine drainage and heavy metal pollution.

DOI: 10.4324/9781003162469-9

Gold mining involves moving around huge quantities of earth, and Johannesburg's mine dumps define its distinctive hilly topography. Initially, proposals to build housing on top of the Top Star dump were refused, but in 1955, permission was given for homes to be built 'all around this toxic pile of waste' (Peša 2022, 392). The opening of the cinema on the top in 1958 shows how mining residues began to be reimagined as a positive part of the city's geography and identity – a reminder and celebration of industrial heritage. By the early twenty-first century, resource depletion had meant that gold mining in the area switched its focus to recovering potentially valuable materials from waste sites. Top Star was closed, and work began to recycle the waste dump, with residues this time transported out of the city. But while patrons of the cinema first protested and then lamented this loss of part of the city's cultural and industrial heritage, 'airborne, soil, and water pollution from the dump site continues unabated, generating health risks for the surrounding [majority Black] residents' (Peša 2022, 392). These residents live mainly in informal and high-density settlements, where unstable tenancy agreements limit their rights to protest, and economic marginalisation hinders their ability to move away. Top Star's neighbours are painfully aware of the toxicity that surrounds them and wonder how such sites can be viewed as having any heritage, cultural, or aesthetic value.

Top Star is a powerful metaphor and exemplar of some of the historical geographies explored in preceding chapters. On top of a hill, wealthy white South Africans enjoyed the products of Anglo-American economic and cultural globalisation from the comfort of their petrol-guzzling automobiles. Yet they sat atop a hill built by the labour of thousands of migrant labourers, toiling as part of a colonial resource extraction economy; a hill which to this day oozes toxic waste into the homes, gardens, farms, and bodies of its impoverished neighbours. One feature of the Anthropocene, however and whenever defined, is how the expanding geography of resource extraction, through the accumulation of ghost acres and the restless development of resource frontiers, has created a cleavage between the world which the biggest consumers live *in*, and the world they live *from* (Latour and Weibel 2020). At Top Star, that cleavage was reflected in a vertical stratification – cinema clientele at the top, victims of toxic pollution below – which crystallised the particular racial politics of the South African state, as well as the broader 'colour line' which so profoundly shapes the global geographies of vulnerability to the ill-effects of the Anthropocene (Yusoff 2024). From the slopes and summit of Top Star, we can see how the geographical transformations of the Anthropocene point to new questions about *justice* – about the distribution of environmental benefits and harms, about who gets a say in decision-making, and about how the violence of social inequality collides with the transformation and destabilisation of environments across spatial scales. Justice is one of the two great 'regions' of morality; the other is how we define and achieve the so-called 'good life'. According to some moral philosophers, justice is the foundation upon which the good life, however defined, must be built. It is therefore crucial to any consideration of a 'good' or 'better' Anthropocene.

In this chapter, we introduce some different ways of thinking about spaces of (in)justice in the Anthropocene. We use international examples of industrial pollution and rare earth mining to illustrate the political geographies of environmental injustice and introduce two concepts – *sacrifice zones* and *slow violence* – which can help us to grasp how environmental injustice plays out across space and time. We then introduce debates about how to retool notions of EJ for the Anthropocene, which have raised issues about justice across borders, across generations, and beyond the human. Conceptualisations of 'planetary justice' in these terms provide a promising new way of integrating EJ scholarship – traditionally concerned with local environmental struggles – with the findings of Earth System Science. Nonetheless, critical geographers, inspired by decolonial and feminist theory, have maintained a vital focus on how injustice is fundamentally embodied and reproduced at the intersections of multiple socially-produced forms of difference, such as race, gender, and class. Geographical approaches to planetary justice therefore need to find a way to think, advocate for, and practice justice across spatial scales, from the human body to the body of the Earth.

In the next section, we introduce ideas on how to conceptualise environmental (in)justice, focusing on questions of distribution, recognition, and procedural justice.

Conceptualising environmental (in)justice

The concept of EJ has always been intensely geographical, 'focused on questions of distribution, inequality and injustice which resonate with long-standing traditions of geographical inquiry and analysis' (Walker and Bulkeley 2006, 655). It is a geographical concept, and a concept with a distinctive geography. Towards the end of Chapter 3, we noted how historical geographies of resistance to the deathly logics of the slave-plantation economy resonated through subsequent decades and centuries, and found expression in the 1991 declaration of the First National People of Color Environmental Leadership Summit (Davis et al. 2019). The quotation on page 87 is the preamble to a document titled *The Principles of Environmental Justice,* which, upon its publication, was hugely influential in knitting together North American communities struggling against things like industrial pollution (e.g. around Love Canal, see Chapter 5), and in helping to crystallise, for the wider environmental movement, the meaning of 'environmental justice'.

If we look closely at the declaration, we can get a sense of how environmental (in)justice was being conceptualised in North America in this period and how conceptualisations have developed in the years since. *The Principles of Environmental Justice* are notable for the kinds of spaces and places referred to. There's little mention of rivers, forests, seas, or skies – the kinds of things we tend to associate with environmentalist calls for the protection of nature. Instead, we are directed to 'healthy work environments', workers' homes, cities, and rural settlements; instead of fragile wilderness, we encounter spaces of military occupation, of Indigenous sovereignty, and of nuclear testing (FNPOCELS 1991). These are thoroughly peopled environments, and environments where the impacts of environmental damage and pollution are felt more acutely by those towards the bottom of hierarchies of power and privilege (Bullard et al. 2008).

These concerns set the EJ movement apart from the wider environmental movement, at least in Western countries. The latter tended to privilege nature conservation and protection over issues of social justice and economic inequality; it was (at the risk of some simplification) about protecting nature from people, on account of nature's inherent moral or aesthetic worth, rather than about protecting people from harms delivered through the medium of their physical environment (Taylor 2016). Nonetheless, that's not to say that the EJ movement wasn't concerned with the inherent value of nonhuman natures and their need for respect and protection. On the contrary, point 1 in the *Principles* states that:

> Environmental Justice affirms the sacredness of Mother Earth, ecological unity and the interdependence of all species, and the right to be free from ecological destruction.

This was certainly a different 'Nature' to that of, for example, early proponents of wilderness protection in the United States (Cronon 1996) – most notably in its insistence on cross-species interdependence rather than Human-Nature separation – but it shares with them an ethic of protecting nature from humans. Where it departed from such mainstream environmentalism (as it existed for much of the twentieth century) is in its positioning of such concerns alongside those of justice among humans.

Justice as fair distribution

This was a conception of EJ concerned foremost with **distribution**. Distributional approaches are a popular way of thinking about justice. Essentially, they ask: 'Who gets what, and are the good (and bad) things in society distributed equally?' (Schlosberg 2018, 630). The influential justice theorist John Rawls argued that unequal distributions could only be justified if the inequities in some way benefited the least well-off, or brought some wider societal benefit. So, for example, if a particular community suffers from toxic exposure due to the siting of a chemical plant in their neighbourhood, distributionists might ask whether that inequality of toxic exposure might be justified by the jobs the facility brings to the neighbourhood, or by the wider economic growth to which it might contribute.

However, despite the *Principles* being clearly concerned with how exposure to environmental bads was distributed unequally, its authors had no time for weighing up such bads against any 'goods' (like jobs or economic growth) that might come with them. Rather:

> Environmental Justice demands the cessation of the production of all toxins, hazardous wastes, and radioactive materials, and that all past and current producers be held strictly accountable to the people for detoxification and containment at the point of production.
>
> (FNPOCELS 1991)

No messing about here – the pollution can't be justified, and must simply be stopped. The wider EJ movement that coalesced around the time of the publication of the *Principles* focused on these kinds of distributional questions, 'with special reference to how environmental harms such as toxic pollution fall disproportionately on poor and marginalised people' (Bullard 2018; Dryzek and Pickering 2018, 67).

Geographical interest in the spatial co-occurrence of polluting industries and toxic waste sites can be traced to the 1982 controversy over the State of North Carolina's plan to dump 120 million pounds of contaminated soil in Warren County, an area with the highest proportion of African Americans in the state. Protestors received support from civil rights groups from all over the United States, and the controversy set in train a new field of academic and activist work to identify whether similar patterns were evident across the country. Sophisticated multi-variate analyses of spatial data proved that they were (e.g. Chavis and Lee 1987; Purifoy 2021), and we are now in possession of hundreds of studies which

> conclude that, in general, ethnic minorities, indigenous persons, people of color, and low-income communities confront a higher burden of environmental exposure from air, water, and soil pollution from industrialization, militarization, and consumer practices.
>
> (Mohai, Pellow, and Timmons Roberts 2009, 406)

Such conclusions have been reached by studies employing Geographic Information Systems and spatial data to analyse either the demographic characteristics of spatial units (like zip-code areas in the United States) that host hazardous sites and those that don't, or analysing the correlations between distance from hazardous sites and neighbourhood characteristics. Numerical models of things like air pollution disposal are also used to identify racial and class disparities in exposure to environmental risks, from both static and mobile sources, such as traffic (Mohai, Pellow, and Timmons Roberts 2009). Historical studies have also been used to examine the question of whether polluting facilities appear first, and poor, marginalised communities gather around them for work, or whether those communities predate the siting of the polluting facilities, thus suggesting discrimination in siting decisions.

Generally, they have found that the communities exist first, and then the toxic industries move in. For example, Saha and Mohai (2005) found a consistent pattern of hazardous waste sources being sited in working-class and non-white communities in Michigan over several decades. However, the pattern became particularly acute after 1970, which they attribute to the public controversy over the Love Canal crisis (see Chapter 5), and the subsequent 'greater success of white communities at keeping noxious land uses from being sited in their neighborhoods' (Mohai, Pellow, and Timmons Roberts 2009, 413).

Such findings refute simple economic explanations for environmental inequalities and injustices, which hold that industries are seeking to keep the costs of doing business as low as possible and locating in areas where land, wages, and raw materials are cheap, with more affluent residents then moving away to more environmentally desirable neighbourhoods. While such variables often also correlate with race in the United States, 'market forces' alone cannot explain geographical patterns in most cases. For example, the ability to move away is determined not just by affluence, but can also be shaped by discrimination in the housing system (such as through access to credit). Looking more broadly, in a classic study of Los Angeles, Pulido, Sidawi, and Vos (1996) showed how historical planning decisions – such as the location of industrial districts in majority Latinx neighbourhoods – continued to produce inequalities in toxic exposure decades later. A tangle of causes can be identified playing out differently in different places – market forces, class inequalities, legacies of racist planning decisions, and direct racial animus in the present, shaped by the same kinds of deep-seated racial prejudices which animated earlier European colonialism (see Chapter 3) and which view non-white bodies as material resources to be exploited, both in the service of production and in the absorption of excess toxicity (Liboiron 2021; Yusoff 2018). This tangle of the economic, the social, the historical, and the cultural emphasises the importance of geographically-sensitive and historically informed scholarship on instances of environmental inequality and injustice.

Box 6.1 Situating the Anthropocene: Environmental injustice in 'Cancer Alley'

One place which has attracted the sustained attention of EJ activists and scholars, including many geographers, is a region of southern Louisiana which has become known as 'Cancer Alley'.

This is a region that was once at the heart of the plantation economy of the US South. Sugar, tobacco, and, most significantly, cotton grown on the slave plantations which lined the middle and lower reaches of the Mississippi river was taken to coastal ports like New Orleans, from where it was dispatched all over the world (including to the revolutionary cotton mills of northern England – see Chapter 4). Towards the end of the nineteenth century, former slaves participated in the agricultural economy as plantation workers, tenant farmers, or sometimes as landowners. Around the middle of the twentieth century, however, with strong government support, former plantation lands began to be sold to petrochemical companies, with the consequence that the 85-mile stretch of the Mississippi River between Baton Rouge and New Orleans now accounts for around one quarter of all petrochemical production in the United States, with around 150 petrochemical plants and oil refineries. This geographical layering of some of the crucial 'detonators' of the Anthropocene (Tsing et al. 2020) – the plantation economy, the oil economy, state-led developmentalism (Holmes, Bolen, and Kirkbride 2021; Woods 2017) – means that the region has attracted renewed attention from Anthropocene scholars interested in how the history and politics of planetary transformation can be narrated and understood through the history and politics of particular places and landscapes (Rosol, Turnbull, and Renn 2021).

For state authorities and oil industry publicists, the region is known as the 'Chemical Corridor'. For many residents, activists, and scholars, it is better known as 'Cancer Alley' (Verdin 2020), so named for its 'toxic inheritance from the historical geographies of slavery' (Davies 2018, 1541). While air quality has generally improved across the United States over recent decades, data suggest it has gotten progressively worse in this corner of Louisiana. The region has well-documented and persistent patterns of ill-health and early mortality from causes which are traceable to its extraordinary concentration of petrochemical facilities, while local communities have seen very little economic benefit from the plants (Baurick 2019; Terrell and St. Julien 2022). The siting of these facilities alongside rural, impoverished, and largely African American communities, many of whose residents can trace their ancestry to slaves who once worked on the plantations, caught the attention of EJ scholars interested in the relationships between social and environmental inequalities.

Figure 6.1 shows how exposure to high-risk toxic pollutants tends to cluster in neighbourhoods with high African American populations. These patterns are well known and are reflected in both official government data and in data and observations made by the myriad 'citizen science' initiatives that have documented the region's toxic geographies over the last few decades, with a view to challenging the activities and expansion of the petrochemical industries. There have been several instances of successful community organising in

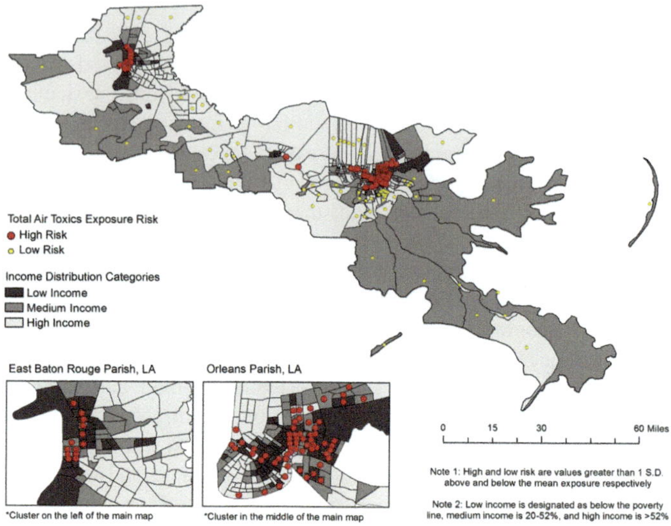

Figure 6.1 Map displaying the relationship between high levels of toxic exposure and race in 'Cancer alley'. (Source: (James et al. 2012, 4373), using US Environmental Protection Agency (EPA) data. CC BY 3.0.)

the region since the 1970s (Berry 2003). Groups have successfully blocked the building of many new plants, but existing plants have proved harder to either close or to halt the expansion of. Sociologist Barbara Allen recounted several successful efforts by communities to challenge industrial interests, and argued that alliances between citizens and activist-experts, which crossed lines of race and class and which linked local and national campaigning organisations, were key to successful resistance to the generation of more pollution (Allen 2003; also Lerner 2006).

However, alliances between citizens and accredited 'experts' can also be fraught (Allen 2004). Even the best atmospheric science can be insufficient to convince state actors to extract compensation or redress from the corporations that run the plants, owing to the inherent uncertainties involved in tracing cause and effect between pollution source and the bodily harms reported by communities (Whitehead 2014, 60–62). And there is evidence of experts using a language of 'corporate responsibility' to convince communities that they are acting in their interest, when in reality they are buttressing the interests of the corporations (Ottinger 2013).

Recently, controversy has raged around Formosa Plastics Corporation's plans to develop a new, 2,500-acre petrochemical and plastics plant in the community of Welcome, which is 98% Black. The plans were fast-tracked by the Trump administration, prompting a coalition of faith and environmental groups to sue the US Government for issuing permits to the company while failing to account for the destruction of wetlands (which are important refuges for biodiversity and which provide protection against hurricanes and flooding) and for the destruction of a cemetery containing the unmarked graves of enslaved people. The legal action succeeded in slowing down the permitting process, but the plans are still being pursued.

The threat to the burial ground prompted Jennifer Blanks and colleagues to examine the exposure of African American cemeteries to other risks, including those associated with climate change such as intensified storm and flooding events. They found that African American cemeteries were more exposed than majority-white cemeteries to a range of such risks (Blanks et al. 2021). These findings emphasise the multi-generational nature of the struggle for social and EJ in 'Cancer Alley', and show how that struggle is not just about proving epidemiological links between petrochemical pollution and bodily ill-health, but about campaigning for basic **recognition** of the presence and value of African American culture and heritage in the region, and for its right to a future.

Recognition and procedure

Alongside market forces, class inequalities, and housing discrimination, EJ scholars have long argued that one possible explanation for the high levels of correlation between race and toxic exposure concerns the ability of communities to fight back (Saha and Mohai 2005). Distribution is not the only way to think about justice, as the authors of the *Principals* well knew. **Recognition** approaches to justice essentially ask how it is that some communities come to be exposed to more harm than others, and which seeks answers in processes of political and cultural marginalisation. Recognition approaches to justice ask who, in societal deliberations over the distribution of goods and bads, is afforded the moral standing to be protected from the bads. They ask how it is that certain communities, whether through cultural domination, prejudice, or ignorance, are essentially devalued as members of the wider community, and therefore rendered vulnerable to an unequal share of environmental bads.

Scholars of North American Indigenous communities' struggles for social and environmental justice have been prominent in advancing recognition theories of justice. For example, Kyle Powys Whyte has shown how the ability of tribal governments to deliver environmental protection for their communities is dependent on equitable relations with US Federal and State governments, and on the provision of adequate funding for them to discharge their duties. In a context of the regular breaking of treaties between the US and tribal governments, the question of whether the most powerful institutions in the equation – organs of the US government – sufficiently *recognise* and respect the political and moral standing of the tribal governments and the communities they represent, is crucial for determining environmental outcomes (Whyte 2011). This becomes particularly stark in negotiations over the protection of Indigenous communities' sacred sites. The protection of such sites (such as burial sites or water bodies) from the incursion of things like oil pipelines (and their attendant pollution risks) is often dependent on whether or not non-tribal authorities recognise the spiritual value of those places, when that valuation belongs to a historically marginalised rather than culturally dominant worldview (Gilio-Whitaker 2019; Whyte 2017).

While some EJ scholars posit recognition as a relatively new feature of the discourse, we would point out that many issues of recognition are prominent in the 1991 *Principles*. For example, it is stated, in point 2, that:

> Environmental Justice demands that public policy be based on mutual respect and justice for all peoples, free from any form of discrimination or bias.

'Mutual respect' here can be read as mutual recognition of the equal moral standing of all communities. More concretely, point 11 states that:

> Environmental Justice must recognize a special legal and natural relationship of Native Peoples to the US government through treaties, agreements, compacts and covenants affirming sovereignty and self-determination.
>
> (FNPOCELS 1991)

Point 11, in its call for recognition of the particular legal position of Indigenous sovereignty in the United States, also points to a closely related theory or element of EJ, that of **procedural justice**. This concerns questions about whether decisions that are made that affect communities have been taken in a manner that is fair and open to the input of those communities. It stems from the 'all-affected' principle: i.e. that everyone affected by a decision has the right to be involved in the making of that decision. Issues of distribution and (mis)recognition can be structural barriers to participation in such decisional procedures: 'If you do not have resources, or recognition, it is difficult to be heard' (Schlosberg 2018, 631). As such,

> Environmental justice activists call for policy-making procedures that encourage active community participation, institutionalize public participation, recognize community knowledge, and utilize cross-cultural

> exchanges to enable the participation of the diversity that exists within a community.
>
> (Schlosberg 2018, 632)

While public participation in decision-making has been institutionalized in recent years in many parts of the world (Cooke and Kothari 2001; Pallett and Chilvers 2013), the struggle to democratise decision-making and environmental governance remains a prominent part of EJ movements (Goh 2020). Calls for procedural justice argue that distributional justice cannot be achieved without, first, recognition, and then procedural justice for those affected. It affirms that authorities – whether public or private – cannot be assumed to fully and fairly account for the interests of others, particularly the poor and the marginalised, and that new forms of democratic practice are required so that those interests might be adequately represented and acted upon. As the *Principles* put it back in 1991:

> Environmental Justice demands the right to participate as equal partners at every level of decision-making, including needs assessment, planning, implementation, enforcement and evaluation.
>
> (FNPOCELS 1991)

However, the ability to participate in political decision-making can itself be shaped by economic and social inequalities (Smith and Rhiney 2016). Sociologist Robert D. Bullard, in his classic work *Dumping in Dixie*, found that 'those communities most capable of mounting effective collective resistance tend to be better educated, have greater levels of income, and fewer people of color' (Mohai, Pellow, and Timmons Roberts 2009, 414). Indeed, in a pattern 'repeated the world over, environmental risks are commonly placed in the path of least resistance, near communities with the smallest reserves of political, economic and social capital' (Davies 2022, 416).

Lower-income and marginalised communities might offer a 'path of least resistance' to companies and governments where siting decisions are less likely to be met with well-resourced resistance from community representatives or grassroots movements. And this isn't always about decisions about where to put polluting factories – emerging research is showing that policy decisions made in response to the threat of climate change can exhibit similar dynamics. For example, geographer Idowu Ajibade (2019) has shown how poor communities have been moved back from the coastal frontlines in Manila and Lagos, ostensibly as an act of adaptation to climate change and rising seas. However, in reality, these 'managed retreats' of the urban poor have opened up desirable seafront real estate to eager property developers, raising urgent questions about power and procedural justice in urban climate change adaptation, and about how 'climate gentrification' is remaking cities around the world (see Taylor and Aalbers 2022).

Cases like these show that there can be very serious material and political barriers to procedural justice, and the EJ movement has devoted significant effort to helping under-resourced communities participate in political and legal processes that they might not otherwise have access to. However, the EJ movement has also been split on whether governments and regulators can be trusted to sincerely want to expand participation in decision making, or whether state institutions should be approached in a more conflictual mode – less as a potential ally to be appealed to, and more as an adversary of the movement, similar to the polluting industries themselves (Pulido 2017).

Epistemic justice

If recognition-based justice is partly about the recognition of different communities' rights, it is often also about recognition of and respect for different communities' forms of environmental knowledge (Parsons 2025). Research has shown how environmental injustice often results from the sidelining of certain forms of knowledge about how certain environmental systems work and how they should be managed (Vaddhanaphuti 2026).[1]

For example, Iokiñe Rodriguez (2017) shows how the Indigenous Pemon people of southeastern Venezuela were consistently marginalised in the management of the complex savannah landscapes of the Canaima National Park. Park managers, informed by Western scientific research, sought to put a halt to Pemon practices of using fire to open up new land to agriculture and to manage the savannah. For decades, the Pemon use of fire was seen by the state and by scientists as destructive to the park's ecosystems and watersheds, and was understood as a product of the Pemon people's ecological ignorance.

However, over recent decades, new forms of socio-ecological research conducted in collaboration with Pemon community members have shown how vital Pemon practices are to maintaining biodiversity in the park, and to managing fuel load accumulation in order to prevent much larger and more destructive fires. As a result, a state of 'cognitive' (Visvanathan 1997) or 'epistemic' justice (Fricker 2007) has started to be achieved, whereby Pemon knowledges and value systems are afforded equal respect and recognition in environmental management processes, to the benefit both of the community and the ecosystems of Canaima National Park (Rodriguez 2017).

Rodriguez's case, and her telling of her own role as an environmental social scientist working in collaboration with the Pemon community, is a reminder that acts of geographical knowledge production are crucial sites 'through which power is deployed with enduring material effects' (Derickson 2022, 841). As geographers of environmental (in)justice, we therefore need to constantly reflect on how our own acts of knowledge production and collaboration serve to bolster or undermine the position of others – simply 'making knowledge *about* people, rather than *for* or *with* people, undermines their own epistemic authority', and can be detrimental to the achievement of epistemic justice (Derickson 2022, emphasis added). Geographers need to cultivate equitable relationships geared around making knowledge *with* and *for* communities – to cultivate epistemic justice for communities' own ways of understanding the world – in order to contribute to struggles against wider forms of injustice (Derickson 2022).

In the next section, we'll work through another example of environmental (in)justice to further illustrate the analytical purchase of distributional, recognition, and procedural theories of justice, and to show how injustices of distribution are tightly bound up with injustices of recognition and procedure in a very different geographical context. It's also a case study of how commodity frontiers (see Chapter 3) and political frontiers intersect to create both local and global geographies of environmental injustice.

Commodity frontiers and sacrifice zones

The rare earths frontier

'[F]irst the animals got sick, then the babies, and then everybody else' (Klinger 2017, 125). So reported an elderly herder in a village some twenty kilometres from the Bayan Obo mining district in the Inner Mongolia Autonomous Region in northern China, in an interview with political geographer Julie Michelle Klinger (2017, 125). Klinger was there to investigate the politics of 'rare earth' mining – the extraction and processing of a family materials used in 'quantities too minute to quantify compellingly' (Klinger 2015, 573), but which are increasingly ubiquitous in twenty-first century life. Rare earth elements such as cerium, neodymium, and yttrium are essential to the infrastructures of technological globalisation and sustainability transitions. From cigarette lighters to smartphones, hard drives, fibre-optic cables, wind turbines and electric cars, rare earth elements are a key part of the material constitution of twenty-first-century modernity and, crucially, of efforts to deal with some of the challenges of the Anthropocene, like developing cleaner energy sources. They are key to the evolution of what some call the 'technosphere' – a new component of the Earth System, made up of human technologies and their residues, which is driving rapid change in the functioning of other elements of the Earth System, like the biosphere (Rosol, Nelson, and Renn 2017).

Despite the name, rare earths aren't actually very scarce.[2] In fact, taken together, rare earths occur between 150 and 200 parts-per-million (ppm) in the planet's crust, while copper is just 55ppm. They're fairly evenly spread over the Earth's surface, with significant reserves in Vietnam, China, Russia, Greenland, India, Australia, and Brazil (USGS 2021). Nonetheless, rare earth mining on any great scale takes place in only a few locations. This is the geographical puzzle Klinger set out to solve – with their widespread geological availability, what political and social factors explain the geographies of their extraction? Why does the physical geography of rare earth prevalence not fully explain the geography of the rare earth commodity frontier? Answers to these questions, it turns out, are rooted in geopolitics and are revealing of some core geographical features of environmental and social justice.

The first 'rare earth' elements were discovered in Sweden in the late eighteenth century, where the term 'earth' was used for any acid-soluble element. Despite much scientific interest, it took until a century later for industrial applications to be found. Carl Auer von Welsbach's cerium-containing gas mantles provided a solution to the problem of cheaply and reliably lighting the urban night, particularly in Western Europe and North America, and by 1930, some 5 billion had been sold, lighting up cities before the later arrival of widespread electrical grids. Welsbach's 'flint stone', which still features as an ignition material in cigarette lighters and weapons, created a further demand for rare earths, and as Scandinavian production began to fall behind demand, new global (and usually colonial) relations emerged. British mining interests began extracting monazite from the beach sands of the Carolinas; later, German and Austrian companies set to work in Brazil and India, while German, Soviet, Japanese, and Chinese geologists surveyed the potential of what would become, in 1947, the Inner Mongolia Autonomous Region in Northern China.

The early Cold War period, with its moves to decolonisation and nuclear arms races, intensified the search for exploitable rare earths. Indian independence in 1947 saw a halt put on exports to the United States, even when food aid was dangled to help the new government deal with a devastating famine. In 1949, the United States discovered significant reserves at Mountain Pass in San Bernardino County, California, but was reluctant to tolerate the already well-known environmental hazards within its own borders. The Indian export embargo, along with the inability of Brazilian mines to keep up with demand, forced US hands, and Mountain Pass became, for the next few decades, the world's biggest source of elements like europium (see Figure 6.2). By 2002, however, local authorities had grown intolerant of the environmental damage being done at the mine, including the regular leaking of toxic and radioactive wastewater, while neoliberal trade reforms in the 1980s made the 'off-shoring' of mineral processing and its environmental consequences easier (Klinger 2015). Meanwhile, the USSR and China had been collaborating on opening up Inner Mongolia to mining and heavy industries. The USSR got materials for its nuclear programme plus an expanded military presence outside of its borders; China in return got Soviet support in the development of domestic industrial capacity in the defence and energy sectors. As Klinger (2017, 53) puts it, 'Mao and Stalin intended to convert the windswept steppes of Inner Mongolia into a military-industrial heartland that could provision both Republics in their struggle against capitalism and Western imperialism'. But China soon felt that the USSR was withholding nuclear expertise, so it eventually resolved to strike out on its own.

Figure 6.2 The Mountain Pass Eare Earth Mine and Processing Facility in San Bernadino County, California. (Source: Wikimedia Commons.)

This story of geopolitical manoeuvring might give the impression that Inner Mongolia was a blank space over which emerging superpowers could simply project their dreams of control. Crucial to the development of rare earth mining in Inner Mongolia was precisely the construction, in scientific and political discourse, of the region as a 'frontier space' – a space in need of conquering, home to a people considered racially inferior whose culture and environment needed to be brought more closely into line with the developing Chinese state. This was a space of nomadic agriculturalists, with strong inclinations towards political autonomy. Through industrial development, the Chinese state reasoned, the region and its inhabitants could be brought more firmly under state control.

The extraction and processing of rare earths come at a great environmental and human cost. These elements are found alongside radioactive materials and heavy metals, and for each tonne of rare earth produced, a tonne of radioactive wastewater is generated, 75m^3 of acid wastewater, up to 12,000m^3 of waste gas containing things like sulfuric acid, and around 8.5kg of fluorine. All the rare earth elements can cause organ damage if inhaled or ingested, many corrode the skin, and some are so toxic they must be handled like radioactive materials (Klinger 2017, 70). Around Bayan Obo, which became the centre of rare earth extraction and processing, the human toll of this toxicity is painfully clear. It is evident in epidemiological studies, in the testimony of people like the herder quoted above, and in the 'skin lesions… malformed bones and decaying teeth' by which, 'heartbreakingly enough', it is possible 'to distinguish true natives' of the region from new arrivals (Klinger 2017, 138).

So why have the residents of Bayan Obo and the surrounding region been subjected to this toxic violence, and not others? What explains this particular **distribution** of environmental injustice?

In Julie Michelle Klinger's analysis, the people of Bayan Obo have suffered because their homeland just so happens to overlap with what has historically been a border region between political powers. The rise of the rare earths frontier at Bayan Obo cannot be explained by geological convenience or economic rationality. Rather, Klinger emphasises how, as with so many major rare earth extraction sites, its positioning close to an international border, and far from the national metropole, is deeply significant. It is a region over which the Chinese state wanted to show it was in control and which, by supplying the raw materials for manufacturing industries elsewhere, would be pulled closer into the orbit of a modernising Chinese state – *territorialised*, as Klinger has it. But its distance from the metropole also means that pollution and toxicity remained invisible for most of the wider population: it was 'out of sight' to those who might be able to make a difference (Nixon 2011).

Klinger's observation that rare earth mining tends to cluster in spaces at the edges of conventional state-based forms of sovereignty – such as Inner Mongolia, parts of the Brazilian Amazon and, if certain companies get their way, the bottom of the sea and the surface of the Moon – invites reflection on how political frontiers tend to enable the extension of extraordinary forms of power; ordinary forms of accountability can be dodged by decisionmakers, rules not followed, environmental standards ignored. Here we get into forms of procedural injustice – the residents of Bayan Obo have not enjoyed the kinds of legal and political protections which residents of Beijing, Brasilia, or San Bernadino might expect, despite their very different political systems. As Klinger (2017, 55) puts it:

> Who is exposed to these hazards, and which landscapes are laid to waste, is seldom a matter of democratic decision making. The spatial allocation of sacrifice illustrates the cartographies of the powerful and the powerless.
>
> (Klinger 2017, 55)

The geography of rare earth extraction is therefore a product of the relationship between their ubiquity and necessity in the infrastructures of Anthropocene modernity, and the capacity to isolate and contain the risks inherent to their production processes. Those risks are both environmental and political – distance and rurality help keep toxic substances away from major population centres, while the politics of the frontier help contain the political risks of distributional injustice. As Klinger argues, the geography of rare earth extraction is 'inseparable from geographies of power and vulnerability: power is exercised in the capacity to make hegemonic claims to the subsoils containing rare earth elements, and power is manifest in the ability to subject some and exempt others from the toxic and radioactive by-products of mining and processing' (Klinger 2015, 754).

The power to subject some to toxicity and radioactivity, while exempting others, points towards issues of **recognition**. As mentioned above, the Indigenous inhabitants of the Bayan Obo region have been constructed in political discourse as racially inferior, and therefore as somehow deserving of fewer protections from the state and from industry. The lack of recognition of these citizens as morally equal members of the political community has gone so far as to see the remaining nomadic pastoralists of the region simply erased from administrative records:

> because rare earth elements are classified as strategic national resources, the Bayan Obo mining district has been re-zoned to legally erase possible conflicts: officially, there are no farmers or pastoralists in Bayan Obo... there is no longer a local census category [for them]. Land is zoned according to three uses: mining, urbanization, or wind energy generation. Livelihoods that do not fit within these zoning codes—whether agropastoralism or artisanal mining—are not counted and therefore exist in precarious relation to the local state.
>
> (Klinger 2017, 30)

This political erasure is about the most extreme form of misrecognition one could imagine, and is a key structural barrier to procedural justice for those who would wish to undo the distributional injustices of rare earth mining and processing.

Box 6.2 Sacrifice zones and slow violence

In Bayan Obo, we have another story of a resource frontier (Moore 2015), a space where the power to know, to extract, and to corral is crucial to processes of accumulation. It is therefore a story of profound environmental injustice, in terms of distribution, recognition, and procedure. Klinger's argument is that resource frontiers and political frontiers often overlap and reinforce each other's status. In this case, political and resource frontiers combined to produce a particular kind of space, familiar to scholars of environmental injustice: a sacrifice zone.

Sacrifice zones are spaces that have borne and contained the environmental and epidemiological costs of the Anthropocene; spaces where the damage inflicted on lands, ecosystems, and human bodies is considered an acceptable price for the 'greater good' of national economic growth and security. The term 'sacrifice zone' originates with the US Government's notion of a 'National Sacrifice Zone' – an official designation for landscapes rendered unhealthy and even inhospitable by the production and processing of nuclear materials and the testing of nuclear weapons. The designation acknowledges the severity of the damage, but frames it as a necessary sacrifice for the greater, national good (O'Lear 2010).

Since then, the term has been adopted and adapted by environmental scholars and activists concerned with places and communities suffering from diverse types of intense pollution. Steve Lerner helped popularise the term in his study of what courts were calling 'environmental high-impact areas'. For Lerner, the term sacrifice zone is better as it:

> dramatizes the fact that low-income and minority populations, living adjacent to heavy industry and military bases, are required to make disproportionate health and economic sacrifices that more affluent people can avoid...this pattern of unequal exposure constitutes a form of environmental racism that is being played out on a large scale.
>
> (Lerner 2012, 3)

The concept has since been used widely by geographers. For example, Marcelo Lopes de Souza, using examples of urban pollution in Rio de Janeiro, shows how social prejudices and 'territorial stigmatisation'[3] function to render some spaces and communities more vulnerable to sacrifice than others (de Souza 2021). The concept is analytically powerful as it highlights how the geographies of environmental injustice are shaped by structural phenomena like economic inequality and racism (Cobs-Muñoz and Westerholt 2025), and by multiple forms of what we might call *slow violence*.

The concept of 'slow violence' was developed by literary critic Rob Nixon to describe:

> a violence that occurs gradually and out of sight, a violence of delayed destruction that is dispersed across time and space, an attritional violence that is typically not viewed as violence at all.
>
> (Nixon 2011, 2)

'Violence' tends to be imagined as immediate, visceral, and spectacular. Nixon draws our attention instead to forms of violence wrought by the slow seepage of toxic chemicals, the gradual inhalation of air pollution, and the unfolding impacts of global climate change. Slow violence is attritional, often detached from its underlying causes, and it can be difficult to see. Political geographer Thom Davies uses the concept to examine the toxic geographies of Cancer Alley, emphasising the lived experiences of those cohabiting with, observing, and feeling the effects of petrochemical pollution (Davies 2018, 2022).

Davies' use of the concept of slow violence urges geographers to 'consider the various scales and speeds at which violence takes place' (Davies 2022, 413). Drawing on ethnographic research and interviews with residents of Cancer Alley, he relates unique attentiveness to the rhythms of this ongoing violence employed by residents of settlements like Freetown, on the banks of the Mississippi. From waking to new smells following overnight gas venting to observations of steadily increasing rates of cancer among neighbours and a feeling of predictable doom – 'if you stay here, you die' (quoted in Davies 2022, 415) – Davies argues, contra Nixon (2011), that far from this slow violence being 'out of sight' to its victims, residents of places like Freetown know the intricacies of this violence well. They know its sources, its temporal rhythms, the geography of its effects. The political challenge is therefore not a lack of knowledge or compelling stories with which to galvanise political action; rather, the problem is that the stories and testimony of some communities don't seem to count as much as those of others when it comes to arresting the 'necropolitics' of 'letting die' (Mbembe 2019; see also Green et al. 2025). The slow violence of toxic pollution in the sacrifice zones of southern Louisiana is shaped and enabled by injustices of recognition, themselves part of historically rooted forms of racial violence and injustice (Pulido 2017).

Planetary justice?

Political geographer Thom Davies offers a sobering diagnosis of the Anthropocene condition as one defined by unavoidable toxicity, which poses challenges for how to think about justice:

> In the Anthropocene, exposure to some form of toxicity is an unavoidable and necessary part of everyday life, yet the discriminatory geographies of pollution ensure that certain populations are subjected to the power of death more readily.
>
> (Davies 2018, 1539)

Davies' argument concerns how technological globalisation – the expansion of the 'technosphere' and its waste products – renders all of us subject to new levels of toxicity. In the twenty-first century, 'humans are chemically-transformed beings' (Murphy, quoted in Thomas 2014, 1599). While the Anthropocene is partly defined by the 70,000-plus different chemicals that make their way into the Earth's ecosystems each year, most of those make their way into human bodies too. Styrene, ethyl phenol, toluene, polychlorinated biphenyls, and other cuddly-sounding substances are found in more than >80% of the US population, for example. Toxins from plastic production inhabit human bodies across the world. Around 80% of the nitrogen in human bodies is industrially produced. It is not just the material make-up of the planet which is being transformed; it is us, too, albeit some more than others.

Would this unavoidable and seemingly 'necessary' toxicity be justifiable if it were more evenly distributed? Do the many conveniences of technological globalisation, and its associated patterns of economic growth, justify the toleration of some level of background toxicity? What about nonhuman creatures who can't enjoy the benefits of owning a smartphone or a streaming subscription? How should their interests be reflected in calculations of allowable toxicity and distributional justice? And what if all those benefits of modern life enjoyed by (some) humans are moot if global capitalism and economic developmentalism are driving us into a place where the global environment, the Earth System, simply won't be able to support us anymore?

These are the kinds of questions animating new debates about the meaning of justice in the Anthropocene. Whatever one thinks are the ethically right answers to the questions above, there's no doubt that the proposed onset of the Anthropocene challenges us to think about justice on new spatial and temporal scales. The EJ movement, which coalesced in the early 1990s in North America took environmental concerns out of the wild and into cities, homes, workplaces, and places of worship. It was a localising move, drawing new attention to the intimate, community and regional geographies of environmental inequality. Since then, the EJ movement – with its core concepts and principals – has expanded into the climate justice movement, which asks similar questions about unequal distributions, misrecognition, and procedural and epistemic injustice in relation to the causes and cascading impacts of climate change (Sultana 2022; Trott et al. 2023). These questions are addressed to wider spatial scales; they inquire into, and challenge, the mismatch in responsibility between those most responsible for the problem (be they governments, companies, or consumers in richer, earlier-industrialising parts of the world) and those already suffering its effects most acutely in some of the world's poorest places (Fisher 2015). Climate justice asks why vulnerable communities aren't adequately represented in global climate policy negotiations, which are dominated by nation-states and industrial lobbyists (O'Lear 2016). And it asks searching questions about recognising the rights of those even more conspicuous in their absence from the policy table – future generations, nonhuman animals, and other ecological beings, like rivers, forests, or waters (Latour 2018; Schmidt 2022).

The climate justice movement has radically stretched the spatiotemporal horizons of thinking around EJ (Schlosberg and Collins 2014). It peers deep into the past of historical greenhouse gas emissions and far into the future to try to give voice to the unborn. It considers unequal distributions of environmental harms on a global scale, and it looks beyond human actors as potential subjects of justice. As such, it points towards new ways of thinking about justice which might rise to the epistemological, ethical, and political challenges of the Anthropocene.

A number of scholars have begun to think about the possible meaning of 'planetary justice' in the Anthropocene. Political scientists Dryzek and Pickering (2018) suggest that planetary justice has three key dimensions that 'reach beyond Holocene thinking' about justice (Dryzek and Pickering 2018, 68): it reaches across *national borders*, across *generations*, and seeks justice for *non-humans*.

Justice across borders

Dryzek and Pickering observe that, until relatively recently, most thinking about justice was conducted at the nation-state scale. The nation-state was (and largely still is) taken as the container within which 'winners' and 'losers' may be identified, and national welfare or legal systems operate (perhaps) to redress those imbalances (also Barnett 2011; Fraser 2009). At the same time, nation-states continue to be the main actors in global politics, which can be problematic for conceiving and addressing injustices across borders. In EJ terms, activism has commonly taken place in relation to local inequities, such as those in parts of southern Louisiana (see above), but has often targeted the levers of the nation-state to seek redress. It is now commonplace to point out that issues like climate change, biodiversity loss, or plastic pollution don't respect national borders: relationships of cause-and-effect span global supply chains and flow through the subcomponents of the Earth System (Parsons, Safra de Campos, and Moncaster 2022); relationships of risk and responsibility therefore connect people, corporations and governments in different parts of the globe (Beck 2016; Okereke 2006). Nonetheless, 'international mechanisms for protecting the vulnerable are far less developed than their domestic equivalents' (Dryzek and Pickering 2018, 70), and of course still rely on the actions of cooperating nation states.

International aid represents one mechanism by which newly recognised responsibilities of citizens of one country to those of others have been put into practice, with varying and contested levels of success. Within international climate change politics, the mismatch between historic responsibilities and present and future vulnerabilities has been a lightning rod for political disagreement and tension (Agarwal and Narain 1991; Okereke and Coventry

2016). The notion of compensation for 'Loss and Damage' has recently come to prominence as a means of recognising unequal responsibilities and vulnerabilities to the impacts of climate change which are already occurring, and of generating flows of money from richer (and more historically-responsible) countries to those which are already feeling climate change's ill-effects. However, critics have argued that the way this has been built into the international climate policy architecture represents less a 'compensation' scheme for historical responsibilities for climate change, and more an insurance scheme premised on complicated and uncertain calculations of cause-and-effect between extreme weather events and global greenhouse emissions; relationships which are hard to prove, least of all for countries which may lack scientific resources (Jackson et al. 2023) – it is a matter, therefore, of epistemic as well as distributional justice.

Political geographer Laurie Parsons, in his work *Carbon Colonialism*, shows how thinking about climate-warming emissions and, by extension, climate justice solely through a national frame obscures important dynamics in the global economy. Parsons develops the notion of 'carbon colonialism' to describe how, through things like the global garment industry, wealth and value flow towards richer countries, while waste and the most severe consequences of greenhouse warming flow towards poorer countries (Parsons 2023; see also Pellow 2007). Complicated global supply chains and production systems mean that the environmental impacts of, for example, producing a cotton t-shirt are obscured and 'off-shored'. When goods are made from raw materials into a finished product in one factory, oversight and regulation is relatively easy. When different parts of the production process take place in different countries, taking advantage, for example, of different wage rates or regulations, complete oversight is impossible with the result that environmental damages (like deforestation or toxic waste) build up in poorer countries with little abatement. What's more, the flow of wealth into the richer parts of the world means that those countries are better equipped to deal with the challenges of a changing climate than those that remain at the margins of the world economy. Echoing the work of world-ecology theorists (Moore 2015), Parsons argues that both researchers and consumers need to consider these distant, denuded environments as part of the economic systems that deliver consumer goods to the global North. Rather than thinking about things like carbon emissions and EJ in national terms, it is increasingly imperative to consider these economic and ecological webs that tie together people, goods, and natures in different corners of the planet (Parsons 2023).

It's these sorts of concerns that proponents of the notion of 'planetary justice' have sought to capture. For political scientist Frank Biermann and colleagues, planetary justice 'encompasses traditional concerns of EJ but foregrounds that the entire human and non-human world is now at stake, not a mere locality'. It is 'concerned with justice among humans as well as between humans and the natural world', and is 'equally concerned with the global and the local, with state and non-state actors, and with individuals and collectives' (Biermann, Dirth, and Kalfagianni 2020, 1). In an earlier publication, Biermann, Bai, and Bondre (2016, 342) argue that the Anthropocene is a useful frame for thinking about environmental politics and justice in so far as it encourages a cross-scalar understanding of local, regional and global change on a planet increasingly understood as an 'interconnected, interdependent social-ecological system' which is nonetheless marked by 'local variation and social inequalities'. To illustrate how considerations of justice are now inseparable from the components of the Earth System, they use the example of nitrogen and global food security.

During the early twentieth century, humans managed to overcome the limited natural availability of nitrogen with the Haber-Bosch process for producing artificial nitrogen fertilizer. Food production ballooned in some countries, leading to new forms of commercialisation, globalised production networks and markets, and the expansion of plantation monocultures (Martín 2017). But this explosion of nitrogen use has disrupted how nitrogen conventionally moves through the Earth System, resulting in local problems of air and water pollution and eutrophication[4] and contributing, through the release of nitrous oxide, to enhancing the greenhouse effect and the problems of global climate change (e.g. Li, Liu, and Yue 2022). At the same time, a significant proportion of the human population is under-nourished and suffering from food insecurity, with many farmers having limited access to sustainable agricultural inputs and technologies, and grappling with changing environmental conditions. Therefore,

> in the Anthropocene we face the problem of an increasing imbalance between 'too low nitrogen areas' and 'too much nitrogen areas'. The global alteration of the nitrogen cycle by humans has thus been highly unequal. Especially wealthy societies have benefited from increased food production, while all people are affected by the negative impacts of nitrogen cycle disruptions, including climate change.
>
> (Biermann, Bai, and Bondre 2016, 343)

Thinking about the geographies of EJ beyond national borders is therefore about more than just thinking about the relations between nation-states. Thinking across supply chains (Parsons 2023) or through the material perturbations of components of the Earth System (Biermann, Bai, and Bondre 2016) provides two means of tracing the inter-scalar and trans-local geographies of planetary (in)justice in the Anthropocene. How could we begin to think about ensuring procedural justice across these geographies? Would it require some kind of global government, or at least a legal system through which the disadvantaged could seek redress? Global environmental organisations, treaties, and agreements exist to try to enable inter-state collective action to redress planetary injustices, but have been critiqued from decolonial perspectives for how their basis in the nation-state system recapitulates – and perhaps even intensifies – the marginalisation of certain voices. For example, Indigenous or minority communities who are marginalised in national politics might find themselves even more firmly shut out of international environmental negotiations, even though they may be bearing the brunt of things like climate change impacts. Geographers and political scientists have therefore studied experiments in different forms and forums of international politics and advocacy, such as networks of cities working for climate action (Castán Broto 2017), or platforms like the International Indigenous Peoples' Forum on Climate Change (IIPFCC), set-up in response to demands for greater Indigenous peoples' involvement with the United Nations climate negotiations process (Claeys and Delgado Pugley 2017).

Achieving procedural justice on a planetary scale is of course an enormous challenge, requiring researchers and practitioners to work together to study and build new kinds of experimental institutions that can begin addressing some of the shortcomings of state-based environmental governance (O'Lear 2016). The scale of the challenge is magnified when we consider that thinking about and enacting justice across borders also demands thinking about justice across time, and across (human) generations.

Intergenerational justice

The Anthropocene is defined by 'cumulative environmental degradation across multiple generations, with far-reaching consequences for the future' (Dryzek and Pickering 2018, 70). The historical geographies we reconstructed in Chapters 3–5 show a clear pattern of unequal responsibility for the onset of things like climate change and toxic chemical pollution, albeit with geographies of responsibility which shift according to which Anthropocene onset point one finds most convincing. But the Anthropocene proposition is also fundamentally about new scientific observations that these environmental changes will reverberate for generations, and potentially for thousands of years to come. Greenhouse gases burnt today might persist in the atmosphere for centuries, and in so doing will affect the life chances of multiple generations to come. But how to even begin thinking about justice for people who might be born several 100 years from now?

The principle of thinking seven generations into the future has become commonly associated with sustainability and futures-thinking, inspired by the worldviews and ethical frameworks of Indigenous communities. The seven-generations concept can be traced to the *Great Law of the Iroquois*, or *Gayanashagowa* – a constitutional framework developed by the Six Nations of the Haudenosaunee, or Iroquois Confederacy, around 800–1,000 years ago (Park et al. 2023). This concept has inspired new approaches to intergenerational justice, and resonates with recent attempts to ensure some measure of *recognition* and *procedural justice* for the as-yet-unborn. In places like the

Netherlands, Chile, and the Philippines, courts have ruled in favour of organisations claiming to represent the interests of future generations, and in so doing have blocked things like logging permits or have forced governments to increase the ambition of their climate change policies. In Finland, a Committee for the Future exists in the national parliament with representation from across political parties, and with a mandate to hold the government accountable on the long-term goals it has set itself.[5] Other proposals have been made for the representation of future generations in parliamentary processes, for example through the reservation of a certain number of seats for people with the special task of representing the as-yet unborn (Gonzalez-Ricoy and Rey 2019). However, questions remain over how such representatives would be selected (or elected), by whom, and on what basis. Would they need to be experts in environmental matters, for example, or rather have life experiences that might be expected to be similar to citizens of the future (Byskov and Hyams 2022)? And how could one guarantee that such representatives were truly acting in the interests of those future people (who, after all, aren't around to hold their representatives accountable), rather than their own interests (e.g. of gaining re-election)?

These profound challenges of political representation have been dodged by another model of trying to ensure intergenerational justice, that of 'independent offices' for future generations – bodies which operate at arm's-length from parliament and government, but which are accountable to them. In 2007, Hungary established a Parliamentary Commissioner for Future Generations. Under environmental lawyer Sándor Fülöp, the office was able to protect the historic Tokaj region from a new power plant, halt the takeover of the national plant gene bank by a multinational corporation, and prevent the privatisation of public water utilities (Vincent 2012). However, the Commissioner post was soon downgraded and subsumed into another office; nonetheless, it provides another road-tested example of an effort to institutionalise considerations of intergenerational justice into policymaking and legislative procedures.

Legislating for intergenerational justice involves combining knowledge of both experienced and anticipated injustices (Kanger and Sovacool 2022). As such, it forms part of an evolving suite of 'anticipatory action' by which governments and other actors make claims about possible futures and take action to modify, avoid, or prepare for them (Anderson 2010). Owing to the challenges of knowing and representing the true interests of the unborn, some scholars of intergenerational justice have argued that the kinds of foresight mechanisms analysed by Anderson (2010) – such as constructing scientific scenarios, forecasts or storylines of possible socioenvironmental futures – might be a more legitimate way of identifying the interests of and risks faced by future generations, and of thereby rendering policymaking 'more farsighted and more respectful of future interests' than direct acts of political or legal representation (Gonzalez-Ricoy and Rey 2019, 10).

However one might try to encourage 'political institutions to outgrow their myopia', rethinking intergenerational justice in the Anthropocene requires 'thinking specifically about what kind of Earth System conditions the present owes to the future' (Dryzek and Pickering 2018, 70). Should the global community aim to pass on the same Earth System conditions as present generations have experienced, or might it be ethically justified to pass on worse conditions if that inheritance includes the tools and knowledge to deal with them? Dryzek and Pickering (2018, 70) arrive at a fundamental principle for thinking about intergenerational justice in the Anthropocene:

> humans living today have a strong responsibility to ensure that future generations can live in a world where the circumstances of justice still prevail, particularly the ecological preconditions necessary for them to pursue justice

That might sound a bit tautological (i.e. 'justice will have prevailed when justice prevails'), but what they mean is that present generations have a responsibility to ensure that future generations at least have a shot at achieving EJ. A background condition of justice is some form of scarcity – if there is lots of everything to go around, then justice would be relatively easy to secure. If there is next to nothing to go round (e.g. extreme scarcity of food), then justice – particularly of the distributional variety – is nearly impossible to secure. Conversations about justice therefore usually take place against a background of *moderate scarcity*, where there is theoretically enough for everyone, but serious obstacles exist to fair distribution. For Dryzek and Pickering then, intergenerational justice in the Anthropocene requires that present generations work to ensure that the Earth System doesn't change so much as to generate material conditions where generalised and extreme scarcity becomes the norm, so that future generations might still be able to strive towards a state of environmental (or rather, planetary) justice.[6]

Multispecies justice

But must the planet, and the vast multitudes of beings other than humans that call it home, be silent in conceptions of planetary justice? Does the onset of the Anthropocene, with all it implies about the distinctiveness of the human species, demand a new anthropocentrism which places humans 'even more firmly at the center of the moral universe' (Dryzek and Pickering 2018, 71)? Or does the 'shock' of the Anthropocene (Bonneuil and Fressoz 2016) constitute not the recognition of humanity's distinctiveness, but rather the recognition of its inherent and ever-deepening entanglement with the rest of nature? Such 'post-Nature' understanding of the Anthropocene (see Chapter 1) would suggest that 'greater entanglements between human and non-human nature point to the need to recognize our *moral* as well as physical commonalities' (Dryzek and Pickering 2018, 71). As humanity (or segments of humanity) creates new vulnerabilities and dangers for non-human natures, perhaps it is time to heed longstanding calls from environmentalists and ethicists to extend the obligation of justice to beings beyond the human: to other species, to ecosystems, perhaps even to the planet itself.

Theorists and activists working with feminist, postcolonial, and posthumanist theory have put forward conceptions of inter- or multi-species justice as a means of correcting the anthropocentrism inherent in mainstream notions of environmental and human justice (Chao, Bolender, and Kirksey 2022). Such perspectives aim not just to foreground how nonhuman beings are largely silent victims of human decisions and actions, but also how human life is inseparable from webs of interconnection and mutual dependence with other species and beings. As such, conceptualisations of justice in this framework have much in common with some variants of the post-nature narrative of the Anthropocene explored in Chapter 1. They seek to enlarge 'the range of obligations and duties vis-à-vis all those whose flourishing is undermined' by the material conditions of the Anthropocene, and to shift the 'focus and subject of justice from the individual and exceptional human being to a wide range of living and non-living entities, and their interactions and processes' (Tschakert, Schlosberg, and Celermajer 2021, 4–5).

Thinking about nonhumans as the 'focus and subject of justice' draws attention to *distributional* injustices – for example, how certain species (like those on the brink of extinction) bear the brunt of anthropogenic habitat destruction. But it is primarily an issue of *recognition* – of recognising nonhuman beings as having 'moral commonalities' with humans, and just as much moral right to existence and protection from harm as humans do.

But is 'existence', or mere survival, enough of a right to afford nonhuman beings like animals and plants? Would that mean, for example, that humans could go on destroying habitats so long as there was space for the displaced animals in a zoo or a gene bank? Or would those animals have a right to a certain *kind* of life? In order to make the case that they do, advocates for multispecies justice have often drawn on another theory of justice, which has become popular in environmental thought, concerned with *capabilities*.

Capabilities-based theories of justice are concerned with whether or not an entity or being – human or nonhuman – cannot just exist, but flourish. Conceived initially as a way of measuring human (as opposed to simply economic) development, the capabilities approach 'begins with a very simply question: What are people actually able to do and to be?' (Nussbaum 2013, x). Rather than asking whether people have the rights and freedoms to do and be certain things, capabilities theorists like Martha Nussbaum inquire into the tapestry of capacities, attachments and resources people require to lead lives that they value, such as bodily health, safety, emotional attachments, and some measure of control over one's political and material environment (Nussbaum 2013). As such, they 'specify that protecting the ability to flourish becomes the central point of justice' (Dryzek and Pickering 2018, 71).

While initially conceived as properties of human lives (albeit with the ability to live in harmonious relations to the rest of nature conceived as a core capability), thinking with capabilities has recently been extended to nonhuman beings themselves. Capabilities can extend across scales – they may be the capabilities of an individual human or a tiger to live full, happy, and healthy lives; or they may concern the capabilities of a herd of reindeer, a forest ecosystem, a wetland, or a border-crossing river system, with all their community members, to flourish. As such, capabilities approaches to justice, for Dryzek and Pickering (2018, 71–72):

> could serve as a basis for understanding planetary justice, grounding obligations to safeguard the capabilities of people living across national borders, and of future generations of non-humans and humans alike.

The importance of safeguarding the rights and capabilities of more-than-human entities to flourish is increasingly being recognised in formal political declarations across local, national, and international scales (Schmidt 2022), and represent attempts to ensure procedural justice for beings who don't usually have a voice in (human) political negotiations. Putzer et al. (2022) identify 409 initiatives to get the 'rights of nature' on the statute books across 39 countries. In Aotearoa/New Zealand, for example, Mount Taranaki and the Whanganui River have been declared as living entities, 'bearing the same legal rights as humans, with human custodians to be appointed to represent them and protect their interests' (Dryzek and Pickering 2018, 72; Charpleix 2018). Earlier initiatives in the United States and Ecuador have been followed by extensions of rights to non- or more-than-human entities like the Ganga and Umunna rivers in India (Kinkaid 2019).

Geographer Petra Tschakert and colleagues point out that expanding conceptualisations of justice beyond the human does not mean smoothing over the category of 'the human' and forgetting, for example, how much environmental change has its origins in the exploitation of human difference and, indeed, in the casting-out from the category of 'the human' of significant numbers of the human population (see Chapter 3). Rather, in their formulation of multispecies justice, Tschakert, Schlosberg, and Celermajer (2021, 5) start from an 'intersectional' account of justice 'that recognizes the simultaneity of identities and categories of difference and inequalities (race, class, gender, age, ability, species, and non-human beings) and their interlocking in structures and processes of injustice and oppression'. Recognising the intersections of these different forms of injustice – for example of climate injustice and housing injustice in Washington DC (see Ranganathan and Bratman 2021) – creates opportunities for building political coalitions with more diverse backgrounds, experiences, and agendas than the climate justice movement has heretofore been able to mobilise (Ducre 2018; Newell et al. 2021).

Central to that diversity must be the recognition and valuing of, and learning from, philosophies and practices of human-nonhuman relationships and entanglements that have been marginalised by Western rationality and anthropocentric thought. Tschakert and colleagues point out that climate justice scholars 'have much to learn and gain from Indigenous frameworks, practices, and protocols' of multi-species justice, although it is vital, of course, that that learning does not itself become a form of neo-colonial appropriation (Tschakert, Schlosberg, and Celermajer 2021, 6). Many of the aforementioned constitutional recognitions of the 'rights of nature' have been driven by the activism of Indigenous communities (Charpleix 2018; Schmidt 2022). And while political scientists have shown how 'the rights of nature' has circulated globally as a new political and legal 'meta-norm' (Kauffman and Martin 2018), geographers have been at pains to point out how articulations and institutionalisations of such ideas need to be understood as part of fundamentally local political and cultural struggles over things such as land, resources, and water (Kinkaid 2019).

Scales and intersections of justice

In debates about the nature of planetary justice, geographers have been concerned with pointing out how struggles for justice for humans or nonhumans always need to be understood as emergent from and embedded within particular political, economic, cultural, and ecological contexts (Avci 2017; Forsyth and McDermott 2022). Geographers have therefore resisted the notion that ideas about things like 'environmental justice' or the 'rights of nature' circulate freely around the globe, progressively converting and inspiring more and more adherents without needing to be adapted to local and regional conditions (Kinkaid 2019; Williams and Mawdsley 2006).

Amidst all the talk of planetary justice, geographers have similarly maintained a focus on thoroughly local sites and scales of justice, including the human body. Gay-Antaki (2023) and Vasudevan et al. (2022), for example, centre the feminized and racialized female body as a key site of environmental violence and injustice, whose exploitation is not a by-product of, but is central to, the workings of racial capitalism (see also Ducre 2018; Pulido and De Lara 2018). Many EJ scholars and aid agencies have noted how women continue to bear the brunt of environmental suffering all over the world. In many settings, for example, women are disproportionately responsible for obtaining things like water, food, and fuel, and in lower-income countries, agriculture is the most important employment sector for women. As such, droughts, extreme weather, deforestation, and pollution tend to impact the lives and livelihoods of women harder, and research has shown that other inequalities, such as those related to race or sexuality, compound vulnerabilities and can multiply exposure to harm (UN Women 2022). EJ research has also shown how female activists have often been central to social movements attempting to end such environmental and intersectional injustices. Examples are legion – we might look to Canadian Inuk activist Sheila Watt-Cloutier, who for many years has provided leadership in political and legal struggles against pollution and climate change across the Arctic region, which she frames as a fight for the 'right to be cold' (Watt-Cloutier 2018); to the work of Sharon Lavigne in fighting for EJ for African American communities in 'Cancer Alley' (Friese 2023; Box 6.1); to the 'practical ecofeminism' of Puerto Rican women leading battles against environmental degradation in the country's impoverished south east (Llorèns and Santiago 2018, 401); or to the leadership of female Indigenous activists in struggles for climate and environmental justice under right-wing authoritarianism in Brazil (Machado et al. 2023).

Scholars working with decolonial and feminist theory have understood many environmental injustices as being a product of enduring colonial and patriarchal social structures, whereby violence towards nonhuman natures proceeds in concert with violence against racialised and gendered human bodies. In geographical thought, the body has increasingly therefore been conceptualised as a key site of geopolitics – a meeting place of intersecting political and environmental inequalities, but also a place of shared vulnerabilities and of interdependence with others, human and nonhuman. The body, particularly the feminized and racialized body, is therefore an important starting point for thinking about justice between humans *and* for thinking about kinship with nonhuman natures (see Vasudevan et al. 2022). But such arguments are not simply a call for a continued localism in geographical engagements with environmental (in)justice, or for the kind of 'thick geographies' we discussed in Chapter 2. Rather, they also call forth radically different ways of thinking about human relationships with the Earth beyond the socioeconomic logics which have arguably paved the way into the Anthropocene. If we follow the narrative of the colonial Anthropocene, for example, 'justice' is not simply about ensuring fair distributions of goods and harms against an assumed backdrop of economic scarcity and liberal democracy, where everyone has a supposedly equal say and equal rights. Rather, such formulations push beyond liberal conceptions of justice, and argue that even basic forms of recognition are impossible to achieve within a system – often described as racial capitalism – which is predicated on the continued marginalisation and exploitation of certain peoples and beings. 'Justice', in any of the forms explored in this chapter, perhaps cannot therefore be obtained without radical *transformation* of the systems and structures that brutalise some bodies – human, nonhuman, more-than-human – to ensure the wealth and comfort of others.

Conclusion

This chapter has explored the evolution of ideas of 'environmental justice' and their implications for the political geographies of the Anthropocene. Much recent scholarship has narrated a progression in EJ scholarship from 'local' concerns with unequally distributed toxicity and pollution to more recent 'global' concerns with climate or even planetary justice, and the complexities of recognition and procedure at such scales. There is no doubt that it is increasingly important to think about EJ in relation to the findings of Earth System science concerning the destabilisation of key components of the Earth System, and to think through what that means for dominant conceptualisations of justice (Biermann, Dirth, and Kalfagianni 2020; Schmidt 2022). Thinking justice beyond national borders, beyond the present, and beyond the human are increasingly urgent tasks. These are tasks that connect to questions about how to give adequate voice and representation to non-human beings (something we'll return to in the next chapter), and about the political mechanisms needed to enact justice in a world that is ever-more materially interconnected, but perhaps increasingly fragmented politically.

Within Geography, feminist and decolonial approaches have built on earlier conceptualisations and observations of environmental injustice to foreground racialised and gendered geographies of inequality, violence, and sacrifice, and have posed challenging questions about the scale of systemic transformations that would be required to redress such injustices. Later in the book, we'll return to some of these more radical proposals for attaining justice in the Anthropocene and look at real-life experiments in building alternative life-worlds where social and environmental justice are placed at the centre of new forms of political organisation.

Study questions

1. Are current notions of distributional and procedural justice adequate to the scale of environmental challenges in the Anthropocene?
2. How well does the notion of 'slow violence' capture the environmental injustices of the Anthropocene?
3. Evaluate the extent to which recent efforts to include non-human beings in environmental decision-making represent the achievement of recognition-based justice.

Notes

1. We can extend this to the epistemic injustices of dominant ways of framing the Anthropocene itself – see Curley and Smith (2024).
2. The term 'rare' was applied because they appear in compounds and are relatively difficult to isolate.
3. 'Territorial stigmatisation' is a term initially developed by urban geographers to describe place-based stigma. Territorial stigmatisation devalues and denigrates certain places, and generates negative stereotypes about their inhabitants. Such stigma is often in turn used to justify 'improvements' and forms of development which prioritise the generation of profits over the material needs of local residents (Butler et al. 2018; Slater and Anderson 2012).

4 Eutrophication occurs in aquatic systems when an excess of nutrients such as nitrogen causes an increase in algal productivity, which in turn can lead to oxygen depletion and suffocation for insects and fish.
5 While recognition and procedural justice has been sought for future generations, controversy continues to rage in many northern European countries about the marginalisation of Indigenous communities and practices of 'green colonialism' – see Keskitalo et al. (2025).
6 This formulation echoes conceptions of 'sustainable development', which posits that economic development must meet the needs of the present without harming the capacities of future generations to meet their own needs.

References

Agarwal, A., and S. Narain. 1991. *Global Warming in an Unequal World: A Case of Environmental Colonialism*. New Delhi: Centre for Science and Environment.

Ajibade, I. 2019. "Planned Retreat in Global South Megacities: Disentangling Policy, Practice, and Environmental Justice." *Climatic Change* 157 (2): 299–317.

Allen, B. L. 2003. *Uneasy Alchemy: Citizens and Experts in Louisiana's Chemical Corridor Disputes*. Cambridge: MIT Press.

Allen, B. L. 2004. "Shifting Boundary Work: Issues and Tensions in Environmental Health Science in the Case of Grand Bois, Louisiana." *Science as Culture* 13 (4): 429–48.

Anderson, B. 2010. "Preemption, Precaution, Preparedness: Anticipatory Action and Future Geographies." *Progress in Human Geography* 34 (6): 777–98.

Avci, D. 2017. "Mining Conflicts and Transformative Politics: A Comparison of Intag (Ecuador) and Mount Ida (Turkey) Environmental Struggles." *Geoforum* 84: 316–25.

Barnett, C. 2011. "Geography and Ethics: Justice Unbound." *Progress in Human Geography* 35 (2): 246–55.

Baurick, T. 2019. "Welcome to 'Cancer Alley,' Where Toxic Air Is About to Get Worse." ProPublica, October 30.

Beck, U. 2016. *The Metamorphosis of the World: How Climate Change Is Transforming Our Concept of the World*. Cambridge: John Wiley & Sons.

Berry, G. R. 2003. "Organizing Against Multinational Corporate Power In Cancer Alley: The Activist Community as Primary Stakeholder." *Organization & Environment* 16 (1): 3–33.

Biermann, F., X. Bai, N. Bondre, et al. 2016. "Down to Earth: Contextualizing the Anthropocene." *Global Environmental Change* 39: 341–50.

Biermann, F., E. Dirth, and A. Kalfagianni. 2020. "Planetary Justice as a Challenge for Earth System Governance: Editorial." *Earth System Governance, Exploring Planetary Justice* 6 (December): 100085.

Blanks, J., A. Abuabara, A. Roberts, and J. Semien. 2021. "Preservation at the Intersections: Patterns of Disproportionate Multihazard Risk and Vulnerability in Louisiana's Historic African American Cemeteries." *Environmental Justice* 14 (1): 1–13.

Bonneuil, C., and J. B. Fressoz. 2016. *The Shock of the Anthropocene*. London: Verso.

Bullard, R. D. 2018. *Dumping In Dixie: Race, Class, And Environmental Quality*. 3rd ed. London: Routledge.

Bullard, R. D., P. Mohai, R. Saha, and B. Wright. 2008. "Toxic Wastes and Race at Twenty: Why Race Still Matters after All of These Years." *Environmental Law* 38 (2): 371–412.

Butler, A., A. Schafran, and G. Carpenter. 2018. "What Does It Mean When People Call a Place a Shithole?" *Transactions of the Institute of British Geographers* 43 (3): 496–510.

Byskov, M. F., and K. Hyams. 2022. "Who Should Represent Future Generations in Climate Planning?" *Ethics & International Affairs* 36 (2): 199–214.

Castán Broto, V. 2017. "Urban Governance and the Politics of Climate Change." *World Development* 93: 1–15.

Chao, S., K. Bolender, and E. Kirksey. 2022. *The Promise of Multispecies Justice*. London: Duke University Press.

Charpleix, L. 2018. "The Whanganui River as Te Awa Tupua: Place-Based Law in a Legally Pluralistic Society." *The Geographical Journal* 184 (1): 19–30.

Chavis, B. F. and L. Charles. 1987. *Toxic Wastes and Race in the United States: A National Report on the Racial and Socio-Economic Characteristics of Communities With Hazardous Waste Sites*. New York: United Christ Church.

Claeys, P., and D. Delgado Pugley. 2017. "Peasant and Indigenous Transnational Social Movements Engaging With Climate Justice." *Canadian Journal of Development Studies*/Revue Canadienne d'études Du Développement 38 (3): 325–40.

Cobs-Muñoz, V., and R. Westerholt. 2025. "Proposing Critical Realism and Second-Generation Environmental Justice for Advancing Sacrifice Zone Analysis." *Progress in Environmental Geography* 4 (1): 44–68.

Cooke, B., and U. Kothari. 2001. *Participation: The New Tyranny?* London: Zed Books.

Cronon, W. 1996. "The Trouble With Wilderness; or, Getting Back to the Wrong Nature." In *Uncommon Ground: Rethinking the Human Place in Nature*, edited by William Cronon, 69–90. London: Norton.

Curley, A., and S. Smith. 2024. "The Cene Scene: Who Gets to Theorize Global Time and How Do We Center Indigenous and Black Futurities?" *Environment and Planning E* 7 (1): 166–88.

Davies, T. 2018. "Toxic Space and Time: Slow Violence, Necropolitics, and Petrochemical Pollution." *Annals of the American Association of Geographers* 108 (6): 1537–53.

Davies, T. 2022. "Slow Violence and Toxic Geographies: "Out of Sight" to Whom?" *Environment and Planning C: Politics and Space* 40 (2): 409–27.

Davis, J., A. A. Moulton, L. Van Sant, and B. Williams. 2019. "Anthropocene, Capitalocene, … Plantationocene?: A Manifesto for Ecological Justice in an Age of Global Crises." *Geography Compass* 13 (5): e12438.

Derickson, K. D. 2022. "Disrupting Displacements: Making Knowledges for Futures Otherwise in Gullah/Geechee Nation." *Annals of the American Association of Geographers* 112 (3): 838–46.

Dryzek, J. S., and J. Pickering. 2018. *The Politics of the Anthropocene*. Oxford: Oxford University Press.

Ducre, K. A. 2018. "The Black Feminist Spatial Imagination and an Intersectional Environmental Justice." *Environmental Sociology* 4 (1): 22–35.

Fisher, S. 2015. "The Emerging Geographies of Climate Justice." *The Geographical Journal* 181 (1): 73–82.

FNPOCELS. 1991. *Principles of Environmental Justice*. Washington: First National People of Color Environmental Leadership Summit.

Forsyth, T., and C. L. McDermott. 2022. "When Climate Justice Goes Wrong: Maladaptation and Deep Co-Production in Transformative Environmental Science and Policy." *Political Geography* 98: 102691.

Fraser, N. 2009. *Scales of Justice: Reimagining Political Space in a Globalizing World.* New York: Columbia University Press.

Fricker, M. 2007. *Epistemic Injustice: Power and the Ethics of Knowing.* Oxford: Clarendon Press.

Friese, Z. 2023. "Fight or Flight: A Story of Survival and Justice in Cancer Alley." *Women Leading Change: Case Studies on Women, Gender, and Feminism* 7 (2): 3–18.

Gay-Antaki, M. 2023. "Embodied Geographies of Environmental Justice: Toward the Sovereign Right to Wholly Inhabit Oneself." *Environment and Planning E: Nature and Space* 6 (2): 25148486231151802.

Gilio-Whitaker, D. 2019. *As Long as Grass Grows: The Indigenous Fight for Environmental Justice, from Colonization to Standing Rock.* Boston: Beacon Press.

Goh, K. 2020. "Planning the Green New Deal: Climate Justice and the Politics of Sites and Scales." *Journal of the American Planning Association* 86 (2): 188–95.

Gonzalez-Ricoy, I., and F. Rey. 2019. "Enfranchising the Future: Climate Justice and the Representation of Future Generations." *Wiley Interdisciplinary Reviews: Climate Change* 10 (1): e598.

Green, L., C. Laing Ebbensgaard, K. Holden, K. Yusoff, and M. Salu. 2025. "Bodies of Evidence, Evidence of Bodies." *E-Flux Journal*, June. https://www.e-flux.com/architecture/insurgent-geologies/667526/bodies-of-evidence-evidence-of-bodies

Holmes, B., J. Bolen, and B. Kirkbride. 2021. "Born Secret (Cash for Kryptonite): A Field Guide to the Anthropocene Mode of Production." *The Anthropocene Review* 8 (2): 183–95.

Jackson, G., et al. 2023. "An Emerging Governmentality of Climate Change Loss and Damage." *Progress in Environmental Geography*, 27539687221148748.

James, W., C. Jia., and S. Kedia. 2012. "Uneven Magnitude of Disparities in Cancer Risks from Air Toxics." *International Journal of Environmental Research and Public Health* 9 (12): 4365–85.

Kanger, L., and B. K. Sovacool. 2022. "Towards a Multi-Scalar and Multi-Horizon Framework of Energy Injustice: A Whole Systems Analysis of Estonian Energy Transition." *Political Geography* 93 (March): 102544.

Kauffman, C. M., and P. L. Martin. 2018. "Constructing Rights of Nature Norms in the US, Ecuador, and New Zealand." *Global Environmental Politics* 18 (4): 43–62.

Keskitalo, A., et al. 2025. *Just Transition Or 'Green Colonialism'?* Amnesty International and the Saami Council. https://www.amnesty.org/en/documents/eur20/8913/2025/en/.

Kinkaid, E. 2019. "Rights of Nature" in Translation: Assemblage Geographies, Boundary Objects, and Translocal Social Movements." *Transactions of the Institute of British Geographers* 44 (3): 555–70.

Klinger, J. M. 2015. "A Historical Geography of Rare Earth Elements: From Discovery to the Atomic Age." *The Extractive Industries and Society* 2 (3): 572–80.

Klinger, J. M. 2017. *Rare Earth Frontiers.* London: Cornell University Press.

Latour, B. 2018. *Down to Earth: Politics in the New Climatic Regime.* Cambridge; Medford: John Wiley & Sons.

Latour, B., and P. Weibel. 2020. *Critical Zones: The Science and Politics of Landing on Earth*, edited by Bruno Latour and Peter Weibel. Cambridge: MIT Press.

Lerner, S. 2006. *Diamond: A Struggle for Environmental Justice in Louisiana's Chemical Corridor.* Cambridge: MIT Press.

Lerner, S. 2012. *Sacrifice Zones: The Front Lines of Toxic Chemical Exposure in the United States*. Cambridge: MIT Press.

Li, S. L., X. Liu, F. J. Yue, et al. 2022. "Nitrogen Dynamics in the Critical Zones of China." *Progress in Physical Geography* 46 (6): 869–88.

Liboiron, M. 2021. *Pollution is Colonialism*. Durham: Duke University Press.

Lloréns, H., and R. Santiago. 2018. "Women Lead Puerto Rico's Recovery." *NACLA Report on the Americas* 50 (4): 398–403.

Machado, M. R. A., et al. 2023. "Indigenous Women Against Bolsonaros Government in Brazil: Resisting Right-Wing Authoritarianism and Demanding Climate Justice." In *Feminist Frontiers in Climate Justice*. Cheltenham: Edward Elgar Publishing.

Martín, W. S. 2017. "Nitrogen, Science, and Environmental Change: The Politics of the Green Revolution in Chile and the Global Nitrogen Challenge." *Journal of Political Ecology* 24 (1): 1.

Mbembe, A. 2019. *Necropolitics*. Durham: Duke University Press.

Mohai, P., D. Pellow, and J. Timmons Roberts. 2009. "Environmental Justice." *Annual Review of Environment and Resources* 34 (1): 405–30.

Moore, J. W. 2015. *Capitalism in the Web of Life: Ecology and the Accumulation of Capital*. London: Verso.

Newell, P., S. Srivastava, L. O. Naess, G. A. Torres Contreras, and R. Price. 2021. "Toward Transformative Climate Justice: An Emerging Research Agenda." *WIREs Climate Change* 12 (6): e733.

Nixon, R. 2011. *Slow Violence and the Environmentalism of the Poor*. Cambridge: Harvard University Press.

Nussbaum, M. C. 2013. *Creating Capabilities: The Human Development Approach*. Cambridge: Belknap Press.

O'Lear, S. 2010. *Environmental Politics: Scale and Power*. Cambridge: Cambridge University Press.

O'Lear, S. 2016. "Climate Science and Slow Violence: A View from Political Geography and STS on Mobilizing Technoscientific Ontologies of Climate Change." *Political Geography* 52 (May): 4–13.

Okereke, C. 2006. "Global Environmental Sustainability: Intragenerational Equity and Conceptions of Justice in Multilateral Environmental Regimes." *Geoforum* 37 (5): 725–38.

Okereke, C., and P. Coventry. 2016. "Climate Justice and the International Regime: Before, During, and after Paris." *WIREs Climate Change* 7 (6): 834–51.

Ottinger, G. 2013. *Refining Expertise: How Responsible Engineers Subvert Environmental Justice Challenges*. New York: NYU Press.

Pallett, H., and J. Chilvers. 2013. "A Decade of Learning About Publics, Participation, and Climate Change: Institutionalising Reflexivity." *Environment and Planning A* 45 (5): 1162–83.

Park, J., et al. 2023. "A World Fit for the Next Seven Generations: Upholding Indigenous Rights for the Foundation of a Sustainable Future." *Canadian Journal of Children's Rights/Revue Canadienne Des Droits Des Enfants* 10 (2): 5–29.

Parsons, L. 2023. *Carbon Colonialism: How Rich Countries Export Climate Breakdown*. Manchester: Manchester University Press.

Parsons, L., R. Safra de Campos, A. Moncaster, et al. 2022. "Trading Disaster: Containers and Container Thinking in the Production of Climate Precarity." *Transactions of the Institute of British Geographers* 47 (4): 990–1008.

Parsons, M. 2025. "Indigenous Knowledge and Politics of Recognition in the Colonialcene." In Handbook of Environmental Political Theory in the Anthropocene, edited by Amanda Machin and Marcel Wissenburg, 330–39. Cheltenham: Elgar.

Pellow, D. N. 2007. *Resisting Global Toxics: Transnational Movements for Environmental Justice*. Cambridge: MIT Press.

Peša, I. 2022. "A Planetary Anthropocene? Views From Africa." *Isis* 113 (2): 386–95.

Pulido, L. 2017. "Geographies of Race and Ethnicity II: Environmental Racism, Racial Capitalism and State-Sanctioned Violence." *Progress in Human Geography* 41 (4): 524–33.

Pulido, L., and J. De Lara. 2018. "Reimagining 'Justice' in Environmental Justice: Radical Ecologies, Decolonial Thought, and the Black Radical Tradition." *Environment and Planning E: Nature and Space* 1 (1–2): 76–98.

Pulido, L., S. Sidawi, and R. O. Vos. 1996. "An Archaeology of Environmental Racism in Los Angeles." *Urban Geography* 17 (5): 419–39.

Purifoy, D. M. 2021. "North Carolina [Un]Incorporated: Place, Race, and Local Environmental Inequity." *American Behavioral Scientist* 65 (8): 1072–103.

Putzer, A., T. Lambooy, R. Jeurissen, and E. Kim. 2022. "Putting the Rights of Nature on the Map. A Quantitative Analysis of Rights of Nature Initiatives Across the World." *Journal of Maps* 18 (1): 89–96.

Ranganathan, M., and E. Bratman. 2021. "From Urban Resilience to Abolitionist Climate Justice in Washington, DC." *Antipode* 53 (1): 115–37.

Rodriguez, I. 2017. "Linking Well-Being With Cultural Revitalization for Greater Cognitive Justice in Conservation: Lessons from Venezuela in Canaima National Park." *Ecology and Society* 22 (4): art24.

Rosol, C., S. Nelson, and J. Renn. 2017. "Introduction: In the Machine Room of the Anthropocene." *Anthropocene Review* 4 (1): 2–8.

Rosol, C., T. Turnbull, and J. Renn. 2021. "Introduction: The Mississippi River Basin—a Model for Studying the Anthropocene in Situ." *The Anthropocene Review* 8 (2): 99–114.

Saha, R., and P. Mohai. 2005. "Historical Context and Hazardous Waste Facility Siting: Understanding Temporal Patterns in Michigan." *Social Problems* 52 (4): 618–48.

Schlosberg, D. 2018. "Environmental (in)Justice." In *Companion to Environmental Studies*, edited by Noel Castree, Mike Hulme, and James D. Proctor, 629–34. Abingdon: Routledge.

Schlosberg, D., and L. B. Collins. 2014. "From Environmental to Climate Justice: Climate Change and the Discourse of Environmental Justice." *Wiley Interdisciplinary Reviews: Climate Change* 5 (3): 359–74.

Schmidt, J. J. 2022. "Of Kin and System: Rights of Nature and the UN Search for Earth Jurisprudence." *Transactions of the Institute of British Geographers* 47 (3): 820–34.

Slater, T., and N. Anderson. 2012. "The Reputational Ghetto: Territorial Stigmatisation in St Paul's, Bristol." *Transactions of the Institute of British Geographers* 37 (4): 530–46.

Smith, R. A. J., and K. Rhiney. 2016. "Climate (in)Justice, Vulnerability and Livelihoods in the Caribbean: The Case of the Indigenous Caribs in Northeastern St. Vincent." *Geoforum* 73 (July): 22–31.

de Souza, M. L. 2021. "'Sacrifice Zone': The Environment–Territory–Place of Disposable Lives." *Community Development Journal* 56 (2): 220–43.

Sultana, F. 2022. "Critical Climate Justice." *The Geographical Journal* 188 (1): 118–24.

Taylor, D. E. 2016. *The Rise of the American Conservation Movement: Power, Privilege, and Environmental Protection*. Durham: Duke University Press.

Taylor, Z. J., and M. B. Aalbers. 2022. "Climate Gentrification: Risk, Rent, and Restructuring in Greater Miami." *Annals of the American Association of Geographers* 112 (6): 1685–701.

Terrell, K. A., and G. St. Julien. 2022. "Air Pollution Is Linked to Higher Cancer Rates Among Black or Impoverished Communities in Louisiana." *Environmental Research Letters* 17 (1): 014033.

Thomas, J. A. 2014. "History and Biology in the Anthropocene." *The American Historical Review* 119 (5): 1587–607.

Trott, C. D., E. S. Gray, S. Lam, R. H. Courtney, J. Roncker, and T. L. Even. 2023. "People's Action for Climate Justice: A Systematic Review." *Local Environment* 28 (9): 1131–52.

Tschakert, P., D. Schlosberg, D. Celermajer, et al. 2021. "Multispecies Justice: Climate-Just Futures With, for and Beyond Humans." *WIREs Climate Change* 12 (2): e699.

Tsing, A., J. Deger, A. Saxena Keleman, and F. Zhou. 2020. *Feral Atlas: The More-Than-Human Anthropocene*. www.feralatlas.org

UN Women. 2022. "Explainer: How Gender Inequality and Climate Change Are Interconnected." February 28.

USGS. 2021. *Rare Earths. Mineral Commodity Summaries*. Reston: U.S. Geological Survey.

Vaddhanaphuti, C. 2026. "Ecological Wisdom in Southeast Asia: Rethinking Resilience, Justice and Sovereignty." *Heinrich Böll Stiftung Regional Dialogue E-Paper Series*. https://th.boell.org/sites/default/files/2026-01/ecological-wisdom-in-southeast-asia-final.pdf

Vasudevan, P., M. M. Ramírez, Y. G. Mendoza, and M. Daigle. 2022. "Storytelling Earth and Body." *Annals of the American Association of Geographers* 113 (7): 1728–44.

Verdin, M. M. 2020. "Cancer Alley: Istrouma to the Gulf of Mexico." *Southern Cultures* 26 (2): 80–95.

Vincent, U. 2012. "Ombudspersons for Future Generations: Bringing Intergenerational Justice into the Heart of Policymaking." *UN Chronicle* No. 1 & 2 Vol. XLIX.

Visvanathan, S. 1997. *A Carnival for Science: Essays on Science, Technology and Development*. New Delhi: Oxford University Press.

Walker, G., and H. Bulkeley. 2006. "Geographies of Environmental Justice." *Geoforum* 37 (5): 655–59.

Watt-Cloutier, S. 2018. *The Right to Be Cold*. Minneapolis: University of Minnesota Press.

Whitehead, Mark. 2014. *Environmental Transformations: A Geography of the Anthropocene*. Routledge.

Whyte, K. P. 2011. "The Recognition Dimensions of Environmental Justice in Indian Country." *Environmental Justice* 4 (4): 199–205.

Whyte, K. P. 2017. "The Dakota Access Pipeline, Environmental Injustice, and U.S. Colonialism." *Red Ink* 19 (1): 154–69.

Williams, G., and E. Mawdsley. 2006. "Postcolonial Environmental Justice: Government and Governance in India." *Geoforum* 37 (5): 660–70.

Woods, C. 2017. *Development Arrested: The Blues and Plantation Power in the Mississippi Delta*. London: Verso.

Yusoff, K. 2018. *A Billion Black Anthropocenes or None*. Minneapolis: University of Minnesota Press.

Yusoff, K. 2024. *Geologic Life: Inhuman Intimacies and the Geophysics of Race*. Durham: Duke University Press.

7 Wild Lives

How Should We Relate to the Non-Human?

Introduction

In Chapter 6, we explored some implications of the new interfusions being caused by anthropogenic Earth System change – namely, the 'collapse' of the former differences between society and nature, between near and far, and between present and future. We did so through the lenses of environmental, climate, and planetary in/justice, making mention of multi- and interspecies justice along the way. Justice is a powerful concept that allows us to see how today's deep and far-reaching changes to Earth's geography raise pressing political and moral questions about who or what is harmed, who or what needs redress or protection, and who's responsible for creating and tackling perceived problems. The questions require local-scale answers, of course, but also transcend national borders and will affect future people and the environments bequeathed to them by us. So, they require large-scale and long-term thinking, as well as thinking (and acting) locally.

In this chapter, we want to focus on the society-nature relationship again, but more specifically on efforts to preserve, restore, conserve, and generally make space for 'wildness'. Wildness is a characteristic of the non-human world, namely its 'self-willed' element. 'Wilderness' areas are ones where wildness is dominant, and the human presence is minimal or absent. The fate of the wild is a question preoccupying very many environmentalists, 'green' businesses, governments, and concerned citizens around the world. It's a key aspect of what the world's present and future geography can and should look like. It's part of the wider question of precisely where geographically, and to what degree, we can seriously rebuild the 'biodiversity' lost to large-scale land clearing, urbanisation, commercial farming, water pollution, and so on. And the question is central to the even bigger issue of the status of 'nature' in our world, something we discussed in Chapter 1 when considering the 'post-nature' narrative. In short, the meaning and future of wildness is a global issue, even as addressing it won't be a one-size-fits-all process. Do we now live in a 'post-wild' world, commensurate with a 'post-natural' Anthropocene era? Is it more that we need to rethink what *counts* as 'the wild', and act

DOI: 10.4324/9781003162469-10

accordingly at different scales? Regardless of how we view and interact with it, 'the wild' is an important political issue underpinned by moral-ethical beliefs and arguments.

As we'll see, different answers to the two questions have been proposed. The present and future of 'the wild' is an issue offering us vivid insight into how different people view what's at stake in the Anthropocene. Once again, we see here how the discipline and wider discourse of G/geography overlap, without being reducible to each other. There are many voices, and the challenge is for them to hear each other in dialogue across a table, not talk past each other in separate corners of a room – even as, in the end, some will tend to win out over others in our far-from-perfect, far-from-egalitarian world where arguments occur within and across multiple fora.

The Anthropocene and the 'de-wilding' of the Earth: three perspectives

What do you think of when you hear the word 'wildlife? Probably some mixture of keystone species (such as polar bears, penguins, pandas, elephants, gorillas, and wolves), iconic ecosystems (such as the Amazon rainforest), and any of those other living entities that aren't domesticated or controlled by people. What do you think of when you hear the word 'wilderness'? Probably an area remote from human settlements in which processes and events are almost entirely natural, such as the highest Himalayan mountains, Antarctica, or the Canadian Arctic[1] (the oceanic equivalent of these is the high seas). This reflects our standard understanding of the world 'wild' in English. Dictionary definitions define the wild as that which is 'uncontrolled', 'un-governed', 'non-domesticated', 'uncultivated, and 'self-willed'. In this light, much of the wildness that emerged through evolutionary processes during the Holocene – against the background of continuous natural variations in life on Earth – has been modified or lost due to human action. One example is the so-called 'Columbian exchange' from the late sixteenth century onwards, caused by Spanish and Portuguese colonisation of central and South America. It reconfigured ecology on both sides of the Atlantic (see Chapter 3).

The human impact on non-human life has been quantified by animal, plant, and insect biologists who frequently talk of the Earth's 'sixth great extinction' event, the novelty being this one is caused by people (Ceballos, Ehrlich, and Dirzo 2017). Stanford University's Rodolfo Dirzo and colleagues have talked of Anthropocene 'defaunation' worldwide (Dirzo et al. 2014). Argentinian ecologist Sandra Díaz and colleagues (2019) adduce evidence of 'pervasive human-driven decline of life on Earth' – a major biospheric loss that helps to distinguish the Anthropocene from the Holocene epoch. Around half the terrestrial surface of the Earth has undergone 'medium' to 'high' human impacts (Riggio et al. 2020). During the period of the Great Acceleration, and without doubt today, 'the wild' inherited from the mid-Holocene period has beaten a forcible retreat. Evidence of this is widely reported in the news – species losses and endangerment; shrinking and fragmented habitats; polluted water and soils that compromise flora and fauna; disturbed ecosystems that were once deforested or farmed; stresses imposed by a warming atmosphere; wild birds, animals and marine life showing-up in unexpected places; invasive (more politely 'introduced') species displacing native species; mass death events of fish, birds and whales: these and related things are familiar to anyone who takes an interest in current environmental affairs. The human footprint is deeper and wider than ever – so much so that American journalist David Biello titled his 2016 best-seller *An Unnatural World*, while more recently the environmental writer Adam Welz (2023) published *The End of Eden*, subtitled *Wild Nature in the Age of Climate Breakdown*.

These contributions echo themes animating earlier publications by green activists, eco-centric scholars, and concerned scientists, such as Bill McKibben's (1989) famous book *The End of Nature*, political theorist John Meyer's (2006) *The End of the Wild*, and John Terborgh's (2019) *Requiem For Nature*. The difference today is that never has so much wildlife been so directly impacted by human activity. Some, like American environmental philosopher Eileen Crist (2019), regard this as nothing less than 'ecocide'. Looking ahead, even if the world population levels off numerically, per capita requirements for food, energy, and other commodities are very likely to grow. It's a sign of the times that great efforts are being made to create 'wildlife arks' before too many more species are lost forever; it's also a sign of the times that extinct species are being bioengineered back into existence in laboratories.

What, then, should our stance be in light of the stark diminution of the Holocene epoch wildness (i.e. wild species and wild areas)? What ethical and moral position should be taken, and what sort of actions follow, depending on the position we take? How can political debate and decision articulate moral-ethical beliefs? To quote legal scholar Jedediah Purdy (2015) from his book *After Nature*, '… in a world we can't help shaping, the question is what will we shape' (p. 3) – to which we can add, 'shape where, to what extent and how'? Broadly speaking, there are three positions on how to respond to an increasingly 'post-wild' world that have been put forward in recent years (see Table 7.1).

First, there are those who decry the outsized human impact on the wild. For various reasons, they value the wildness bequeathed to us by millennia of evolution by natural selection. While they recognise that it will be impossible to turn back the hands of time, they argue for concerted efforts to preserve and conserve what remains of wild nature. Some are environmentalists, such as McKibben and philosopher Ned Hettinger (2021). Others also favour a massive reduction in the human footprint so as to give space back to wildlife. Some are Left-leaning activists and scholars who favour a post-capitalist 'de-growth' economy, such as Andreas Malm (e.g. 2018) and critics Rob Fletcher and Bram Büscher in their book *The Conservation*

Table 7.1 Wildlife and wildness in the twenty-first century: three paradigms

Approach	*Rationale*	*Advocates*	*Moral and political ethos*
Protect, preserve, and restore	Wildness is fast-disappearing and urgently needs defending	'Deep green' critics of capitalism; some socialist, communitarian, and feminist critics of capitalist instrumentalism, inequality, and exploitation; some anticolonial critics	*Radical*: strong commitment to equality for non-humans and for humans. Either strong ecocentrism or a weak anthropocentrism. A post-growth economy. Ideas of 'ecological justice', within which sits 'inter-species justice', i.e. just treatment of non-humans for their own sake
Actively manage and intervene strongly where necessary, using advanced technology	To retain and increase wildness, human intervention is needed rather than a 'hands-off' approach	'Ecomodernists', such as Erle Ellis and Michael Shellenberger	*Radical*: strong commitment to deepening capitalist modernisation of human life and the material environment. Strong and 'benevolent' anthropocentrism. 'Decoupling' of economic growth from environmental destruction. Reduction of environmental injustice for vulnerable human communities.
Balance creation of 'anthropogenic' environments with autonomy for multiple wild ones	Ensuring a good quality of life for people requires further development but it should be spatially concentrated development, giving plenty of space back to wild things	Social democrats with a 'medium green' sensibility, such as Emma Marris; some anticolonial, socialist, feminist, and other critics seeking better forms of human 'development'	Moderate: favours strong reform of current society-environment arrangements to meet the goals of better human welfare and stronger environmental protections. Strong anthropocentrism but tempered by moderate ecocentrism. Managed economy, strong national government role in protecting people and planet. Reducing environmental injustice, and fostering ecological and inter-species justice by reducing the human 'footprint' in many areas.

Revolution (2020). For them, a rewilded world is one where people live more local lives that are far less resource-intensive, more in touch with nature, and less economically unequal than present-day ones in the West and in so-called 'developing' countries. Still others are scientists, such as the late entomologist Edward O. Wilson. In his best seller *Half Earth: Our Planet's Fight for Life*, he famously proposed a massive expansion in protected land areas that would be largely off-limits to human activity. Then, finally, there are those Indigenous peoples who lament the destructiveness of Western-style development and push for a kinder, more respectful approach to the non-human realm. Some of their ideas have made it into a global movement of activists and scholars to enshrine 'rights of nature' in different national legal systems (as we mentioned in passing in Chapter 6). Many of these ideas are core to the 'decolonial' movement, whose arguments are prominent in large parts of Anglo-Western social science, the humanities, and parts of the arts.

Second, as we noted in chapter one, there are so-called 'ecomodernists' who argue that humans are now so powerful they need to take full responsibility for a world they shape so very much, if not exactly control. The American biogeographer Erle Ellis is one of these. In a set of scientific and more popular publications, he's for some time argued that a great deal of the planet was partially or significantly domesticated and modified *hundreds of years ago* (Ellis et al. 2021). This is linked to the proposition that the Anthropocene, in fact, began a long time ago. For Ellis, recent decades are but an *intensification* of a now old process of us creating different 'anthromes' (humanly altered ecosystems). The same argument is implied by physical geographers Simon Lewis and Mark Maslin's (2016) research into the Columbian exchange from the 1600s onwards. Given the enormous extent of human development, and the sheer number of people who need to be supported on Earth, Ellis favours intelligent intervention to exploit nature efficiently, while also creating space for other (admittedly 'unnatural' or 'semi-natural') life forms to thrive. 'The question now', he once put it, 'is not how nature can continue to sustain humanity, but how humanity can continue to sustain nature' (2015, 29). In Ellis's view, it's largely too late to try to restore, preserve, or conserve wild creatures and locations on our 'used planet'. Moving forward, wildness will need, paradoxically, plenty of human intervention to persist – and will thus be wild in ways that aren't particularly natural. For instance, we might bring back extinct species and relocate them outside their historical habitats and ranges, be they wolves or beavers. Advanced technologies will be part of the way we rewild the planet, while also ensuring high levels of human welfare. Ellis advocates for much greener development using all the tools of cutting-edge science.

Finally, there's a third position that blends elements of the first and second, being at once pragmatic and yet very committed to fostering the 'autonomy' and 'otherness' of a myriad of wild entities. This is sometimes called 'the new nature conservation' for two reasons. First, it recognises the important role that humans can play in protecting valuable ecosystems and species – like

Ellis, it sees no necessary contradiction between human intervention and non-human wildness flourishing. Secondly, though, it does not fixate on the global patchwork of 'protected areas', such as Yellowstone or Kruger national parks (in the United States and South Africa, respectively). Instead, it sees the best chance for wildlife lying in a much broader conception of when and where wild entities can exist – even welcoming their proximity to dense human settlements in some cases, as we'll see later in this chapter. Among the most famous statements of this third position is the book *Rambunctious Garden* (subtitled *Saving Nature in a Post-Wild World*) by science writer Emma Marris (2011). More recently, Marris published *Wild Souls* (2021), which recapitulates earlier arguments. In academic geography, her position finds some echoes in research by 'new materialist', 'post-humanist' and 'more-than-human' geographers (such as Jamie Lorimer) – some examples will be highlighted in section four below. This third stance, broadly, is both 'anthropocentric' – recognising the necessary role of human values and needs – but also 'ecocentric' in that there's concern for wildness in-and-for itself, to the extent people can ascertain its nature. It takes a more cautious approach to using advanced technology to rewild the Earth, tempering the Promethean confidence of stance number two. See Box 7.1 for more about 'anthropocentric' and 'ecocentric' attitudes to the wild.

Box 7.1 'Anthropocentric' and 'ecocentric' ways of valuing the wild

In Anglo-Western cultures, a distinction is often made between 'society' and 'nature', 'domesticated' and 'wild, and 'human' and 'animal'. We attach values to other people and various elements of the material world we inhabit, seeing that world as Other to us, even as we're connected to it. Anthropocentric values are those that place human rights, needs, and wants above the rights, needs, and wants of non-humans in many or most situations. Strong anthropocentrism is hierarchical, seeing the material world as but a means to the end of satisfying the requirements of human activities. Morally, not too much is off-limits. Weak anthropocentrism is more egalitarian. It accepts human entitlement to appropriate the non-human world for various uses, but is prudent and respectful, placing certain limits on use for our own good. By contrast, 'ecocentrism' assigns considerable value to the non-human realm. Strong ecocentrism finds reason to place the rights, needs, and wants of non-human entities above human requirements in many cases. Weak ecocentrism is more egalitarian, aiming to ensure the voice of nature is heard properly when humans propose to do things like clear-cut a forest or build housing near the habitat of rare butterfly species. The middle ground between anthropo- and ecocentrism is often labelled 'light green', whereas strong ecocentrism amounts to a worldview called 'ecologism'. Strong anthropocentrism is characteristic of modern capitalist societies, notwithstanding pockets of opposition and efforts to moderate large-scale environmental harm. Note that the environmental justice advocates discussed in Chapter 6, can be strong or weak anthropocentrists, or weak ecocentrists, depending. Meanwhile, advocates of interspecies justice tend to sit somewhere on the ecocentric spectrum.

From the perspective of strong ecocentrism, strong anthropocentrism spells the 'death', 'domination', and 'control' of nature. In her best-selling book *This Changes Everything*, Canadian critic Naomi Klein (2014) called it 'extractivist'. From the reverse perspective, strong ecocentrism comes across as misanthropic: it risks overriding the imperative of ensuring that all humans have their capabilities realised through good jobs, enough food, proper shelter, rapid transportation, and so on. For instance, some ecocentric critics call for controls on human reproduction worldwide.

In Anglo-Western cultures, ecocentric views extend back to the seventeenth century. However, they received their fullest and most morally radical expression from the early 1970s onwards. In that period, scholars such as Peter Singer, Christopher Stone, and Arne Naess codified the ecocentric position for both 'wild nature' and also things like farm animals and domestic pets. They made the case for the fulsome protection of the non-human. In the academic world, the position achieved its strongest support in the disciplines of philosophy, history, and politics. Meanwhile, geographers rather ignored the ecocentric worldview (we don't have space to explain why here, though do note the signal exceptions of Tim O'Riordan and David Pepper, who published the fine books *Environmentalism* [1976] and *The Roots of Modern Environmentalism* [1984]). Only in recent years have geographers paid closer attention to the affordances, capabilities, and rights of non-humans, as we explain in Box 7.3 below. Even so, few are strong ecocentrists, even as many are critics of strong anthropocentrism.

Clearly, all three positions are moral-ethical through and through, feeding into political debate and decision. The second and third are more tolerant of sustained and significant human involvement in the fate of wildness than the first. They each define and value wildness in rather different ways, and propose rather different actions so as to preserve, conserve, restore, and even enlarge the domain of 'the wild'. Arguments for social and/or interspecies justice can be aligned with all three positions, according to the context. 'Environmental justice' is about ensuring healthy, liveable environments for all people, especially the poor and vulnerable. 'Ecological justice' aims to respect the needs and rights of the non-human realm and is the ultimate goal of an 'ecocentric' worldview. The differences among the positions rest, in part, on how the facts of species losses, extinctions, habitat shrinkage, and so on are perceived in light of particular ideas about 'nature' and its present-day status. That is to say, the facts of society-nature interfusion do not, in themselves, dictate how we should think about them. Before we turn to these ideas, though, we need to set the three positions in the context of earlier approaches to the wild prevalent during the post-1945 period.

The nature-society dualism, protected areas, and 'fortress conservation': time for a paradigm shift?

Efforts to protect wild animals, forests, grasslands, lakes, wetlands, estuaries, waterfalls, gorges, canyons, beaches, and so on go back well over a century. Long before there was any serious talk about the Anthropocene, many individuals and groups had realised that the non-human world needed protecting from unchecked human use. For instance, the poorly regulated killing of bison, pigeons, and fur seals in North America 150 years ago led to initial attempts to protect nature. Some protection was about nature conservation on land and in water (i.e. sustainable use over time), some of it about preserving special land- and seascapes, including the various 'charismatic' species within them. Efforts at 'restoration' and recovery when a species or ecosystem has been badly over-exploited can be geared to either conservation or preservation, depending. For decades, efforts to eradicate 'invasive species' introduced by humans have illustrated both these goals of habitat restoration.

The area of the Earth's surface devoted to nature protection grew during the twentieth century. Today, around 10% of the globe is designated as a national park, a nature reserve, a national sanctuary, a wildlife reserve, a biological reserve, a wildlife park, and so on (see Figure 7.1). While initially focussed on land – for instance, Yosemite National Park was created back in 1890 in the United States – a growing area of ocean has received some level of (admittedly often nominal) government protection by being designated as a marine reserve. Some areas benefit from UNESCO World Heritage status, such as the Galapagos Islands and the Great Barrier Reef. There have long been organisations that advocate for wildlife and the non-biotic aspects of nature – for instance, the Royal Society for the Protection of Birds (RSPB), the International Union for the Conservation of Nature (IUCN), the National Audubon Society, the World Wide Fund for Nature (WWF), the Sea Shepherds, and the Nature Conservancy. There have also long been laws and (inter-)government regulations that aim to conserve or preserve certain species and natural areas – for instance, the Endangered Species Act in America

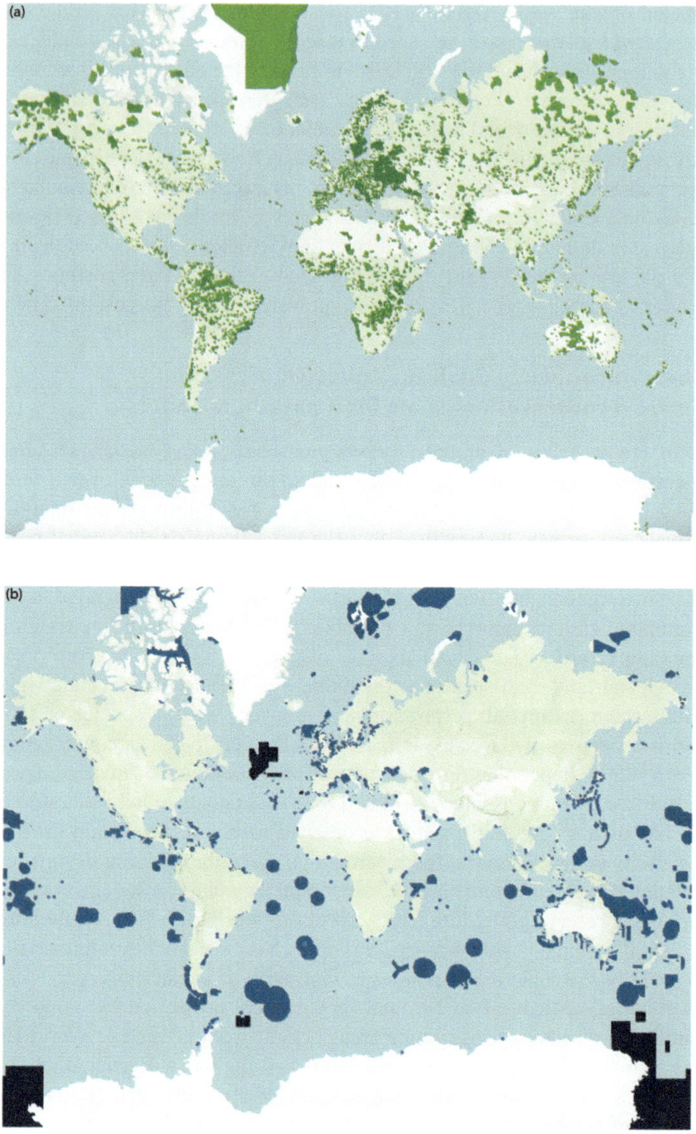

Figure 7.1 Global terrestrial (a) and marine (b) protected areas as of March 2026. (Source: UNEP-WCMC and IUCN (2026), Protected Planet: the world database on protected and conserved areas (WDPCA) [Online], March 2026, Cambridge, UK: UNEP-WCMC and IUCN. Available at: www.protectedplanet.net.)

and the international Convention on Trade in Endangered Species (CITES). Correspondingly, there are numerous government departments and agencies that seek to enforce laws and rules. In short, for decades, significant energy and resources have gone into circumscribing human access to, and use of, certain creatures and areas that are predominantly wild and natural.

To generalise, this led to the creation of a distinctive geography of wildness on the face of the Earth – at least up until the turn of the millennium. On the one hand, there was a patchwork of 'natural' and 'semi-natural' areas – some very large, some quite small, some well protected, others less so. Many were public, communal, or government-owned; others privately owned. These retained some or much of their historic character, ecologically speaking. On the other hand, there was a vast area where wildlife and other natural phenomena were, variously, able to be extinguished, exploited, reconfigured, or looked after, depending on people's values and priorities. This vast area comprised a patchwork of towns, cities, farmland, managed forests, roads, common land, railways lines, artificial lakes and reservoirs, canal networks, dammed rivers and streams, mining facilities, industrial districts, port complexes, military training and test sites, and so on. It also covered the high seas worldwide.

In a geographical sense, this expressed an imagined dualism between the realm of 'nature' and the realm of 'society'. Nature was variously excluded, simplified, and managed in the realm of society, but given much freer reign in areas designated as parks, reserves, refuges, or common (open access) lands and waters. Its 'purest' and fullest expression was in designated or otherwise remote wilderness areas such as much of Alaska and the Franklin-Gordon Wild Rivers National Park in Tasmania, Australia. Conversely, as the world population has ballooned, in the context of an expanded capitalist economic system (exemplified by the rise of China), the extent of modified – that is to say, damaged, decomplexified or otherwise altered – environments has grown enormously. This is captured graphically in those Great Acceleration graphs shown earlier in this book. In Western countries, the front or back garden of the typical home is, perhaps, representative of the massively modified – far from 'wild' – environments characteristic of modern life. In these heavily altered environments, much wildlife is seen as out of place (for instance, urban foxes are often seen as pests in the UK, as are urban raccoons in Canada). See Box 7.2.

Box 7.2 Situating the Anthropocene: the manicured garden in modern western societies

How wild is a garden?

In many industrialised and urbanised cultures around the world, a domestic garden has been a space to connect with elements of the non-human world – sunshine, a gentle breeze, pretty flowers, perhaps some visiting birds. But gardens have conventionally been highly regulated and carefully maintained spaces, meticulously designed according to the wishes and value systems of their owners. They are spaces whose boundaries are carefully policed. Take a walk around any garden centre in a country like the UK and see bird-feeders designed to cater to certain species and not others, especially 'pests' like squirrels, rats or pigeons; chemicals used to kills plants that are considered 'weeds' and to make space for more desirable plants; fences, mesh and even spikes to keep your outside space free of unwanted visitors like pigeons, foxes or cats, or to keep ornamental fish safe from attack from the skies.

A classic work of political ecology, geographer Paul Robbins' 2007 book *Lawn People* explores the culture of the manicured garden in modern North American society. He shows how the perfectly clipped, bright green lawn became an aspirational part of suburban life, and documents the incredible amount of inputs – human and machine labour, water, and chemical feed, weed killers and pesticides – that lawn ecosystems require if they are to be kept in that young, emergent state of lush greenness throughout the year. The manicured lawn might therefore be seen as iconic of a conceptualisation of nature as something to be dominated, controlled, and subdued for human pleasure – wildness thoroughly tamed.

However, Robbins shows how the lawn is more complex than that. He reveals that many people who pour their time, money, and chemicals into maintaining a perfect lawn are often painfully aware of the environmental costs of things like pesticide use – often more so than 'non-lawn people'. They frequently profess environmental values, even as they engage in a gardening practice many would see as anti-environmental. Robbins' analysis provides further evidence for our argument above that the facts of the matter don't determine how people think or act. In Robbins' case, many other things are at play too – the pull of consumerism, the marketing prowess of lawn-care companies, cultural pressures, aesthetic principles, and so on.

More recently, we can see how such pressures and principles might be changing. If you move from the pesticide aisle to the magazine rack in the aforementioned garden centre and pick up a gardening magazine, you'll likely see the new prominence of ideas about 'wildlife-friendly gardening'. This shift has encouraged many gardeners to embrace diversity over uniformity, pollinator-friendly flowering plants over those which might have more aesthetic value, and a measure of wild untidiness over crisp lines and tidy flower beds (Wignall, Alton, and Ratnieks 2019). Australian geographer Leslie Head has documented this emerging trend for 're-wilding' the backyard, and shows how different gardening practices reveal very different ideas about the boundaries of 'nature', and about the human role in preserving, conserving, restoring, or even creating it (Head 2016).

For some Anthropocene thinkers then, if gardens can be spaces for the 'guided flourishing' of diverse beings (White 2019, 85), and not just grassy monocultures (Aloi 2025), then they perhaps offer a model for how the Anthropocene's paradoxes of human domination and nonhuman rights and agency can be resolved at a planetary scale, in creating a 'Garden Earth' (Cannon and Kua 2017) or a planetary 'Rambunctious Garden' (Marris 2011) in which human-wildlife proximity is celebrated and embraced and environments are designed to enable multi-species flourishing (Nastold 2025). Such ideas draw inspiration not only from the shifting norms of Western suburban gardens, but also from the gardening practices of marginalised and Indigenous communities, many of which embody fundamentally different attitudes to the non-human world, centred on ideas of reciprocity, respect, diversity and care (see e.g. Carney 2021; Davis et al. 2019; Tillett 2020). Driven by a whole range of cultural influences, Anthropocene gardens certainly seem to be getting wilder.

Figure 7.2 A carefully manicured lawn. (Source: Shutterstock.)

This relentless 'defaunation' of the Earth may seem to put a premium on doubling down on efforts to ring-fence wild areas and protect endangered wildlife. It seemingly points to a need to strengthen the guardrails around existing areas, such as the Sundarbans National Park on the India-Bangladesh border (a 'biodiversity hotspot'). Indeed, it apparently indicates the need to *expand* the area of land and water that's robustly protected in law from things like settlement, agriculture, fishing, and deforestation (as with ecologist E. O. Wilson's idea of 'half Earth'). Part of this could involve expanding privately owned areas reserved for wildlife, which has become a growing feature of nature conservation and preservation since around 1990. Increasingly, it could also involve stronger punishments for trespassers and poachers who raid wild areas, along with a more militarised approach to defending wildlife in some situations.

The rationale for all this seems watertight at first glance. First, wildlife needs protecting for its own sake, before it's lost forever; this is the so-called 'ecocentric' view that nature has intrinsic or 'existence value' (refer back to Box 7.1), and 'ecological justice' is geared to respecting this value to varying extents. Second, wildlife is fascinating and inspiring, such that its protection brings many people emotional, aesthetic, and even spiritual benefits when controlled access is ensured (witness the enduring popularity of safaris). Third, many remarkable products have originated from basic research into the chemical and mechanical properties of wild entities, so we need to preserve a storehouse of wild things in anticipation of future useful discoveries. Fourth, weakly regulated incursions into what remains of wild nature pose real risks to human health, as we've seen with various new viruses and pathogens entering human populations over the last 30 years. Finally, many people earn money through 'ecotourism' and so the protection of wild areas and wildlife helps to ensure their future prosperity in countries such as Costa Rica, Malaysia, and New Zealand. In short, for several reasons, nature needs space to survive, to bounce back, and even to thrive. This is best achieved, the argument goes, by minimising and tightly regulating human interaction with wildlife and wild areas.

However, things aren't so simple. In the first place, many ostensibly 'natural' areas where wildlife flourishes were, in fact, once *occupied landscapes* where traditional societies lived in settlements, or else nomadically. For instance, this is true in America, Australia, and Canada, all settler-colonies where Europeans gradually displaced Indigenous peoples, usually denying them access to land, water, and natural resources. This displacement created the false impression of 'natural areas' that colonists could visit and enjoy – to hike, to camp, to hunt, or to see things like grizzly bears and wolves first-hand. In his study of Tanzania, the British geographer Dan Brockington (2002) coined the term 'fortress conservation' to describe this approach. It largely restricted access to local people and enabled access for colonisers and wealthy overseas tourists, all the while fencing in wildlife that had, historically, co-existed with humans. This 'coercive conservation', as it's sometimes also been called, has created a sense of injustice, and sparked calls for greater access to – even full control of – things like wildlife reserves and parks by Indigenous peoples. In settler-colonies such as Canada, it's linked to wider efforts among these peoples to secede from the nations they've been forced to become part of, or at least enjoy some real political autonomy. The critique of fortress conservation is that it's helped to create a 'two domain' geography of nature that's socially unjust. While the appearance of a relatively 'pure nature' full of wildlife may appeal to ecotourists, it rests upon forced exclusions and evictions going back decades and centuries. It has imposed an Anglo-Western dualism of 'society vs nature' and literally written it into the landscape. In many cases, the protection of wild nature has involved violence (Duffy 2010). Efforts to expand the number of terrestrial wildlife parks and reserves have raised the spectre of so-called 'green grabbing', where low-income people in rural areas are denied access to adjacent forests, grasslands, and so on (Fairhead, Leach, and Scoones 2012).

Secondly, there's plenty of evidence, both historically and today, that there's no necessary antithesis between human action and the flourishing of wildlife. Erle Ellis and Emma Marris have a point, and they're not alone in making it. While species extinctions and biodiversity loss at human hands get a lot of news media coverage, during much of the Holocene, there have been numerous cases of successful co-existence between people, wild animals, natural grasslands, and so on. For instance, in their influential research into desertification in West Africa, anthropologists Melissa Leach and James Fairhead (1996) found evidence that pre-colonial shifting cultivators had helped to afforest a largely dry grassland landscape. This served to *increase* biodiversity on the savannah-forest fringe. More recently, in his book *Inheritors of the Earth*, British ecologist Chris Thomas (2017) argues that human action can help to significantly *enhance* biodiversity, harnessing the living world's propensity to sift-and-sort itself into new species, niches, and communities. For Thomas, adaptation and selection are constant; life goes on.

Third, related to this is an argument made by British environmental and social justice campaigner George Monbiot. In his best-selling book *Feral* (2013), he noted that most modern people, living as they do in cities, are cut off from direct engagements with wildlife. This diminishes our lives kinaesthetically and emotionally. If we allowed wild things more space in our towns, our urban fabric and our suburban landscapes, he argued, we'd develop a new sense of self and a heightened appreciation of the wonders (and rigours) of nature. We need, he argued, to 'rewild' *ourselves* by rewilding the world we inhabit. We might thereby challenge the deep anthropocentrism of Anglo-Western culture today, and develop a new, wider appreciation of the value of wild things. Ecocentrism can flourish if we re-immerse ourselves in the non-human realm.

Fourthly, the advent of anthropogenic climate change – with atmospheric carbon dioxide concentrations now a troubling 50% higher than 200 years ago – sees the spatial patterning of wildlife changing as weather patterns alter. That is, both *within* areas reserved for nature restoration, preservation, and conservation and *all other areas too*, wildlife is now incrementally 'on the move' as temperature, wind patterns, ocean temperatures, and moisture levels change. Insects, birds, fish, mammals, trees, shrubs, plants, mosses, reptiles, amphibians … the habitats and ranges of a great deal of wild nature are shifting and spill-out beyond the system of reserves, parks, and refuges currently in place to protect them. Marris noted this in her influential book *Rambunctious Garden*, as did leading environmentalist journalist Fred Pearce in his widely read text *The New Wild* (2015). The older geography of wildness is no longer stable as the atmosphere and oceans warm up. This arguably requires active management not passive acceptance.

Fifth, even prior to the advent of anthropogenic climate change, the large-scale introduction by humans of species far outside their natural ranges has altered ecosystems irreversibly, creating practical and moral problems that can't be side-stepped. What is deemed 'alien' is now very much *a part of* 'native' ecosystems in many parts of the world. Aside from the practical impossibility of restoration, there are moral questions about the rights of invasive species – such as wild horses ('brumbies') in eastern Australia whose ancestors were farm animals brought by the British from the 1780s onwards. They arguably now have the right to exist in rural New South Wales and other states, so long as the overall numbers are managed. Again, this is an argument made by Marris and Pearce, among many others. As Pearce put it, even putting aside climate change, 'We've altered our planet too much, and nature never goes backwards' (2015, 2). For this reason, there are now many international 'biodiversity' offset policies. They enable 'compensation' and 'substitution' to be achieved for biodiversity loss in one or other place – far-distant wetlands, grasslands, meadows, lagoons, woods, and so on can be created or biologically enriched so as to make up for biodiversity losses in other parts of the world. Here, wildlife becomes 'commensurated' and is treated as something that can be built back (so-called 'net gain') among otherwise unique locations on Earth.

This links to a final point. Many years ago, the historical geographer William Cronon identified a key problem with the Western habit of placing 'nature' and 'the wild' on something of a pedestal. In a now classic chapter called 'The trouble with wilderness', in what became an influential book titled *Uncommon Ground* (Cronon 1996), he challenged lovers of the natural environment to pay greater attention to the supposedly 'unnatural' landscapes they inhabited – the manicured gardens, the orderly public parks, the farm fields, the grazed hillsides, the motorway verges, the brownfield sites, the disused railway lines, the canal-side walkways, and so on. Why, he asked, are these far less deserving of affection and attention than places like the Grand Canyon? Are they really bereft of wildness, at best being but ersatz versions of 'real nature'? Isn't it perverse to pay so much attention to 'extraordinary' places, such as Nigaloo (in Western Australia), while being indifferent or even negative towards the familiar living landscapes on our doorsteps? Why do we need to travel away from home to 'experience nature'? Why do we seek to exclude and tame the wildness that *could* flourish nearby, if only we'd allow it to? Cronon's intervention sparked a sometimes acrimonious debate about the future of wild(er)ness (e.g. see Nelson and Callicott 2008).

Because of these challenges to efforts to delimit 'truly' or significantly wild places, the perceived and real geographies of wildness have begun to change in recent years. There's a much greater recognition that we should both (i) widen our sense of where wild things could and should exist, and (ii) attach more value to entities that, while they bear a human thumbprint, can still in significant respects be considered wild. Nature *preservation* is thus seen by many as rather futile, whereas 'the new conservation' (as it's sometimes called) is valuing the non-human world in novel ways, such that many restoration efforts are no longer geared to recreating the past 'authentically' (Usher 2022). The advent of the Anthropocene, we might say, is obliging us to rethink what counts as the 'wild', despite some resistance to this in certain quarters. New geographies of wildness will emerge accordingly, bringing new benefits and challenges.

But there's more to this than the fact of deepening society-nature entanglement telling us just exactly what to think. As society-environment geographers and other scholars note, we still have freedom to respond to the 'fact' in different ways – hence those three perspectives on wildness summarised in Table 7.1, two of which take an expansive view of wildness, one a stricter view. As we'll now explain, at base, the freedom pivots on different senses of when 'nature' is no longer sufficiently natural to be so described. For, despite what some say, while the world is undoubtedly changing, it's not now 'obviously' a '*post*-natural' world. There's a debate to be had about where 'nature' ends and 'society' begins, and what it signifies morally and practically – as we'll now see. The debate reflects different ways of valuing the living world we inhabit.

Which 'nature' are we post? Thinking about what counts as wildness and how it's valued

As we noted in Chapter 1, some have rather dramatically pronounced 'the end of nature', such is the stupendous magnitude, scale, and scope of the human impact. For instance, Paul Crutzen – writing with Christian Schwägerl (2011, np) – once opined that '… it's no longer us against 'Nature.' Instead, it's we who decide what nature is and what it will be'. To make sense of such a claim, we need to establish what the word 'nature' means in Anglophone, and much Western, culture. The meanings of 'wild', 'wildness', and 'wilderness' are nestled within this larger concept.

The idea that there's a realm of process and objects that are 'natural' took hold from the Middle Ages onwards in Europe. As historians of ideas (e.g. Glacken 1967) and cultural anthropologists (e.g. Descola 2013) have shown, perceptions of this realm became progressively 'disenchanted' and 'secularised'. Nature was seen as something that could be understood objectively, and either left alone, adapted to, managed, or exploited accordingly – as a material realm beyond the realm of symbolism, social relationships, and human institutions. Over time, nature was instrumentalised – seen as a 'resource' to be rendered intelligible and used – even as some pointed to the beauty and majesty of certain species and untouched places, seeing them as antidotes to the ills of industrialisation and urbanisation. All this contrasted with cultures where the concept of 'nature' does not (did not) really exist, and where the non-human realm is perceived to have certain powers, rights, intentions, and even 'character' that affects societies. By the early twentieth century, in 'developed' countries 'nature' was regarded as the antonym of 'society', and a whole vocabulary of familiar dualities reflected this, such as 'biology', 'ecology', 'instinct', 'rural', 'beast', and 'wild' versus 'culture', 'civilisation', 'cultivation', 'reason', 'man' and 'urban' (Castree 2014). Today, we still routinely refer to nature as that which is Other to society or (in the case of our own bodies) other to our volition, imagination, creativity, and so on. But, this general point having been made, we still need to ask: what *exactly* does the word 'nature' signify in Anglo-Western cultures?

Four meanings have endured for several centuries (Daston 2019). Most will perhaps be familiar to readers of this book, but are worth spelling out even so.

- *The non-human world* (**'external nature'**): Colloquially called 'the environment', external nature spans areas and entities considered to be wild ('first nature') through to various cultivated or otherwise altered phenomena, like the English Lake District, the Mississippi River delta, and secondary West African rain forest. In other words, this is nature seen either to be independent of humans (i.e. 'pure') or else nature modified – possibly to a considerable degree – by human action ('second nature'). In either case, nature is seen as that domain that exists *regardless of how we characterise it or influence it*, and which is, wholly or in part, *not of our own making*. This is why some people don't regard the environment as shaped by people as 'nature' – certainly not the built environment, which is by definition a human construction.
- *The defining character of specific phenomena or groups thereof* (**'essential or intrinsic nature'**): Pretty much everything, we in Anglo-Western societies are wont to say, has its own particular nature. To say that something is 'in its nature' is to comment on a phenomena's distinctive and enduring qualities, '… that which makes it what it is and not something else, its ontological identity card' (Daston 2019, 7). For instance, it's in the nature of puffer fish to swim in water, eagles to fly through the air and humans to be land-dwelling omnivores who communicate using elaborate linguistic systems. The nature of something, in this sense, can refer to either or

all of **(i)** its appearance and qualities (size, shape, colour, weight, density etc), **(ii)** its functional specificities (e.g. a capacity to swim but not fly), **(iii)** what it does (or can do) in a material sense (e.g. evaporate at 90°C), and also **(iv)** the fundamental physical laws and processes that animate it or created it. The notion of essential or intrinsic nature gets applied to the tiniest phenomena in existence, such as molecules, right the way up to the Earth System and beyond. Artefacts are also said to have a nature, though one designed rather than involuntary or autopoetic.

- *The autonomous, automatic or irresistible force* (conatus or entelechy) *that operates in and through material phenomena* (**'superordinate nature'**): Often, though by no means only, at scales larger than an individual entity (e.g. a kangaroo) or a single site (e.g. the grassy railway verge near your home), we're often apt to talk about nature as a large and/or enduring domain of processes that can be very difficult to resist, which enables life (e.g. via photosynthesis in plants), but which can often cause us serious harm. For instance, El Niño events are natural in this sense, as are tsunamis like the one that destroyed Japan's Fukushima nuclear power station in 2011. In some situations, it could well be that human actions have triggered such superordinate processes, though without, of course, creating them. Poor land management, for instance, can contribute significantly to wildfires being earlier in the season and much larger than they might otherwise be. Likewise, anthropogenic climate change threatens to release vast methane stores from the northern hemisphere permafrost, with feedback effects on the whole Earth System. In other cases, human action can deliberately alter 'superordinate nature' at a specific level, as with the genetic engineering of mice, crops, and various insects.
- *The totality of biotic and abiotic processes and forms, including humans as biological organisms* (**'universal nature'**): We often refer to almost everything there is as 'nature' – or what we sometimes call Nature, with an upper-case n. We say almost because, as per 'external nature' above, Anglo-Western societies dichotomise humans into **(i)** their biological ('natural') elements and **(ii)** the other attributes that are seen to mark us out as different from all other lifeforms. We regard ourselves both as 'of nature' – fleshy, mortal – and 'beyond nature' insofar as, in the last 200 years especially, we've exerted massive control over our lived environments via science, technology, and various other forms of inventiveness. In Andreas Malm's words, 'Nature is not reducible to humans, who are part to it; humans are not reducible to nature, which is part of them' (2018, 163–4). This is why it's rather glib, to the point of being meaningless, to say that all humans do and have ever done is simply 'natural' because we, too, are part of nature.

In light of this reminder about what 'nature' means, we can draw two important lessons. First, 'wildness' is necessarily differentiated according to the conventions of the word 'nature', which it subtends. For instance, a caged chicken grown for its meat or eggs in France is not at all 'wild' compared to its forest-dwelling east Asian forebears, but in another sense it retains some 'self-willed' aspects irreducible to human designs and meddling. It's artificial, yes, but only to a certain degree. Or take a tornado making landfall in Florida – its intensity may be affected by anthropogenic climate change, but it's hardly 'unnatural' and its intrinsic properties remain intact, expressing as they do fundamental physical laws. Or, if we want to conserve wild animals, is the salient issue the whole species, a local population, a sub-species, their ecological role relative to other species, how long they've been part of an ecosystem …? Just as there's no one 'nature', there's no one domain of *the* 'wild' to which we can point to vouchsafe our values, wishes and plans (Arias 2024).

But, secondly, this reminds us that it's *we people who decide when and where to assign labels such as 'wilderness' or 'unnatural'.* There are choices being made, and choices to be made. This is a key insight furnished by human geographers, sociologists, and many others starting some 40 years ago. As English philosopher Kate Soper phrased it, in her classic book *What Is Nature?* (1995), 'There can be no adequate attempt … to explore 'what nature is' that's not centrally concerned with what it's … *said to be* …' (1995, 21). This echoes a famous observation made by Welsh cultural critic Raymond Williams (1980), that when we talk about nature we are, at some level, talking about *ourselves* – whoever 'we' happen to be. By implication, the same applies to all claims about wildness, wildlife, and wild areas: these claims are not 'objectively' imprinted on our minds from the outside.

Given this, we can return to the Crutzen and Schwägerl claim that 'it's we who decide what nature is and what it will be'. As noted above, they lodged this as a *reality claim, a truth claim,* or what philosophers call an *ontological claim.* As a scientist, Crutzen was saying that the realm of nature has been eclipsed in a material sense due to the outsized human footprint on Earth. This claim undergirds the 'post-nature' narrative summarised in Chapter 1, and related laments by environmentalists, like McKibben, that the Otherness of nature has been fatally compromised. However, there are two problems here, relating to the two lessons identified above.

First, it's not clear which 'nature' is being said to have been superseded in light of the four itemised meanings of the word. Nature could well be *eclipsed* in one sense but necessarily *endure* in other senses, according to these meanings. Which 'nature' do we value most, and can and should we restore what's been damaged or lost? Secondly, added to this complexity is that claims about the end of 'nature' – including 'the wild' – tend to forget that it is *we people* who are speaking for a living world that *cannot speak for itself.* As Soper and Williams argued – later echoed by geographers (e.g. in Castree and Braun's [2001] edited book *Social Nature*) – there's no 'pure' representation of what we call 'nature' that eliminates the contingencies of what we happen to notice, and how we happen to place value on what we notice, in any given situation.

We don't speak as neutral conduits as if reality imprints itself on our minds, like waves sculpting a rocky shoreline. 'The meaning of nature', legal theorist Jed Purdy (2015, 20) writes, 'is ultimately a human question'.

Accordingly, the three stances on present-day wildlife identified in Table 7.1 cannot all be 'right' in a factual sense, but neither is any one of them 'wrong' in a factual sense either. The point is that they view 'the facts' of our current world through lenses shaped by certain cognitive and normative preferences. The debate, then, is not just about 'the truth' of wildlife in a scientific sense. It's also about how the protagonists determine what *counts as wildness and how they value it* in a world undergoing enormous change. The judgements made are not simply 'mirrors' held up to the biosphere, hydrosphere, cryosphere, atmosphere, and pedosphere (Arias-Maldonado 2020). They are cognitive and normative claims emerging from how situated human actors perceive their reality. See Box 7.3 for a classic study of this 'social construction of belief' about the material world.

Box 7.3 The social construction of reality: 'denaturalising' the wild

In Anglophone human geography, along with several social science and humanities disciplines, the 1990s and 2000s were marked by a major focus on 'discourse' – i.e. on how people talk about the world, often conflating their representations of that world with the world itself, as if language acts as an unblemished 'mirror'. Key thinkers included cultural critic Edward Said. His focus was on how the world's 'real geography' was seen via all manner of 'imaginative geographies'. The Canadian geographer Bruce Braun's much-cited book *The Intemperate Rainforest* (2002) was a notable example of research in this vein. Even after 20-plus years, it contains lessons for how we can understand the rather dissonant discourse about wildlife today. Influenced by Said, Raymond Willians and Bill Cronon, among others, Braun focussed on a dispute over how to manage an area of old-growth temperate rainforest in British Columbia, in a location called Clayoquot Sound. Environmentalists sought to protect the area from commercial logging. They strongly emphasised its rarity, its special character, and its wildness against a background of decades of logging in other parts of British Columbia, which had destroyed primary (or 'climax') forest. Meanwhile, the forest products multinational Macmillan Bloedel wanted to log the area. It emphasised the economic value of the timber, the jobs that would be created, and the utility of the wood once processed for the benefit of all manner of manufacturers and consumers. Despite the differences, Braun showed that the antagonists both unthinkingly projected an Anglophone idea(l) that 'wild nature' is a people-free zone, to be appreciated or exploited, depending. This idea, in British Columbia and elsewhere, rationalises the previous forced removal of widespread, often itinerant, Indigenous communities who'd lived harmoniously in 'wild' territories for thousands of years before European colonists arrived. British geographer Sarah Whatmore (2002) made a similar point about the 'purification' of nature in her influential book *Hybrid Geographies*, published the same year as *The Intemperate Rainforest*. In short, representing 'nature' is a political act. It involves inviting (obliging?) others to see nature in certain ways, what another Canadian geographer – Stephanie Rutherford – called 'eco-governmentality' in her monograph *Governing the Wild* (2011).

Braun's research was one example of the 'denaturalising' approach that showed that what we call 'nature' and the 'wild' are not merely real domains but also domains coloured by the 'epistemic' lenses we wear – the lenses are far from natural because different cultures see the world somewhat differently. The value of the approach was, and remains, two-fold. First, it exposed the social content of reality-claims made by various people about the world - in Braun's case, claims about the beauty or economic utility of a 'wild' Clayoquot Sound. Secondly, it thereby lent a voice to (or spoke up for) people whose lives may be affected negatively when more powerful groups of people purport to speak for nature and for wild places. This is especially important when claims about the 'nature' of the world are employed in overtly discriminatory ways, as has occurred time and again in history (e.g. when unqualified references to 'race', 'genes' or the biology of men and women have justified unequal treatment of certain groups of people). As scholars like Braun have shown time and again, powerful groups often seek to avoid political discussion by 'naturalising' certain things as if they're 'common sense,' i.e. so obvious that no one would bother to challenge them.

The lessons today are clear. When anyone makes claims about 'what we need to do' in response to problems thrown up by the Anthropocene – including what to do about 'the wild' – we need to ask what values and goals they're projecting onto the world, and whose interests they're seeking to uphold. The claims both draw inferences from the declared 'facts' of Anthropocenic change, and will often imply that the facts are more-or-less beyond question. In their own way, each of the five narratives about the Anthropocene outlined in Chapter 1 has a propensity to naturalise elements of their story.

Despite appearances, this isn't an 'anti-realist' argument about nature and wildlife. Clearly, there wouldn't be so much discussion of the Anthropocene and planetary boundaries if there weren't very large, very real, and ultimately undeniable alterations to the Earth System. Decades of research into the 'social construction of nature', and the comparative anthropology undertaken by Philippe Descola and others, simply remind us that the world we perceive is rendered intelligible through culturally specific 'webs of meaning'. As Kate Soper once aptly put it, '... 'nature' is a kind of self-denying concept through which what's culturally ordained is presented as a pre-discursive external determination upon that culture' (1995, 34). 'No reality with representation', one might say (with certain caveats) – this has become axiomatic across many Anglo-Western social science and humanities disciplines. Furthermore, it reminds us that our categories and modes of valuing reality are as material, in their own ways, as the myriad of phenomena we study, use, destroy, or seek to protect. They shape action and, in turn, our actions are increasingly world-changing. In the context of this chapter, how the wild is defined and valued by different people matters enormously for its future on Earth. To riff-off Raymond Williams, when people talk about wildlife and wild places, yes they're talking about living entities, but they're *also* discussing them via their own cultural frameworks. Consider creating an artificial wetland, using native wild plant, insect, and fish species suitable for the task. Is the wetland 'fake' because it wouldn't otherwise be there, or is it in fact 'natural' because it's self-sustaining once created? The wetland itself cannot provide us with the answer. Similarly, neither can an extinct woolly mammoth tell us if its 'right' to use its frozen DNA, 'recreate' it in the 2020s, and let it roam in a remote part of northern Europe. The fact that we could 'de-extinct' it does not tell us it's right to do such a thing.

The way the wild is represented is, therefore, consequential even as what we call 'the wild' exceeds our various representations of it. We thus need to be realists about the material world *and* the various means by which it's depicted. This is the lesson of decades of research into the linguistic construction of reality, and the associated effects of our constructions on the social and non-human worlds. To repeat: the advent of the Anthropocene doesn't mean we now somehow get outside of our 'worldviews' to achieve a pellucid grasp of the once-and-for-all truths of our fast-changing reality (Arias-Maldonado 2020). As geographers Paul Robbins and Sarah Moore (2013) showed in a study of rival camps in the field of conservation biology, even scientists apprehend reality via certain cultural filters that aren't shared, or even recognised, by all people alive today. They don't stand above the fray on some Archimedean promontory but are, instead, shapers of the action.

Living in a world that's (still) wild, semi-wild and post-wild, all at once: pathways, policies, experiments, and challenges

Let's take stock. The very fabric of wildlife and wild areas has undeniably been altered profoundly in the last 80 years in particular, but much further back in time too in certain regions. Even so, there's no purely factual answer to the question of whether we now live in a 'post-wild' world. It's a question of values and applying those values to pass judgement on the ongoing alterations to living entities. What we can say, though, with some certainty, is that the second and third perspectives on wildlife summarised earlier are gaining considerable traction. One indication of this is that former Chief Scientist of the prominent organisation The Nature Conservancy, Peter Kareiva, was a prominent critic of the 'fortress' approach to conservation that excluded communities and circumscribed human intervention. He embraced 'the new conservation', acknowledged the limits of preservation, and saw nature restoration as a pragmatic endeavour, not an effort to turn back the hands of time.

This doesn't mean that efforts to protect what's perceived as the 'truly wild' are pointless or likely to cease in the near future. But it does mean – and we're already seeing this around the world – that all sorts of other initiatives are gaining support that challenge older understandings of where wildlife can and should exist. Some have a strong social and environmental justice element, some are fairly ecocentric, some are designed to yield broad civic benefits, some are linked to nation-building, others to ethno-cultural survival (or revival), and still others provide income streams (variously) to governments, business owners, and communities. We're today seeing urban and rural 'rewilding' initiatives, species reintroductions to rural and peri-urban areas, 'de-extinction' efforts involving highly advanced biotechnology and ex-situ conservation programs, and new management approaches to once heavily utilised areas of land and water so as to create biodiverse 'novel ecosystems'.

From the perspective of position one (the 'doubling-down' stance), initiatives expressive of position two amount to an arrogant effort to *dominate* the wild, while position three will, at best, yield *semi*-wildness. From the perspective of positions two and three, meanwhile, position one is idealistic, backward-looking, lacking in pragmatic good sense, and often contributory to injustice towards communities in Africa, Asia, the Middle East, and elsewhere. If we look globally today, we see a number of policies and initiatives enacted in the name of all three positions, with their various supporters and critics. But, as we've just noted, overall it's the new conservation and restoration that's on the rise, with older style wildlife preservation rather on the back foot. The result is that there's what geographer Jamie Lorimer (2015) calls a 'multi-natural' sense of the wild abroad in the 2020s: different definitions, and different modes of valuing, everything from captive-bred ground parrots to Nile perch to Tasmanian devils to Tamil lacewings to the bacterium Burkholderia pseudomallei, now prevail. There's no consensus on what the wild is and where it should rightly be, but there's an openness to broader visions. A patchwork of initiatives is emerging, including ones employing advanced technology.

Geographers and many other scholars have both studied, and also at times advocated for, policies and initiatives consistent with positions two and three.

They've done so in the belief that the overall human footprint can be better managed, but not radically reduced anytime soon. Let's consider some key topics, with examples (listed in no particular order of priority). Throughout, the emphasis is on *diverse relationships*, both material and affective, between people and wild things.

- *Wild experiments in unlikely locations:* As noted earlier, there are these days efforts being made to 'rewild' areas that have suffered significant ecological loss, and to foster existing wildlife in heavily inhabited areas such as Calcutta or Vancouver. A noted example of the former is Oostvaarderplassen in the Netherlands. Created in the mid-1980s, it's a nature reserve on reclaimed land along the North Sea coast, very close to densely populated Amsterdam. The ecosystem is an approximation of a pre-human landscape, with red deer, Konik ponies, and Heck cattle introduced and left largely to fend for themselves in the early years. These herbivores have helped to increase the diversity of plant, insect, and bird life through their grazing habits. Yet, of course, Oostvaarderplassen has been created through human decisions, with an ecologist called Frans Vera the key player early on. The site is now managed so as to ensure the large animals do not suffer unduly and biodiversity is maximised. There are no natural predators of the animals in the reserve. People can walk and cycle through Oostvaarderplassen, and there are several lookout points. While clearly not 'the same' as a wild pre-human ecosystem, Oostvaarderplassen is nonetheless significantly wilder than what was in this area after it was reclaimed from the sea in the 1950s. It's uninhabited, but doesn't conform to a strict notion of 'restoration' of a lost ecosystem. British geographer Jamie Lorimer is among those who've studied Oostvaarderplassen carefully. In his book *Wildlife in the Anthropocene* (2015), he commended the experimental ethos of those who manage the reserve. He argued that, in allowing for surprise, in being adaptable, and in being flexible about the species mix, the Dutch managers of the reserve have set a globally relevant example of how wildness can be reinvigorated in unexpected locations, bringing all manner of benefits for nature and for people.
- In-situ *wildness often needs* ex-situ *wildness:* Historically, places such as zoos and aquariums have been *ex-situ* locations where, in effect, people have wild things brought to them, so that they can marvel at things like tigers and sharks – but knowing full well that the wildness in them is contained in highly artificial settings. However, the global loss of *in situ* wildlife in recent decades has caused a change to how many zoos, aquariums, and hatcheries operate. Many are now sites for creating small populations of endangered species with a view to reintroducing them into remaining areas of their native habitats. Here, then, historical roles are reversed: the 'artificial' settings become supply sources for wild populations. In her detailed study of these altered linkages between wildlife *in-situ* and *ex-situ*,

legal geographer Irus Braverman (2015) has pointed to the enormous care shown by all those involved in this effort to save wild communities of black-footed ferrets, Sumatran rhinos, Gila trout, and far more besides. Her book *Wild Life* demonstrates just how necessary *ex-situ* work is, where wild creatures are actively reared by people in artificial environments, if *in-situ* wild populations are to survive.

- *Slippery species and baselines:* Even in situations where there's a desire to conserve, preserve or restore wildlife in a notionally people-free ('truly wild') state, contestable judgements need to be made about what the desired state of a species or wild area is going to be. For, as naturalist Charles Darwin famously showed, life evolves: incremental change is a constant. And, more recently, as ecologists discovered during the 1960s and 1970s, the natural world doesn't always settle into some sort of 'equilibrium' state for decades at a time; instead, it can cross thresholds quickly, even without human intervention, and flip into new states (Botkin 1990). Clearly, such 'flipping' is now more common due to human interference in natural systems. So, which 'baseline' should be used as the criteria for 'successful' preservation, conservation, or restoration? Manchester University geographer Aurora Frederiksen (2015) is one of several researchers who's asked this question. She examines efforts to protect Scottish wildcats from breeding with feral cats descended from domestic cats (pets). Crossbreeding between wild and feral cats is ongoing and producing healthy offspring. The focus on wildcat species 'purity', Frederiksen argues, is based on a cultural preference for a piece of Scotland's biogeographic history. If captive breeding and animal release plans are to succeed, she shows, a large culling of often healthy feral cats will ensue, many containing wildcat genes. There is thus a certain arbitrariness at work here that draws lines between wild animals, despite their natural propensity to reproduce and intermingle. Is it thus cruel to pursue 'purity'?
- *Co-existence with companion species:* Domesticated animals have been a feature of many cultures for millennia, whether it be as working animals, as food sources, or as pets. Likewise, we've domesticated all manner of plants. The total 'unnatural' biomass on Earth has grown enormously. However, both native and introduced wild creatures are common across the towns, cities, and settled rural areas of the planet – as Cambridge University geographer Maan Barua shows in his book *Lively Cities: Reconfiguring Urban Ecology* (2023). Where once often seen as 'pests', there are now plenty of examples of where people develop relationships with these creatures. There's now a body of research into these 'companion species' and the material and affective ties that bind people to them (e.g. Wilson, 2023). Understanding, and learning from, these ties may help to normalise positive relationships between people and the wildness on their doorsteps.

  One of many examples is a study of people adjusting to 'wildlife out of place', undertaken by English geographer Helen Wilson (2022). It focuses on how people in the city of Newcastle, in north east England, have responded to the recent appearance of a colony of Tyne kittiwakes – birds whose breeding habitat is coastal cliffs some 15km away. The kittiwakes have spontaneously nested on buildings and bridges in Newcastle, with some locals seeing them as a noisy, messy intrusion, but others pointing out that accommodating them may help to protect the bird population over time. Wilson shows how debates about the 'proper place' of the birds are beginning to shift conventional understandings of wildlife in Newcastle, altering people's sense of urban companionship and belonging. Some now see them as an icon for Newcastle. Meanwhile, in London Barua (2023) shows how many locals feel emotionally attached to sub-tropical parakeets, an introduced species that's adapted to its temperate, urban environment. These two studies echo Sarah Whatmore's argument, in *Hybrid Geographies* (2002), that species don't have entirely fixed natures – we need, she argued, to attend to the differentiations that emerge as wild creatures shift, or are shifted, to new circumstances (see also Turnbull, Fry, and Lorimer 2025). The new social and ecological relations they enter into alter them, and may alter us too.

- *Learning from non-Western cosmologies*: The studies cited above are both examinations of, and encouragements to foster, different perceptions of the wild, with all that entails. But different perceptions from Anglo-Western ones are integral to many other cultures. A case in point is Australian Aboriginal culture, in which what people call Country (with a capital C) is seen as animated, communicative, and very much integral to human lifeways through time and across land and sea. Along with many anthropologists, in recent years some human geographers in the Anglosphere, such as Sarah Wright (of the University of Newcastle, New South Wales), have sought to give voice to Country in different parts of Australia – i.e. to make it seen and heard so that non-Indigenous people might appreciate the caring Aboriginal sensibility towards what non-Indigenous people call 'wildlife', and 'nature' more broadly. Innovatively, this has involved writing *as* Country, not merely *on behalf of it*, along with Aboriginal people (e.g. Bawaka Country et al. 2018). Country is a 'congress' of life and spirits where time flows in all directions. This is a graphic example of how thought and practice regarding the 'wild' can be decolonised in the world's many former settler-colonies. Quite simply, it is unthinkingly ethnocentric to presume that 'we' in the Anglo-Western world have the right or only outlook on wildness.
- *Conflicts in contact zones, risks elsewhere*: Despite the retreat of 'wildness' wholly untouched by humans, wild entities still possess considerable 'agency' and power. They have a capacity to harm people or to be harnessed in ways that enable social power to be exercised. This is obviously the case with large animals or predators, such as elephants and wolves when they wander out of protected areas into settled areas. Clearly, there's

a need to balance people's safety against the rights of wild creatures in these situations. This theme has been explored widely, e.g. in Rosemary Collard's (2012) analysis of cougars on Vancouver Island, Canada. There are also questions of farmer's rights in rural areas where wild animals intrude (see, for instance, Vaccaro and Beltran 2009; Wenborn et al. 2025). And there are questions of how to reintroduce species valued by minority cultures but where the majority culture sees them as undesirable (see Beggen and York 2025).

Large wild creatures aside, there are health concerns with managing infectious wild creatures, such as mosquitoes. There are also biosecurity concerns relating to illegal or uncontrolled transfers across political borders of insects, microbes, and other small wild things. New diseases, for instance, are a major risk when people begin to enter hitherto uninhabited areas, become unknowingly infected, and spread a new disease (Barker 2015; Hinchliffe et al. 2013). Stringent regulation and forensic environmental monitoring are important for safety reasons, though often lacking. Meanwhile, in other situations, powerful social actors may use conservation and restoration for political reasons. A fascinating case is the Israel-Palestine border. In her monograph *Settling Nature* (2022), Irus Braverman (whose research we've already mentioned) shows how Israel has deliberately shaped the non-human landscape in its ethnic image: fallow deer, gazelles, wild asses, griffon vultures, pine trees, and cows have been introduced on the Israeli side of the border, as against goats, camels, olive trees, hybrid goldfinches and akkoub (a thorny plant) which are characteristic of the Palestinian side. Here, then, wildlife becomes a vector for a political power play. But in other cases, wildness *countermands* the political goals of national governments, as Juanita Sundberg (2011) showed on the semi-arid US–Mexico border. She recounts how efforts to conserve wildcats along the border frustrated attempts to stop illegal migrants coming from Mexico and pitted different US government departments against each other.

- *Inhuman nature and the big wild world*: Unlike what's been discussed in the last couple of pages, the final theme in the research isn't that consistent with positions two and three, though it does have loose filiations to position one. It's the theme of wildness as an occasional force of such stupendous magnitude that any ideas of 'entering into relations' with it are made a mockery of. This theme is strongly highlighted in research by UK–New Zealand geographer Nigel Clark (2011). His book *Inhuman Nature* pointed to the rapid, catastrophic events that can occur as we collectively exit the Holocene epoch. Here 'wildness' is abiotic and biotic (as with COVID-19) – it's the uncontrollable, 'self-willed' forces of fire, flooding, a pandemic, and so on whose contemplation, in his view, should not only humble us, but spur us to collective action to prevent the worst impacts. Here the wild is that which eludes our control and threatens a great many of us. Clark is more pessimistic than Erle Ellis about the prospects of a new wave of 'green' development, wherein people are able to orchestrate the wild with all the tools of science and advanced technology.

A number of the geographers whose research we've just cited sit within intellectual camps called 'post-humanist', 'more-than-human', 'non-representational', and 'new materialist'. They employ concepts such as 'actor-networks', 'hybridity', 'rhizomes', and 'assemblages' to describe the various entanglements, both up-close and more impersonal, between people and wild entities (see Box 7.4). Others, as we just saw, are less interested in these linkages and more focussed on the enduring power of wildness to shock and overwhelm us (for instance, a collaborator with Nigel Clark, the British geographer Kathryn Yusoff, uses the term '*in*human geography' in her self-selected professorial title). They focus on large-scale 'natural forces' and the outsized 'agency' possessed by the non-human realm, which are often actualised harmfully for humans.

Box 7.4 Post-human, more-than-human, non-representational, and new materialist approaches to wildlife

From the late 1990s onwards, a number of scholars in the Anglo-Western social sciences and humanities began to criticise aspects of the then-influential 'social constructionist' approach to nature, wildness, and reality more broadly. The criticism was that the 'materiality' of the non-human world was inadvertently being ignored in the desire of critics like Bruce Braun to 'denaturalise' references to the world (see Box 7.3). This was not an out-and-out ecocentric critique in an ontological and moral sense. It was more an acknowledgement of just how reliant people are, in a corporeal and practical sense, on all manner of non-human entities. Leading proponents of this 'post-constructivist' perspective have included anthropologist-philosopher Bruno Latour, science studies scholars Donna Haraway, John Law, and Steve Woolgar, political theorist Jane Bennett, feminist philosophers Judith Butler, Elizabeth Grosz and Elspeth Probyn, British geographers Sarah Whatmore, Jonathan Murdoch, Steve Hinchliffe, and Nigel Thrift, anthropologist Tim Ingold, and eco-critics Timothy Morton and Cary Wolfe. Between them, their research focussed on both non-human 'nature' and also the human body (our fleshy existence as organisms).

While this perspective isn't homogeneous, the different names chosen to signify it have strongly overlapping meanings. 'Post-humanism', in this context, refers to the fact that what makes people distinctively human relies entirely on a plethora of physical laws, processes, and specific material entities irreducible to their symbols and actions. The term 'more-than-human' also captures this, and is the title of another book in the series to which the present book belongs (by Lorimer and Hodgetts 2024). Then, Nigel Thrift's neologism 'non-representational' denoted all that which exceeds our verbal or pictorial depictions of it. Meanwhile, the term 'new materialism' refers to studies of how material phenomena come to matter by way of the relations they have with other things, including people.

Conceptually, all this has been expressed by way of widely used ideas such as 'actor-networks', 'assemblages, 'rhizomes', 'cyborgs', 'meshwork', and 'hybrids'. In Geography, these ideas have been integral to the rise of two related sub-fields called 'animal geographies' and 'plant geographies'. Neither are closely linked with biogeography in a pure science sense, and are more a product of human geographers now paying close attention to people's connections with domesticated, otherwise altered, and wild non-humans. A number of studies rightly pay very close attention to how economic relationships are having world-changing impacts on people, animals and plants alike – such as Canadian geographer Rosemary-Claire Collard's monograph *Animal Traffic* (2020, which is about the global trade in exotic pets) and the book *Ingrained*, authored by Australian geographers Lesley Head, Jenny Atchison and Alison Gates (2012), which is about wild and cultivated wheat.

Clearly, the themes above – seen in the context of the three stances towards wildness and its future in the Anthropocene – raise very tricky and important issues of ethics and policy. The modalities and geographies of wildlife are already plural – 'multi-natural' – and may well become more complex and fluid over time. Researchers who ponder the fate of the wild increasingly recognise that there can be no single solution to questions of conservation, preservation, or restoration. The problem is that approaches to wildness that seem legitimate in one place may well cause outrage in other places. Given that the Anthropocene is an 'implicate order' where everything affects everything else at some level, some way needs to be found of opening up ostensibly local and regional scale decisions to wider input. At a minimum, the need to revegetate the Earth in order to sequester greenhouse gases, often using 'off-setting' as a policy tool, will enable some international coordination around approaches to wildlife. The challenge is not to subordinate biodiversity policy to the needs of climate policy.

Conclusion

The fabric of the wild has shrunk, partly unravelled but also been made anew in the last century or more. Many people are horrified at the damage caused by a careless human footprint. Looking ahead, wildlife in the Anthropocene will be influenced as much, if not more, by human perceptions and associated actions. In this chapter, we've surveyed the major schools of thought about the current and preferred future state of the Earth's wildness. We've outlined the reasons why new geographies of the wild are emerging, be it through indirect or direct human causes. How one assesses these geographies is partly a question of perception (what *counts* as 'wild' these days?). We've outlined some of the key issues concerning the 'new wild' that need to be wrestled with in different places, even as some sort of coordinated (if not necessarily unified) approach among places needs to be taken in the medium-to-long term. The future of the wild is a topic relevant to all people on Earth, raising questions of how to balance environmental, ecological, and interspecies justice against all manner of pressures and competing priorities.

In closing, we want to do two things. The first is to return to the distinction made in Chapter 2 between 'thin' and 'thick' understandings of the world's geography. The wild entities visible and invisible to us shape our world functionally and affect us in psychosomatic ways. The nature, density, and distribution of wild things matter profoundly to us – it affects the quality of our places and regions, it shapes the long-distance ecological links shaping people's lives, it involves new fusions of the social and the environmental, it requires multi-scalar management, and it affects our political and administrative partitioning of land and water. The Anthropocene opens the door to the epic task of deciding how and where to rekindle or protect wildness at *all* points of the compass. It invites us to consider how to make a habitable world for ourselves, our descendants, and a vast array of non-human others. The wild is a key vector for geography, and history, in the making. How it's defined and discussed matters every bit as much as the insects, animals, plants, fish, and so on that constitute the wild world's ecologies.

But, secondly, the wild is also a vector for those five Anthropocene narratives summarised in Chapter 1. As attentive readers may have noticed, there are a range of ways the three general outlooks on the wild presented in this chapter intersect with those five narratives (recall: the narratives are the 'standard' one, the post-natural one, the catastrophist one, the eco-Marxist one, and the decolonial one). The key is how 'anthropocentric' the narratives are, and how politically and morally ethically radical they are. Only the eco-catastrophist narrative sees little or no room for meaningful choice about how to relate to wild phenomena, at least not until after some sort of cataclysm. The other narratives have different visions of how far to reconfigure the relations between people and the biophysical world, and different senses of what is just, good, and necessary for people and the non-human realm. In the

Anthropocene, much rides on which narratives get enacted in different places and how they will shape what transpires elsewhere.

Study questions

1 If 'nature' is on the back foot in the Anthropocene, on what basis can decisions to protect major parts of the non-human world be justified in our 'post-natural' age?
2 If 'the wild' and 'wildness' are many things, not one thing, what future relationships with wild phenomena could and should humanity aim for?
3 Try to recall the last time you encountered 'wildness' of some kind. Where was it, and why did it occur? How did it make you feel? Would you value more opportunities to encounter wildness, and if so, why?

Notes

1 Common imaginaries of the Arctic, including environmentalist ones, depict the region as an un-peopled place. However, the Canadian Arctic is home to around 150,000 people, and the wider Arctic region is home to around 4 million.

References

Aloi, G. 2025. *Lawn*. London: Bloomsbury.
Arias, S. 2024. "The Abandonment of the Ideal of Wilderness: Rewilding as the Consequence of the Anthropocene Metaphysics on Restoration Ecology." *The Anthropocene Review* 12 (1): 35–50.
Arias-Maldonado, M. 2020. "Bedrock or Social Construction? What Anthropocene Science Means for Political Theory." *The Anthropocene Review* 7 (2): 97–112.
Barker, K. 2015. "Biosecurity: Securing Circulations from the Microbe to the macrocosm." *The Geographical Journal* 181 (4): 357–65.
Barua, M. 2023. *Lively Cities: Reconfiguring Urban Ecology*. Minneapolis: University of Minnesota Press.
Beggen, K., and R. York. 2025. "Colonizing Canis Lupus: Wolf Management as a Settler Colonial Project." *Rural Sociology* 90 (3): e70009.
Botkin, D. 1990. *Discordant Harmonies*. Oxford: Oxford University Press.
Braun, B. 2002. *The Intemperate Rainforest*. Minneapolis: University of Minnesota Press.
Braverman, I. 2015. *Wild Life*. Stanford: Stanford University Press.
Braverman, I. 2022. *Settling Nature: The Conservation Regime in Palestine-Israel*. Minneapolis: Minnesota University Press.
Brockington, D. 2002. *Fortress Conservation: The Preservation of the Mkomazi Game Reserve, Tanzania*. Bloomington: Indiana University Press.
Buscher, B., and R. Fletcher. 2020. *The Conservation Revolution: Radical Ideas for Saving Nature Beyond the Anthropocene*. London: Verso.
Cannon, C. H., and C. S. Kua. 2017. "Botanic Gardens Should Lead the Way to Create a "Garden Earth" in the Anthropocene." *Plant Diversity* 39 (6): 331–37.

Carney, J. A. 2021. "Subsistence in the Plantationocene: Dooryard Gardens, Agrobiodiversity, and the Subaltern Economies of Slavery." *The Journal of Peasant Studies* 48 (5): 1075–99.

Castree, N. 2014. *Making Sense of Nature*. London: Routledge.

Castree, N., and B. Braun, eds. 2001. *Social Nature: Theory, Policy and Practice*. Oxford: Blackwell.

Ceballos, G., P. R. Ehrlich, and R. Dirzo. 2017. "Biological Annihilation via the Ongoing Sixth Mass Extinction Signalled by Vertebrate Population Losses and Declines." *Proceedings of the National Academy of Sciences* 114 (30): E6089–96.

Clark, N. 2011. *Inhuman Nature*. London: Sage.

Collard, R. C. 2012. "Cougar-Human Entanglements and the Biopolitical Un/Making of Safe Space." *Environment and Planning D* 30 (1): 23–42.

Collard, R. C. 2020. *Animal Traffic: Lively Capital in the Global Exotic Pet Trade*. Durham: Duke University Press.

Country, B., et al. 2018. "Goŋ Gurtha: Enacting Response-Abilities as Situated Co-becoming." *Environment and Planning D* 37 (4): 682–702.

Crist, E. 2019. *Abundant Earth*. Chicago: Chicago University Press.

Cronon, W. 1996. "The Trouble With Wilderness, or Getting Back to the Wrong Nature." In *Uncommon Ground*, edited by William Cronon, 69–90. New York: W.W. Norton.

Crutzen, P. J., and C. Schwägerl. 2011. "Living in the Anthropocene: Toward a New Global Ethos." *Yale Environment* 360. http://e360.yale.edu/feature/living_in_the_anthropocene_toward_a_new_global_ethos/2363/. Accessed 13 May 2025.

Daston, L. 2019. *Against Nature*. Cambridge, Mass: MIT Press.

Davis, J., A. A. Moulton, L. Van Sant, and B. Williams. 2019. "Anthropocene, Capitalocene, … Plantationocene?: A Manifesto for Ecological Justice in an Age of Global Crises." *Geography Compass* 13 (5): e12438.

Descola, P. 2013. *Beyond Nature and Culture*. Translated by Janet Lloyd. Chicago: Chicago University Press.

Díaz, S., et al. 2019. "Pervasive Human-Driven Decline of Life on Earth Points to the Need for Transformative Change." *Science* 366: eaax3100.

Dirzo, R., H. S. Young, M. Galetti, G. Ceballos, N. J. Isaac, and B. Collen. 2014. "Defaunation in the Anthropocene." *Science* 345 (6195): 401–6.

Duffy, R. 2010. *Nature Crime: How We Are Getting Conservation Wrong*. New Haven: Yale University Press.

Ellis, E. 2015. "Too Big for Nature." In *After Preservation*, edited by Ben Minteer and Stephen Pyne, 24–31. Chicago: Chicago University Press.

Ellis, E., et al. 2021. "People Have Shaped Most of Terrestrial Nature for at Least 12,000 Years." *Proceedings of the National Academy of Sciences* 118 (17): e2023483118.

Fairhead, J., and M. Leach. 1996. *Misreading the African Landscape*. Cambridge: Cambridge University Press.

Fairhead, J., M. Leach, and I. Scoones. 2012. "Green Grabbing: a New Appropriation of Nature?" *The Journal of Peasant Studies* 39 (2): 237–61.

Frederiksen, A. 2015. "Of Wildcats and Wild Cats: Troubling Species-Based Conservation in the Anthropocene." *Environment and Planning D* 34 (4): 689–705.

Glacken, C. 1967. *Traces on the Rhodian Shore*. Berkeley: University of California Press.

Head, L. 2016. *Hope and Grief in the Anthropocene: Re-Conceptualising Human-Nature Relations*. Abingdon: Routledge.

Head, L., J. Atchison, and A. Gates. 2012. *Ingrained: A Human Biogeography of Wheat*. Farnham: Ashgate.

Hettinger, N. 2021. "Age of Man Environmentalism and Respect for an Independent nature." *Ethics, Policy and Environment* 24 (1): 75–87.

Hinchliffe, S. 2013. "Remaking Emerging Infectious Diseases." In *Biosecurity: The Socio-Politics of Invasive Species and Infectious Diseases*, edited by Andy Dobson et al., 199–213. London: Routledge.

Klein, N. 2014. *This Changes Everything: Capitalism vs Climate*. New York: Simon and Schuster.

Lorimer, J. 2015. *Wildlife in the Anthropocene*. Minneapolis: Minnesota University Press.

Lorimer, J. and Hodgetts, T. 2024. *More Than Human*. London: Routledge.

Malm, A. 2018. *The Progress of This Storm*. London: Verso.

Marris, E. 2011. *Rambunctious Garden*. New York: Bloomsbury.

Marris, E. 2021. *Wild Souls*. New York: Bloomsbury.

McKibben, B. 1989. *The End of Nature*. NYC: Vintage.

Meyer, S. 2006. *The End of the Wild*. Cambridge, Mass: MIT Press.

Monbiot, G. 2013. *Feral: Rewilding the Land, Sea and Human Life*. London: Allen Lane.

Nastold, F. 2025. The Garden as Scene of the Anthropocene: Heterotopia, Queer Ecologies, and Green Archives. *Insert* 8 (November 30). https://doi.org/10.5281/ZENODO.17771825

Nelson, M., and J. B. Callicott, eds. 2008. *The Wilderness Debate Rages On*. Athens: University of Georgia Press.

O'Riordan, T. 1976. *Environmentalism*. Brondesbury: Pion.

Pearce, F. 2015. *The New Wild*. London: Icon Books.

Pepper, D. 1984. *The Roots of Modern Environmentalism*. London: Croom Helm.

Purdy, J. 2015. *After Nature*. Cambridge, Mass: Harvard University Press.

Riggio, J., et al. 2020. "Global Human Influence Maps Reveal Clear Opportunities in Conserving Earth's Remaining Intact Terrestrial Ecosystems." *Global Change Biology* 26 (8): 4344–56.

Robbins, P. 2007. *Lawn People: How Grasses, Weeds, and Chemicals Make Us Who We Are*. Philadelphia: Temple University Press.

Robbins, P., and S. A. Moore. 2013. "Ecological Anxiety Disorder: Diagnosing the Politics of the Anthropocene." *Cultural Geographies* 20 (1): 3–19.

Rutherford, S. 2011. *Governing the Wild: Ecotours of Power*. Minneapolis: University of Minnesota Press.

Soper, K. 1995. *What Is Nature?* Oxford: Blackwell.

Sundberg, J. 2011. "Diabolic Caminos in the Desert and Cat Fights on the Rio: a Posthumanist Political Ecology of Boundary Enforcement in the United States–Mexico Borderlands." *Annals of the Association of American Geographers* 101 (2): 318–36.

Terborgh, J. 2019. *Requiem For Nature*. Washington DC: Island Press.

Thomas, C. D. 2017. *Inheritors of the Earth*. London: Allen Lane.

Tillett, R. 2020. "The Necessity of Lived Resistance: Reading Leslie Marmon Silko's "Gardens in the Dunes" in an Era of Rapid Climate Change." *Studies in American Indian Literatures* 32 (1–2): 188.

Turnbull, J., T. Fry, and J. Lorimer. 2025. "(Re)Wilding London." *Transactions of the Institute of British Geographers* 50 (3): e12748.

UNEP-WCMC and IUCN. 2026. *Protected Planet: The World Database on Protected and Conserved Areas (WDPCA)* [Online], March 2026. Cambridge: UNEP-WCMC and IUCN. Available at: www.protectedplanet.net

Usher, M. 2022. "Restoration as World-Making and Repair: A Pragmatist Agenda." *Environment and Planning E* 6 (2): 1252–77.

Vaccaro, I., and O. Beltran. 2009. "Livestock Versus 'Wild'." *Geographical Review* 99 (4): 499–516.

Wenborn, M. J., et al. 2025. "Assessment of Mitigation of Human–Elephant Conflict in the Highlands of Northwest Namibia." *The Journal of Environment and Development* 34 (3): 684–706.

Whatmore, S. 2002. *Hybrid Geographies*. London: SAGE.

White, J. 2019. "Gardening in the Anthropocene." In *The Poetics and Politics of Gardening in Hard Times*, edited by Naomi Milthorpe, 71–86. London: Rowman and Littlefield.

Wignall, V. R., K. Alton, and F. L. W. Ratnieks. 2019. "Garden Centre Customer Attitudes to Pollinators and Pollinator-Friendly Planting." *PeerJ* 7 (June): e7088.

Williams, R. 1980. *Problems in Materialism and Culture*. London: Verso.

Wilson, H. F. 2022. "Seabirds in the city: Urban Futures and Fraught Coexistence." *Transactions of the Institute of British Geographers* 47 (4): 1137–51. https://doi.org/10.1111/tran.12525

8 Governing *Geo*

Geographies of Security and Resilience

Introduction: politics on a lively planet

How to deal with the increasing 'liveliness' of planetary systems, spurred into potentially erratic and unpredictable new behaviours by the combined actions of humanity? How might geophysical and ecosystem turbulence shape the turbulence of human politics, and vice versa, from local to global scales? These are fundamental and timely questions about the politics of the Anthropocene, and political thinkers and actors are already responding to them with new models and proposals for how politics should be done. As such, a key task for political geographers and like-minded scholars is to interrogate these emerging governing logics of the Anthropocene, to pick at their assumptions and interrogate their consequences – both likely or actual – for communities and environments across spatial scales. We therefore round out this 'Geo-Politics' section of the book with a close look at how politics is being rethought for the Anthropocene. To begin, we need to start with some consideration of how the relationship between physical environments and human politics has conventionally been thought about in Geography and beyond.

Conceptualising geopolitics

As we noted in Chapter 1, when we think about 'politics', we tend to imagine a decidedly human endeavour where people cooperate and compete to realise their aims. In democratic societies, the term might conjure images of elected representatives sitting in a parliamentary chamber debating affairs of state; deciding on ways to manage things like market economies, systems of welfare, justice, public health and education, armies and navies, and debating relationships with other countries. Politics, in this vein, is the system by which, in any given place and situation, human communities determine the nature and trajectory of their collective affairs and relationships through discussion and compromise.

When we think about 'geopolitics' too, we might similarly imagine a very human drama of war and conflict, international rivalries, and the hard work

DOI: 10.4324/9781003162469-11

of diplomacy. Classically, 'geopolitics' has meant the study of relationships between states or empires, and their struggles for power, hegemony, and control over territory and resources (Toal 2024). In the early twentieth century when, for example, Halford Mackinder – one of the first university-appointed geographers in the UK – was busy defining the scope of the study of geopolitics, the great state rivalries were understood to be those of the major imperial powers. Mackinder himself theorised that imperial control of the 'heartland' region of central Eurasia was the key to global hegemony (Mackinder 1904). Much like in the popular boardgame *Risk*, winning the 'great game' (Becker 2012) of geopolitics was about occupying and controlling key chunks of the Earth's surface. The Earth itself, in this theory of geopolitics, was largely static, even passive. It was an inert storehouse of resources to be claimed and a terrain to be managed. While 'environmental determinism' also loomed large in these theories, granting the physical environment agency in shaping peoples and cultures (e.g. through theories of climatic determinism – see Livingstone 2024), and in shaping political and economic fortunes (e.g. through the distribution of resources and types of landscape), these ideas similarly regarded the physical environment as a static thing. The environment's influence on human affairs was the result of a set of unchanging characteristics that could, perhaps, be overcome through military strategy or technological innovation (Carey 2011; Peet 1985).

The discipline of Geography, and particularly Human Geography, has spent several decades trying to rid itself of the spectre of environmental determinism, with all its nineteenth-century racism and crude understandings of complex socio-ecological realities (Radcliffe et al. 2010). This has meant that subfields like political geography and geopolitics have devoted less attention to the role and agency of physical environments in the workings and effects of political power. However, in the Anthropocene, it is no longer tenable to conceive of the Earth as a static backdrop to human affairs (Clark and Szerszynski 2020; Clark and Yusoff 2017). As a result of the geopolitical exercise of state power over the preceding few centuries, and of the associated transformations of economic production and consumption (see Chapters 3–5), the earthly playing board of the 'great game' is undergoing rapid transformation. As a consequence, the disruptive agencies of biophysical phenomena are inserting themselves into geopolitical dramas in new ways (Dalby 2020a), while communities the world over are grappling with how to live on the surface of an increasingly unruly planet (Westgate 2020) and working out what it means to position the Earth System as both an object and agent of politics (Biermann 2021).

In the rest of this chapter, we explore some of the ways in which political geographers are reconceptualising the scope of their field in response to the challenges of the Anthropocene. In the next section, we examine some of the new theoretical terms which are being used to conceptualise the liveliness of the Earth as a political actor, in contrast to twentieth century, Holocene conceptualisations of the Earth as a fairly static and predictable backdrop (with some exceptions) to human affairs. In subsequent sections, we examine critical geographical engagements with two emerging 'governing logics' of the Anthropocene, which we conceive of as observable political discourses and practices that have emerged as means of governing societies and their relations to an increasingly turbulent physical environment. Logics of 'geo-management' and 'resilience' are both explicit responses to the challenges of the Anthropocene, albeit based on very different ideas about human agency and security, their entanglement with nonhuman natures, and about the rightful place of science in making Anthropocene futures. As such, they exhibit very different geographies and have the potential to produce very different forms of spatial order.

Anthropocene assemblages

As mentioned briefly in the preceding chapter, one increasingly influential way in which geographers and researchers in related fields (such as International Relations) have been conceptualising nonhuman agency in politics is via so-called assemblage theory. This is a cluster of thinking inspired by thinkers such as Gilles Deleuze, Felix Guattari, Manuel De Landa, and Bruno Latour, which directs attention to the 'contingent, non-necessary coming together of various phenomena, actors, institutions, or processes to form a temporarily stable bundle of relationships and capacities' (Castree, Kitchin, and Rogers 2013, np.). Contrary to classic philosophical approaches which understand the world as comprised of discrete entities which have external relations with each other (following Cartesian philosophies) or, following Marx, as a totality of relations governed by defining logics (such as capitalism or class conflict), assemblage theorists 'see the world as a complex, changing, and emergent product of interactions that are only partly planned, ordered, or controlled' (Castree, Kitchin, and Rogers 2013).

The roots of assemblage thinking can be traced back to 1970s complexity theory, and thus share an origin story with the Earth System Science which, as discussed in Chapter 1, was so central to the emergence of the Anthropocene hypothesis (Steffen, Richardson, and Rockström 2020; Thrift 1999). Increasingly, as human geographers, sociologists, and other social scientists gaze at the human world, they see similar things to Earth System scientists gazing at planetary systems: complex relations between different entities, whose interactions can create novel, 'emergent' effects and nonlinear change which may well not be predictable. These modes of seeing and styles of thought are also helping scholars to look across the human-nonhuman divide, which has historically structured academic disciplines.

For example, assemblage theory has enabled political geographers to think in new ways about how earthly forces – like volcanoes, earthquakes, tropical storms, or wildfires – interact with and shape political relationships.[1]

Studies of volcanoes are particularly instructive (e.g. Bobbette 2023). Take the example of Montserrat, a relatively small Caribbean island that has been a British dependency, in one form or another, relatively consistently since 1632 (it was originally named by Christopher Columbus in 1493, after a monastery in the mountains above Barcelona). McConnell and Dittmer (2018) conceive of the polity of Montserrat (i.e. its people and government) as an assemblage, consisting of an ever-shifting set of relations between entities like the island's citizens, their economy, the physical environment, and sets of political ideas about sovereignty and independence from the UK. They argue that British Overseas Territories like Montserrat represent 'liminal' polities, existing in a strange equilibrium which is always poised somewhere between independence and dependence, but which is vulnerable to being shunted in one direction or another by the interactions of other elements of far-reaching geopolitical assemblages (Dittmer 2014).

In 1961, Montserrat had opted to remain a dependency of the UK, but subsequent economic successes meant that by the early 1990s, despite setbacks caused by Hurricane Hugo in 1989, Montserrat was increasingly seen as a 'viable' independent polity. However, from 1995 onwards, 'the dynamism of the earth and its systems … worked to undercut the perceived viability of Montserrat' (McConnell and Dittmer 2018, 151). That earthly dynamism was expressed through the Soufrière Hills volcano, which became active in 1995 and fully erupted in 1997. By that point, most of the southern two-thirds of the island had been evacuated, and the area remains an exclusion zone at the time of writing. Nineteen people were killed in the July 1997 eruption, the capital Plymouth was buried under ash, and more than half of the population went on to emigrate to neighbouring Caribbean islands or to the UK or the US. The devastating effects of the eruption led Montserratian politicians to reaffirm their dependency on the UK, and even to call for direct rule. Criticism was levelled at the slow, hesitant response of the UK Government, with various actors suggesting that a similar disaster within the British Isles would have generated a much more concerted and well-resourced response from London. For McConnell and Dittmer (2018, 152):

> The paradox here – of a colonial power reluctant to wield executive power, and of a colony reiterating its continued dependence – indicates how much the agency of the volcano has materially reshaped the processes by which political subjectivities were emergent in the Eastern Caribbean. In short, the 'viability' of Montserrat had been decimated, and the subsequent relations between the UK and Montserrat would come to be defined by long-term dependency.

Work like this which emphasises the agency of disruptive geophysical events within geopolitical assemblages (e.g. McGowran and Donovan 2021) shares some similarities with earlier observations made by hazard and disaster scholars about how so-called 'natural disasters' can create 'tipping points' within political systems, for example by giving marginalised communities an opportunity to mobilise around social and political grievances that may get magnified in the aftermath of a disaster, for example through inequalities in the distribution of disaster aid (Pelling and Dill 2010). Disasters have also been shown to enable powerful actors to further entrench their interests, for instance through inequitable urban redevelopment and gentrification under the guise of 'reconstruction', or through the imposition of punitive economic policies to pay off debts accrued during the recovery period (Rivera 2022). Interestingly, while hazard geographers like Pelling and Dill (2010) don't talk in terms of assemblages, they do use a related language of complex systems and 'tipping points' to describe the interaction of human and nonhuman entities in shaping political outcomes.

With all the talk of 'tipping points' in the Anthropocene, and of the potential of the Earth System to change quite rapidly into a new state, perhaps the 'Anthropocene event' we described in Chapter 1 is one of recognition of shared liminality in the planetary assemblage of human and more-than-human life. From the scale of anthropogenically altered Earth Systems, and down to the scales of local environmental crises – of floods, droughts, or storms – many human societies arguably exist in a new state of profound liminality:

> a state in which a person, place or thing is unsettled, teetering on a threshold with micro-scaled dynamics threatening to tip them/it into a different state of being.
>
> (McConnell and Dittmer 2018, 142)

Geophysical hazards like earthquakes and volcanoes might not seem obvious products of the anthropogenic environmental transformations that have motivated the heralding of the Anthropocene. Nonetheless, physical scientists are starting to piece together connections. For example, more and more data is being collected about 'induced earthquakes' – earthquakes set off by things like nuclear explosions, and extractive industries (Foulger et al. 2018). Relatedly, Liu, Linde, and Sacks (2009) report that typhoons travelling over Taiwan – themselves perhaps strengthened by anthropogenic climate change – might trigger small earthquakes underground, as the intense low barometric pressures of the typhoon systems can, they contend, be sufficient to allow crustal tension building up in faults deep below the island to be released. Observations and models of volcanic systems have also suggested that intense precipitation might trigger 'seismic events' on the surface of volcanoes, which could, for example accelerate processes of dome collapse and eruption, as observed in Montserrat (Farquharson and Amelung 2020; Matthews, Barclay, and Johnstone 2009). And previous periods of climatic warming, for example at the dawn of the Holocene, seemingly saw increased levels of volcanic and seismic activity, as it is thought that the melting of huge ice sheets released seismic forces that had been building up over millennia. While earthquakes and volcanoes continue to be largely controlled by geological processes, it is increasingly recognised that they, too exist in a kind of 'liminal' state, with complex relations to the changing physical attributes of the land surface and atmosphere, which can help push systems into new states (Fan et al. 2019; McGuire 2024).

Volcanic eruptions triggered by rain, and earthquakes set off by melting ice, are further illustrations of the interlinkages and connections between components of the Earth System, and of the potential for human influence to cascade through them. They might be interpreted as further support for 'post-nature' arguments (see Chapter 1) that there now exists no biophysical realm that is untouched by human influence; nature-culture hybrids and 'assemblages' proliferate, even reaching into the faults and magma chambers of the geosphere. Yet these examples perhaps also add credence to the warnings of geographer Nigel Clark, who contends that such 'end of nature' arguments retain a problematic anthropocentrism which denies the existence of a 'functionally intact nature enduring beyond' the sphere of these 'new hybridised environments'. This is a denial of a 'geo-physical materiality with an autonomy and integrity of its own' (Clark 2011, 11). Scientists' cautious, hesitant, and uncertain ascriptions of human agency in helping to set-off volcanoes and earthquakes emphasise how, in many respects, the agency of humans in shaping environmental processes is still dwarfed by the powers of the Earth itself (Clark and Szerszynski 2020). Viewed at a planetary scale – taking into account entities like the Earth's molten core and its upper atmosphere – 'human' and 'natural' agency still exist in a very asymmetrical relationship. For thinkers like Clark and Szerszynski, we must practice humility in the face of these dynamic earthly forces, and recognise the planet's capacity for profound rearrangement of its subsystems in ways which may, or may not, be amenable to continued human flourishing (Clark and Szerszynski 2021).

Nonetheless, if the material conditions of the Anthropocene are likely to increase the severity and frequency of what we might now term post-natural hazards (cf. McGowran and Donovan 2021), then we might expect the coming decades to be a period of significant geopolitical change, as material agencies transform geopolitical assemblages, creating new political crises, new geopolitical relationships and conflicts, and even new political identities and subjectivities (e.g. Farbotko, Stratford, and Lazrus 2016). The question for us, then, is how are human societies, in different places and operating at different scales, trying to govern these dynamic assemblages? How are the challenges of asymmetrical forms of human and nonhuman agency, of interconnection, emergence, and surprise, reshaping the conduct of politics? How are the spatialities of political power being remade by these assemblages? Is power being wielded in the spirit of humility called for by the likes of Nigel Clark, or in the continuation of a twentieth-century spirit of geopolitical hubris?

In the following sections, we analyse two emergent 'governing logics' of the Anthropocene, and their political geographies: *geo-management* and *resilience*. We suggest that while both are united by a concern with security amid collapsing boundaries – between nature and culture, science and politics, and between the 'insides' and 'outsides' of modernity – they differ in their conceptualisation of the Earth System and its amenability to control. As such, the logics of *geo-management* and *resilience* represent different 'geopolitical formations'; that is to say, different ways of conceiving the relationship between human and nonhuman natures, between science and politics, and between space and power.[2] The two logics intersect in different ways with the narratives of the Anthropocene we explored in Chapter 1, and in many respects are in competition to become a defining logic of Anthropocene political geographies. Their relative success in this regard will be hugely consequential for how communities across spatial scales feel the effects of the Anthropocene and navigate a path through it.

Geo-management for 'whole-earth security'

In his book *Rethinking Environmental Security*, Canadian political geographer Simon Dalby (2022) examines how the onset of the Anthropocene challenges conventional notions of 'security' that have shaped geopolitics and international relations for decades. The desire for security has conventionally led states to use large amounts of firepower to dominate other states and environments, culminating in the development of nuclear weapons and all the security dilemmas and existential angst they brought with them. Now, 'firepower' in its broadest sense – the power to manipulate and control fire to create energy, usually from fossil fuels – is creating a new existential crisis for humanity and the living world. As such, *security* needs to be fundamentally rethought along ecological lines, with the security of ecosystems and their life-supporting functions elevated to the highest levels of diplomatic wrangling and state strategy.

Thinking about environmental security on a planetary scale is not altogether new; Dalby cites papers from the 1980s which argued that 'the planetary system itself was what needed to be secured' (Dalby 2022, 134), in contrast to traditional formulations of geopolitics that focused on state strategies of territorial containment and control. In subsequent years, interest in 'geopolitics' and 'ecopolitics' went in separate directions, but more recently, environmental concerns have risen back up the agenda of academic geopolitics. Gerard Toal has recently revisited Halford Mackinder to theorise how geopolitics drives the climate crisis, citing inter-state economic rivalry, territorial expansionism, and competing technological ambitions as being anathema to effective global decarbonisation efforts (Toal 2024).[3] Meanwhile, Gerry Kearns has argued that scholars of geopolitics need to pay more attention to Marxist arguments (and what we're calling narratives; Chapter 1) about the dynamics of capitalism being the root of the ecological crisis, and therefore being central (but so far largely absent from) attempts to characterise the geopolitics of the Anthropocene (Kearns 2025). Kearns' arguments align with Dalby's insistence that thinking geopolitics simply in terms of state interests and security strategies is no longer tenable – now, global supply chains, energy infrastructures and circuits of capital are equally significant units of geopolitical analysis, entities whose 'security' is as much a function of ecosystem and Earth System changes as it is about military might and state strategy (Dalby 2020a).

Tuning the geo-engine

For Toal, geopolitics has always, at its root, been about different groups trying to secure their 'living space'. With the heralding of the Anthropocene that threatened 'living space' can now be regarded as the whole planet, and the 'security' and stability of that living space might demand radical new geopolitical strategies. One set of strategies aiming for 'whole-earth security' (Dalby 2022), and which is increasingly attracting the attention of national governments and international organisations, is 'geoengineering' (Fleming 2012; Kornbech, Corry, and McLaren. 2024). As the originator of the term 'Anthropocene' put it:

> A daunting task lies ahead for scientists and engineers to guide society towards environmentally sustainable management during the era of the Anthropocene. This will require appropriate human behaviour at all scales, and may well involve internationally accepted, large-scale geo-engineering projects, for instance to 'optimize' climate.
>
> (Crutzen 2002, 23)

This quote, which you may recall from Chapter 1, is exemplary of a stance towards the challenges of the Anthropocene that emphasises the need for technical experts to 'guide' and 'manage' society and its relations to the physical environment. Often reproduced, it is perhaps a founding statement of an emerging Anthropocene logic of 'geo-management'.

By 'geo-management', we mean an emerging and increasingly influential approach to governing human-environment relations in the Anthropocene which involves deliberate intervention in the physical workings of the Earth System. As aspects of the Earth System become increasingly lively participants in global politics (owing to previous and ongoing human interference with them), several scientists, experts, policymakers, and activists have proposed that we need to further manipulate those systems, to bring them back from the brink of 'tipping points' or 'planetary boundaries'. In the case of climate change, technoscientific means of doing so are often bracketed together as 'geoengineering', but can be more helpfully divided into two families: (1) technologies to re-engineer the global carbon cycle by pulling carbon dioxide out of the air and sequestering it elsewhere; and (2) technologies to manipulate the energy balance of the atmosphere, thus offsetting the global warming caused by greenhouse gas emissions (Buck 2019). The former is certainly not without its geopolitical controversies, particularly in relation to, for example, efforts to develop fast-growing energy crops in parts of the formerly colonised world to feed the energy requirements and offset the carbon emissions of former imperial powers (Hickel and Slamersak 2022); or to fertilise international ocean waters with iron to increase CO_2 sequestration (Siegert et al. 2025). However, in the section below, we focus on the latter – technologies of 'solar radiation management' (SRM) – as they are arguably the most expressive of the logic of geo-management.

SRM involves strategies to manipulate the radiative balance of the Earth and atmosphere, in order to offset the warming effect of increased atmospheric greenhouse gas concentrations (Figure 8.1). Techniques range from the mundane to the Promethean. Changing the albedo – or reflectiveness – of human settlements (e.g. by repainting roads and roofs) can reduce the amount of heat that is absorbed. At the other end of the spectrum, mirrors arranged carefully in outer space, in between the Earth and the sun, could help keep the planet cool. A popular and commonly discussed set of proposals involves injecting sulphur aerosols into the stratosphere, mimicking the cooling effect of atmospheric aerosols released by large volcanic eruptions.

The quotation on page *259* is illustrative of how, in the early days of Anthropocene discourse, geoengineering and SRM loomed large in emerging

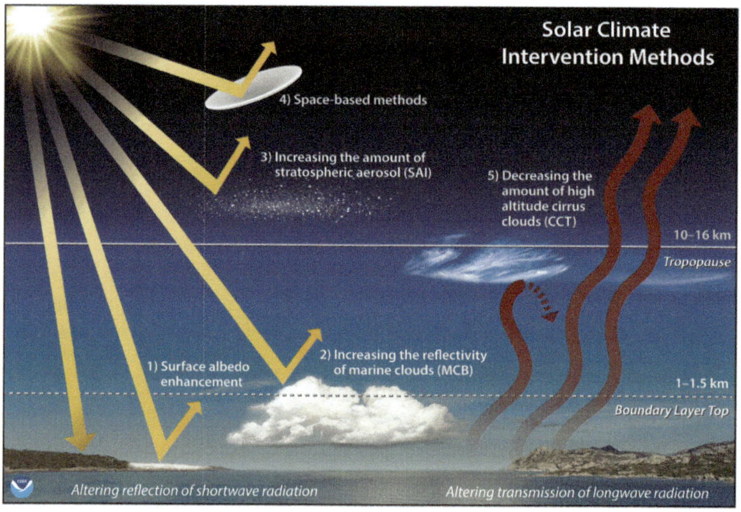

Figure 8.1 Illustration of different SRM techniques. (Source: US National Oceanographic and Atmospheric Administration, via Wikimedia Commons.)

narratives of Earth System perturbation and of the need to harness science and technology to put things right (Baskin 2019). In the 'standard narrative' of the Anthropocene, scientists often appear as heroic protagonists warning society of the dangers, while also being the source of high-tech solutions (Moore 2025).

> Given the grossly disappointing international political response to the required greenhouse gas emissions, and further considering some drastic results of recent studies…research on the feasibility and environmental consequences of climate engineering…which might need to be deployed in future, should not be tabooed.
>
> (Crutzen 2006, 214)

Yet despite the emergence of this narrative in the early 2000s, the notion that humanity might start to *deliberately* intervene in the climate system remained taboo, and highly unattractive to many climate scientists and policymakers. The quote above is from a 2006 editorial essay Paul Crutzen penned for the influential journal *Climatic Change*, which analysts of geoengineering discourse have pinpointed as the moment the 'taboo' around solar geoengineering began to be lifted.

In his book *Geoengineering, the Anthropocene and the End of the Nature,* South African STS scholar Jeremy Baskin (2019) outlines how, at the time of writing at least, geoengineering continues to be a highly controversial subject, with many social scientists in particular warning that we shouldn't even be talking about it, let alone conducting research and experiments (Biermann, Oomen, and Gupta 2022). Most of the climate science community is wary of it, and most assessments of SRM's potential by authoritative institutions conclude by calling for more research before deployment could even begin to be considered. This is because the potential risks of SRM are profound. While techniques are designed around the goal of lowering average global temperature, uncertainties abound about the effects of something like Stratospheric Aerosol Injection (SAI) would have on key weather systems, like the South Asian monsoon (Ricke et al. 2023). And while deployment of something like SAI might create unquantifiable climatic risks, the 'termination effect' of halting SAI could be devastating too – models show that climate systems would quickly revert to how they would have been without SAI, with rapid changes of temperature and precipitation creating serious adaptation challenges for societies and ecosystems (Bhowmick et al. 2021).

Nonetheless, as SRM has moved up the political agenda, political geographers have begun to analyse the geopolitical implications of both the discourse and potential deployment of SRM technologies. Geopolitically powerful states are prominent actors in geoengineering discourse. The UK recently became the biggest funder of geoengineering research and experimentation, supporting experiments in cloud seeding and investigating what volcanoes like Montserrat's might tell us about adjusting the atmosphere's radiation balance (O'Callaghan 2025). Much official interest in SRM in the United States has been motivated by geopolitical concerns about whether other countries might unilaterally pursue SRM strategies, whether such moves would be detectable, and how they might impact the climatic security of other nations (Schubert 2021). But the spectre of geoengineering raises wider geopolitical questions about the potential for international conflict and cooperation (Dalby 2015). How would countries decide which global temperature was desirable one? How would disagreement be managed, or unilateral action dealt with or punished? How would countries collectively guarantee the maintenance of any intervention in the Earth System, to avoid negative 'termination effects'? Scholars have wondered whether geoengineering has a basic incompatibility with democracy (Horton, Reynolds, and Buck 2018; Szerszynski et al. 2013), and whether it would simply serve to further concentrate power in the hands of already powerful states and institutions, who would now have a hand on the 'global thermostat' (Cairns and Stirling 2014; Hulme 2014; Surprise 2019). Proponents have conversely wondered whether geoengineering could, on the contrary, be an effective way for richer countries acting to protect poorer countries from the more extreme effects of climate change, and of 'buying time' while mitigation and adaptation measures are put in place (Keith 2013). In terms of assemblage thinking, geoengineering figures here as a technology of liminality, concerned with delaying transitions from one climatic regime to another, while holding open the door to other solutions to climate change mitigation and adaptation that might emerge in the meantime.

Geographer Kathryn Yusoff points out that geoengineering, and particularly SRM, involves a profound remaking of the conceptual architecture of geopolitics, most notably the *geo* epithet. Geoengineering 'involves the conceptualisation of an abstract, nonlinear biodynamic model of the earth in which atmospheres and earth systems … are imagined as 'engines' open to forms of making'. These engines are figured as sufficiently predictable to be amenable to deliberate control (see also Zee 2020). This represents a profound conceptual shift 'from the earth as a given' – as was common in geopolitics until recently – 'to the earth as characterised by ideal forms of output, such as climate' (Yusoff 2013, 2806). As such, geoengineering 'represents a distinctly new type of geopolitical formation, informed by a reconceptualization of geophysical processes rather than distinct geographical territories' (Yusoff 2013, 2800). That is to say, geoengineering could remake geopolitics, so it is less about (the study of) competition between states to control territory, and instead (or at least also) about struggles to control elements of the Earth System. The *geo* of geopolitics shifts from meaning the Earth's surface, to denoting Earth System components and their complex interactions (Clark 2013). These components and interactions necessarily cross-cut different scales of space and time, as well as existing borders between national territories and 'common resources' such as the high seas and polar ice caps. They illustrate the sort of 'thick' geographies we discussed in Chapter 2.

Nonetheless, work on the geopolitics of geoengineering has emphasised that emerging struggles to 'control' elements of the Earth System may play out through some familiar geopolitical images of state rivalry, international hierarchies, and global inequalities. SRM proponents sometimes present the idea as part of what Baskin (2019) calls a 'new climate realism', which emphasises the doomed complexity of multi-lateral climate change agreements and which favours actions that can be taken by a small number of technologically-advanced states on behalf of the rest of the world (e.g. Victor 2011). Critics in civil society and academia – including many in Geography – have used the language of colonialism and imperialism to criticise this logic, which variously positions some countries as the key knowledge-brokers and decision-makers who will manage the atmosphere on an assumed trusteeship basis for the rest of the planet (Group 2017), or which positions low-cost SRM options as a market-friendly way of allowing capitalist business-as-usual while masking some of the effects of continued greenhouse gas emissions (Surprise 2019).

Box 8.1 War-gaming geoengineering

Scenarios or stories abound of possible futures marked by the deployment of geoengineering technologies. These scenarios play an important role in shaping concerns and expectations about emerging technologies. For instance, concerns about scenarios of unilateral deployment of SRM technologies have seemingly motivated much US military interest in geoengineering (Baskin 2019). Yet these scenarios rarely arise from quantitative modelling, but rather from qualitative, speculative accounts of possible technological and diplomatic developments.

For example, US Government security analyst William Morrisey outlines four scenarios of SRM deployment, which each raise the spectre of different forms of geopolitical conflict:

1. A powerful state (such as the United States) deploys geoengineering unilaterally. That state then uses its control over the global climate, and the risk of termination effects, as leverage in other geopolitical struggles.
2. A powerful state deploys unilaterally, but this time other states engage in counter-geoengineering measures, perhaps motivated by a 'degree of asymmetry in temperature preferences between countries' (Heyen, Horton, and Moreno-Cruz 2019, 169). Climatological

outcomes that are unfavourable for certain countries could see SRM activities deemed a deliberately hostile act by one country towards another.

3 A global SRM regime is established, perhaps as in scenario 1, but unlike scenario 2, there is no capacity for counter-deployment. Instead, weaker states accelerate local weather-modification efforts to try to offset negative local effects of either the original climate change or of the SRM. Unintended cross-border effects (such as changing rainfall patterns) could ensue, prompting hostility between states.
4 Experimentation in SRM leads to technological innovation and a greater confidence in the 'controllability' of the atmosphere, leading states to make more focused, targeted interventions. As such, 'regional hegemons begin engaging in limited solar geoengineering efforts intended to elicit specific climatological modifications [such as suppressing tropical cyclones] intended to benefit themselves and allied regional partners'. Regions like the Middle East could feasibly see competing geoengineering regimes, with similar cross-border tensions as scenario 3 if shifting weather patterns begin to affect rainfall and agricultural productivity (Morrissey 2024, 307).

While critical political geographers may be sceptical of the straightforward links assumed here between environmental conditions and geopolitical conflict, the emergence of scenarios like this (and the serious role they seem to be playing in state considerations of future security threats), bears out the arguments presented above about geoengineering representing a new kind of 'geopolitical formation', where control over elements of the Earth System – in the form of global and regional temperatures and hydrometeorological systems – becomes a new object of geopolitical calculation and intrigue. Geoengineering here figures as a liminal technology which may be capable of pushing geopolitical assemblages from a condition of relative equilibrium into one of hostility and conflict, in the process perhaps even becoming a means and target of military conflict itself (Sovacool, Baum, and Low 2023).

In many ways, geoengineering is being thought about in similar ways to how nuclear weapons were conceived in the second half of the twentieth century. Both are technologies which reveal 'a belief in humankind's right to exercise total mastery over nature' (Hamilton 2014, 17), and which have ushered in new kinds of strategic thinking about risk, geopolitics, and conflict on a planetary scale (Furuhata 2022). If nuclear weapons were one of the defining 'triggers' of the Anthropocene (see Chapter 5), then perhaps solar geoengineering will become a defining technology of Anthropocene geopolitics (Dalby 2015).

Much security and geoengineering discourse – from calls for experimentation to scenarios of conflictual deployment (see Box 8.1) – represent 'the world imagined from the standpoint of the geopolitically powerful' (Baskin 2019, 135). Political scientists Frank Biermann and Ina Möller (2019) show how, despite the significant consequences and risks that SRM would imply for developing countries, research and policy debate around geoengineering has been notably dominated by institutions and actors from the global North (also Sovacool et al. 2024). While Horton and Keith (2016) offer distributional justice arguments (see Chapter 6) in favour of SRM to mitigate the suffering of poorer countries under climate change, Hourdequin (2020) counters that procedural and recognitional justice demands that voices from the Global South be given far more space and prominence in geoengineering research and debates.

It is significant then that in the sphere of international diplomacy, country representatives from the Global South (or Majority World) have been increasingly vocal in opposing geoengineering experimentation. The 6th United Nations Environment Assembly, held in Nairobi in February 2024, saw a proposal put forward by Switzerland that the UN Environment Program should establish an expert group to assess emerging knowledge about geoengineering options. Highlighting the fact that scientific assessments are rarely neutral exercises but can instead be lightning rods for environmental politics (Castree, Bellamy, and Osaka 2021), vigorous arguments ensued. On the one hand, the United States – with frequent support from Saudi Arabia – argued that scientific research should be allowed to run its course first, and that the Swiss proposal risked putting 'policy' before 'science'. On the other hand, a group of African countries led by diplomats from Djibouti argued strongly that considerations of ethics, governance, and power should be at the forefront of geoengineering assessments, and that deliberations on geoengineering's potential to ameliorate climate change should happen in the political arena, not in the narrow domain of (largely Northern) technical experts. The debate not only revolved around disagreements over the distinction and proper relations between 'science' and 'policy', but also pointed to how issues like SRM transcend the conventional boundaries of 'environmental' policymaking:

> Potential scenarios of unilateral development and control of geoengineering technologies by a few powerful countries suggest that solar geoengineering surpasses traditional "environmental" policies. Instead, solar geoengineering implicates unequal and shifting global power dynamics, mitigation delay for the wealthy, geopolitical conflicts and, for some, a novel threat of neocolonial imposition.
>
> (Biermann and Gupta 2024, 3)

Practical efforts to enact 'open-air' experiments in SRM have also brought forth conflicts between scientists, state actors, and Indigenous communities, and have highlighted conflicting conceptualisations of proper human-environment relationships (Schubert 2021). In February 2021, it became known to members of the Indigenous Sámi community in northern Sweden that researchers from Harvard University planned to test an aerosol deployment mechanism in the skies above their homeland. The Sámi Council sent objections to Harvard, and the test was suspended. The researchers tried to develop new ways of consulting with communities like the Sámi, but for the latter, it was too little, too late. The plans to conduct research on and above their lands without consultation simply re-enacted centuries-long colonial practices (Bronen and Cochran 2021; Whyte 2018), while the whole notion of geoengineering was anathema to Sámi ways of thinking about human-nature relations. As Sámi Council vice president Åsa Larsson Blind put it:

> This goes against our worldview that we as humans should live and adapt to nature… The way of thinking that humans are entitled to change and manipulate our surroundings has actually brought us into the climate crisis in the first place.
>
> (quoted in Dunleavy 2021, np.)

Other Indigenous groups rallied to the cause (Oksanen 2023), and eventually the entire research project was shut down, with Harvard researchers promising to focus less on open-air experiments and more on the ethical and legal aspects of geoengineering research and deployment. Among critical social scientists, attention has increasingly turned towards thinking about both carbon dioxide removal and SRM proposals through the lens of planetary justice (Sovacool, Baum, and Low 2022), while others have argued forcefully that global bodies should develop an international non-use agreement, on the basis both of geoengineering's seeming incompatibility with democratic norms of governance, and its potential to distract from the urgent task of reducing emissions (Biermann, Oomen, and Gupta 2022).

So, despite increasing scientific research on the topic, for many social scientists, human geographers, activists, and diplomats, geoengineering remains something of a taboo subject, and the logic of geo-management is yet to be fully embedded in national and international institutions. Might that change? Advocates argue that if climate change is not halted, geoengineering might become increasingly attractive and important as a policy leaver and that societies should prepare for that eventuality by conducting further research and small-scale experiments so that SRM interventions can be planned and their outcomes predicted.[4] Others suggest that the cascading impacts of climate change (such as rising food prices, deepening inequalities, and perhaps mass migration) might push some societies towards forms of authoritarian nationalist politics, under which unilateral geoengineering might become a favourable option (Morrissey 2024). Jeremy Baskin warns that acceptance of the concept of the Anthropocene itself might be what it takes for geoengineering to finally become 'normalized' as a policy option. In suggesting the end of a completely autonomous or external 'nature', the hierarchical domination of humanity over the nonhuman world, the normalisation of the exceptional recent history of human-environment relations, and in seemingly affirming the responsibility of humanity to better steward and control the environment, the Anthropocene provides 'fertile ground' for the normalisation of geoengineering (see also Collomb 2024; Moore 2025). Baskin argues that 'the "Anthropocene" naturalises a new ordering of the world in which [solar geoengineering] might be understood as natural, inevitable, legitimate and unavoidable' (Baskin 2019, 234). If geoengineering becomes unmoored from the science and politics of climate change, and gets attached instead to a 'positive vision of a good and modern "Anthropocenic" world', then some of the cultural and political hurdles to deployment might be overcome.

It is certainly true that particular narratives of the Anthropocene have helped geoengineering rise up the agenda. Some (but certainly not all) proponents of the 'standard narrative', such as Paul Crutzen, have positioned geoengineering as part of humanity's new geo-managerial responsibilities in this new epoch. For Harvard geoengineering researcher David Keith, 'post-nature' versions of the Anthropocene offer equally compelling justifications for humanity taking ever-greater control of its physical environment (Keith 2013); humanity has always built and engineered its 'ecological niches', so SRM is just the next logical step. But there are many other narratives and versions of the Anthropocene, which make quite different descriptive and normative claims about the world (see Chapter 1). And these are reflected in efforts to, for example, imagine a world where certain geoengineering technologies are deployed in ways that are democratic, decentralised, and placed in the service of social justice (Buck 2019; Parenti 2020). Can we imagine an explicitly decolonial form of geoengineering, for example, which seeks to protect and enhance the sovereignty of Indigenous and colonised peoples (Fish 2023)?[5] Could geoengineering, if deployed and governed in such ways, even contribute to the project of creating a 'charming Anthropocene' (Buck 2015)?

These are the kinds of questions we'll return to in Chapter 10. For the rest of this chapter, we investigate another emerging governing logic of the Anthropocene, which differs in some important ways from the 'geo-management' logics embedded in geoengineering proposals.

Resilience

If liminality is 'a characteristic of any assemblage that has the potential for rapid transformation', then its opposite is arguably 'resilience' (McConnell and Dittmer 2018, 154). 'Resilience' has become the 'go-to crisis-governance concept' in domains as diverse as climate policy, finance, economic development, and disaster management (Davies and Arrieta 2024, 1). Yet it is a term which, like *assemblage*, has its origins in some rather arcane debates in ecology. In this section, we briefly trace the trajectory and use of resilience from its ecological origins to its role as a prominent governing logic in the Anthropocene, stopping along the way to explore how geographers have used, critiqued, and adapted the term to describe the evolving political geographies of life in the Anthropocene.

The concept of resilience first emerged in the 1970s to describe the ways in which ecosystems could absorb changes from the outside, while remaining cohesive. The classic paper by ecologist C.S. Holling (Holling 1973) drew from the same well of 'complexity science' as the emerging Earth System Science, which would eventually birth the notion of the Anthropocene (see Chapter 1), and the concepts have since followed entangled paths (Wakefield et al. 2020).[6] Holling's paper contended that existing understandings of ecological systems were too mechanistic, and that they propped up management ideas like 'Maximum Sustained Yield' which, in fields like forestry and fisheries, maintained that large and predictable amounts of 'surplus' could be taken from a system without affecting its basic functioning. For Holling:

> Acting on assumptions of a permanent resource yield may undermine the complex interdependencies comprising the resilience of the system as a whole, rendering its organisation ever more fragile and vulnerable to collapse … Holling's perspective reflects emerging critical voices which, in the early 1970s, insisted that intensive maximisation of agricultural and industrial production would at some point meet inherent limits, resulting in mass extinctions and intolerable over-pollution.
>
> (Walker and Cooper 2018, 91)

As Holling put it, 'resilience is concerned with probabilities of extinction' (Holling 1973, 20). His arguments were part of the pushback against the rapid economic expansions which characterised the Great Acceleration (see Chapter 5), particularly the intensification of agriculture, industrial fishing, and forestry. Holling's warnings about how intensified production processes were damaging the resilience of ecosystems, and making them ever more fragile and prone to sudden collapse, were influential in emerging ideas and narratives of a coming eco-catastrophe, which have more recently motivated waves of activism in the name of reducing the 'probabilities of extinction' (see Chapter 1). Nonetheless, Holling's notion of resilience as a quite radical conceptual critique of the ecological metabolism of industrial modernity soon transmuted into something else, as the concept travelled from ecology into other spheres of thought and action.

Also latent in Holling's text was a prescription for a form of 'crisis management' which would abandon pretences of being able to predict the future, and which would instead be oriented around 'sustaining productivity even under conditions of extreme instability' (Walker and Cooper 2018, 91). The ongoing productivity of systems depended not on stocks of components like fish or trees, but on stocks of a systemic capacity – *resilience*. As users of the term – Holling included – began to apply it not just to ecological systems, but to coupled socio-ecological systems (Gunderson and Holling 2002) and even to more obviously 'human' systems like businesses and markets, it evolved into an 'approach to risk management which foregrounds the limits to predictive knowledge and preventative action, insisting on the prevalence of the unexpected', and on the importance of cultivating a stance of absorbing and accommodating surprises. 'Resilience' has thus become an 'alternative theory of growth in far-from equilibrium conditions, in which the capacity of systems to spontaneously re-organise through catastrophic events is denoted as 'capital'' (Walker and Cooper 2018, 92; also Dalby 2020b).

Walker and Cooper (2018) attribute the subsequent popularity of the concept in governmental and business circles to its resonance with the ideas of one of the architects of economic 'neoliberalism', Friedrich Hayek. He depicted markets as complex, spontaneous, and largely unpredictable forms of socioeconomic order which were best left to their own devices rather than be made subject to governmental oversight or to regulation in the service of social objectives. In 1973, the year Holling's influential article was published, a global oil price shock heralded the end of the period of large-scale state direction of Western economies – the 'Anthropocene mode of production' which, we argue in Chapter 5, was so crucial to the trajectory of the Great Acceleration. The end of the decade saw the election of Ronald Reagan to the US Presidency and Margaret Thatcher's Conservatives in the UK – two radical politicians who were deeply influenced (directly or indirectly) by the ideas of thinkers like Hayek, and who would help promote varieties of 'neoliberalism' as a new governing logic and mode of regulation.

Much has been written by critical geographers and others on the impacts of neoliberalism on environmental politics, particularly in relation to policies such as marketisation and privatisation (see e.g. Apostolopoulou et al. 2021; Castree 2010). Geographers concerned with the effects of neoliberalism on the environment were quick to point out how the notion of 'resilience' seemed to sit very easily with a style of politics that emphasised the reduction of state welfare functions, individualisation, and the prioritisation of the health of markets over the wellbeing of societies and biophysical environments. As resilience thinking made its way into fields like disaster management, it drew critique for seeming to provide a quasi-scientific smokescreen for the retreat of the state from meeting basic welfare needs in the aftermath of disasters and socioecological disruptions (MacKinnon and Driscoll Derickson 2013). As resilience moved from being an observable capacity of ecological systems to being a quality that seemingly needed to be cultivated within social systems, it met with resistance from scholars and activists, who viewed resilience initiatives as facilitating:

> the spread and consolidation of neoliberal socio-ecological relations – that is, they normalize vulnerability, seek to make individuals responsible for

managing the effects of social and ecological insecurities and disruptions, and thus provide discursive justification for further cuts to social welfare provisioning.

(Grove 2018, 6)

Box 8.2 Situating the Anthropocene: New Orleans, the resilient city?

> New Orleans has unique experience in dealing with and recovering from major urban emergencies. Hurricane Katrina, Hurricane Isaac, the Deepwater Horizon Oil Spill, the city's frequent boil water advisories – New Orleans has learned important lessons about what it takes to become a vibrant, resilient city that serves all its residents, particularly the most vulnerable.
>
> (Resilient Cities Network 2026)

Despite this notable recent history of 'major urban emergencies', not everyone in New Orleans accepts the label 'resilient'. In 2015, posters started appearing around the city (Figure 8.2) in response to the city's new 'resilience strategy'. The posters quoted Tracie Washington of the Louisiana Justice Institute, who objected to how victims of Hurricane Katrina (2005) and the Deepwater Horizon oil spill (2010) were constantly being labelled as 'resilient'.

This labelling of populations as resilient implied, for urban geographer Maria Kaika,

> that they can take more suffering, deprivation or environmental degradation in the future. If we took [Washington's] statement seriously, we'd need to focus instead on identifying the actors and processes that produce the need to build resilience in the first place. And we would try to change these factors instead.
>
> (Kaika 2017, 95)

The city's new resilience strategy formed part of a broader move to resilience thinking in the region, most notably in the context of coastal protection and restoration in response to the mounting threats of sea-level rise, land subsidence, wetland destruction, and extreme weather. Between 2013 and 2019, New Orleans featured prominently in the Rockefeller Foundation's Resilient Cities Network, with money funnelled into resilience projects and the city championed as a leading 'laboratory' for urban resilience. While the city initially joined the

Network with a holistic view of infrastructural and social resilience, the following years saw efforts increasingly focus on funding major infrastructural projects, rather than addressing issues of social inequality (Naef 2022). In 2017, the Louisiana Coastal Protection and Restoration Authority (CPRA) published a resilience master plan to be funded in large part by the same industries that were causing the environmental problems. While this might seem like a sensible mobilisation of the 'polluter pays' principle, critics have argued that the focus on securing resilience to external shocks distracts governmental attention from the underlying causes of those shocks, such as oil and gas exploration and drilling, and industrial fishing. Max Woods therefore asks whether, 'under the rubric of resiliency, do oil and gas companies still reap the benefits of environmental exploitation even when they are forced to fund restoration projects'? Paraphrasing Tracie Washington, he contends that 'Rather than challenging environmentally destructive practices and industrial operations, policymakers opt to prepare Louisiana residents for the inevitable, to make them resilient so that these industries "can do something else to them"' (Woods 2017, np).

This sense of 'resilience' increasingly being a concept and tool of the powerful is reflected in interviews conducted by urban geographer Patrick Naef. For example, a charity worker in the Lower Ninth Ward, the area of the city most devastated by Katrina, asked forcefully:

> Instead of asking the question [of] why are we demanding that people be resilient and what are the circumstances in which they have to be resilient … Why don't we just fix that shit so that nobody has to be resilient? Congratulating people for their resilience, there's something so patronising about it.
>
> (quoted in Naef 2022, 5)

The example of New Orleans highlights a point made by Simon and Randalls (2016), that we need to see resilience as fundamentally 'multiple' – the concept will mean different things in different places, and what is done in its name will be largely shaped by local circumstances. The resistance to the notion in New Orleans is tied to long histories of racial segregation, inequality, and violence, and to the stark socioeconomic iniquities in disaster impacts and responses that Hurricane Katrina laid so bare. The reduction of official uses of the term to hard infrastructure projects perhaps represents a wish to side-step the difficult politics of urban inequalities, and to focus instead on what may, to engineers at least, appear more tractable problems – measuring rising waters and designing new flood defences.

Figure 8.2 The New Orleans poster reproduced as a mural in the Lower Shankhill area of Belfast, Northern Ireland, where communities similarly wish for a more a collaborative approach to solving social problems. Photo by virtual Belfast mural tour. 'Stop calling me resilient – virtual Belfast mural tour' (https://www.virtualbelfastmuraltour.com/stop-calling-me-resilient-mural.htm).

Urban geographer and resilience theorist Kevin Grove is partially sympathetic to these critiques of resilience being fundamentally neoliberal, but suggests that in the notion's growing popularity in fields like urban policy, something else is going on beyond the smuggling-in of neoliberal norms and policies under the guise of resilience. He identifies resilience as 'a distinct *style of thought* about social and environmental change, a certain way of posing problems and devising solutions to these problems' which he calls 'a will to design' (Grove 2018, 8). He points out that much of the history of things like urban planning and disaster management has revolved around what Michel Foucault dubbed a *will to truth*: 'the notion that truth should be objective and lead to reliable prediction, unsullied by subjective values and tastes; predictive knowledge will enable humans to bend nature to their will' (Grove 2018, 4). This notion of scientific knowledge enabling humans 'to bend nature to their will' is evident in much geo-management and pro-geoengineering discourse. By contrast, resilience's *will to design* is based on the premise that knowledge of physical and social processes will always be partial and incomplete, that the world is 'too complex for an individual to know, understand and manage' (Grove 2018, 13), and that different forms of knowledge need to be brought together and synthesised in order to develop practical solutions to immediate problems, which must then be monitored and perhaps adapted over time as conditions and requirements change. As such:

> Resilience thinking's underlying will to design does not displace a will to truth, but rather mobilizes truth towards a new end: neither prediction nor total control, but rather designing solutions that are necessarily partial and limited, and that will have to be revisited and revised as new problems arise.
>
> (Grove 2018, 22)

In contrast to the hubris of command-and-control approaches to socio-environmental management, resilience thinking offers 'modesty and pragmatism' (Grove 2018) and a more open-ended, 'experimental' approach to policymaking (Wakefield 2020). On this reading, the rise of resilience thinking should not be attributed solely to its apparent 'fit' with neoliberal political economy, but rather to a growing erosion of some of the basic assumptions about modernity and security, particularly in the western world:

> [Resilience] began to gain prominence within policymaking circles during the late 1990s and early 2000s, as scholars and practitioners alike grappled with a series of social, geopolitical, technical and political economic events that exceeded modernist technologies of security premised on boundaries, prediction, stability, linear temporality and control.
>
> (Wakefield et al. 2020, 2)

Just as the heralding of the Anthropocene rendered assumptions about the seemingly 'predictable environment' of the Holocene untenable, so did these realisations begin to undercut assumptions about 'the stable, predictable world of European modernity' (Wakefield et al. 2020, 3). Financial crises, global terrorism and climatic shocks began to rattle the cage of modernity, Wakefield and colleagues suggest, forcing realisation of 'humanity's embeddedness within complex social, environmental and technical systems' that are deeply complex, defined by surprising, 'emergent' phenomena, prone to cascading instabilities, and which cannot be contained within either conventional metaphysical borders (like 'society' and 'environment') or within the political borders of modern nation-states. Therefore, 'at the moment the Anthropocene annihilates modernity's metaphysical fantasies of security-as-stasis, resilience arrives to reconstitute security as a problem of affirming rather than rejecting worldly connectivity and emergence' (Wakefield et al. 2020, 3).

So, for Wakefield and colleagues, the prominence of resilience thinking is not reducible to shifting economic orthodoxies (such as neoliberalism; see also Anderson 2015) but is related to a profound unsettling of the philosophical foundations and governing logics of modernity (see also Pickering 2025). The Anthropocene, on this reading, represents the end of modernity, and 'resilience' is the new way to think about governing life in this 'post-modern' age.

Resilient cities: Anthropocene urbanism

But what gets done in the name of 'resilience'? Cities have become key sites for this new 'experimental' ethos of urban administration, in what Kate Derickson (2018) terms a new form of 'Anthropocene urbanism' (see also Bulkeley 2021). The Rockefeller Foundation's Resilient Cities Network, for example, boasts one-hundred members worldwide and has supported the appointment of a Chief Resilience Officer in each one. In cities like Miami and New York, major investments have been made in initiatives to improve the resilience of the cities' physical and 'social' infrastructure. For example, in New York, the challenge of maintaining urban systems in the face of rising seas and more frequent storms has generated ideas like installing reefs and oyster beds offshore, in the hope of softening the blows of storm waves. Unlike conventional sea walls, built upon modernist convictions of sure boundaries, reliable prediction, and structural permanence, the plan is that these reefs will rise with the seas, providing an ongoing adaptive response to changing and unpredictable conditions. At the same time, disaster preparedness has become a key topic of education and public outreach, with super-heroine character 'Ready Girl', for example being a key means by which the authorities seek to cultivate a disaster-ready citizenry:

> neighbors with backyard gardens and disaster go-bags are called on to shovel out, clean up, and rebuild, taught to be knowledgeable in first aid

and search and rescue—and, as importantly, to be ready to transmit data to city agencies.

(Wakefield 2020, 42–43)

Building resilience means new ways of building, but it also means cultivating new social responsibilities and subjectivities. But 'resilience' can manifest differently in different places (Grove, Reid, and Molinari 2024). Further south down the United States's Atlantic seaboard in Miami, resilience policies have seen new kinds of hard engineering, such as raising roads above sea level and installing industrial water pumps to try and alleviate the increasingly frequent 'sunny-day flooding' when, even in the absence of heavy rains, water seeps into the city's highways and buildings. Despite seemingly being less about flexibility and adaptability than New York's policies, these initiatives in Miami nonetheless still represent a departure from conventional, modernist urban design. With respect to sea level rise and flooding, the aim is not to keep water and city apart, but to find new ways for the two to coexist (Wakefield 2025).

In New York, Miami, and elsewhere, including a number of major cities in the global South, the notion of resilience has been met with storms of enthusiasm from city authorities and from communities of architects, designers, academics, and artists. Enthused by the possibilities of abandoning modernist, managerial pretences and embracing new ways of working creatively across conventional siloes and disciplines, of working in an ecologically-attuned way 'with nature', and of rethinking the fundamental assumptions of urban design, the concept of resilience has become a 'boundary object' (Grove 2018) around which multiple communities invested in urban futures have gathered. Nonetheless, this mobilisation of resilience still attracts critique. For Stephanie Wakefield (2020), its enrolment by powerful actors and its emphasis on maintaining urban systems in the face of exogenous shocks reveal how 'resilience' continues to be a means of working out how to sustain the socioeconomic status quo in the face of a new range of threats to its longevity. In Cape Town, for example, activists have mobilised in opposition to resilience planning, which they see as being top-down, technocratic, and reinforcing of existing urban inequalities. In contrast to the reductionist maps and quantitative analyses of urban systems mobilised by city authorities in the service of fostering resilience, activists have urged for greater consideration of qualitative data about different groups' actual experiences of urban space (Borie et al. 2019). In Miami, resilience is most often talked about in economic terms of maintaining land and property values, tourism, and investment. The frequent irony of course is that it is the socioeconomic status-quo which is creating many of the threats in the first place. Those reefs off New York, for example, were only permitted if they don't interfere with the continued flow of shipping – much of it carrying fossil fuels – in and out of regional ports. In the UK meanwhile, new infrastructural experiments in 'working with nature' to ensure resilient coastlines have clustered around (and been heavily funded by) the very same hydrocarbon infrastructures which are helping drive coastal erosion in the first place (Lorenzoni et al. 2024). As Wakefield puts it:

> Though resilience recognizes deep-seated problems, its key tenet and goal is to attenuate and govern disruption in order to maintain the identity of the system… Resilient infrastructures are designed to manage destruction and disaster caused by the ongoing function of [climate change-causing] infrastructures like pipelines, mines, and power plants. Put otherwise, the role of resilient infrastructures is to manage and adapt to changing conditions of catastrophe at sea – rising seas and storm surge – to secure and manage an unchanging urban order on land.
>
> (Wakefield 2020, 48–50)

Nonetheless, both Grove (2018) and Wakefield (2020) see potential in 'resilience'. While present mobilisations of the concept might be geared towards ensuring the endurance of existing socioeconomic orders, it also contains more radical potentials. As we summarise in Table 8.1, resilience discourse

Table 8.1 The logics of geo-management and resilience compared.

	Geo-management	*Resilience*
Conceptualisation of human-environment assemblages	Earth System properties – and human influences on them – are generally predictable and amenable to control, at least up to certain 'thresholds' after which emergent behaviour can occur	Characterised by emergent properties, uncertainty, and surprise
Whither modernity?	An extension of the modernist project of exerting human control over nonhuman systems	Recognition of the limits/end of modernity's notions of control and security
Epistemology	Objective science can make the Earth System known with sufficient detail and certainty to enable technological interventions	Complex socio-ecological systems exceed our capacities for scientific understanding and prediction
Scales addressed	The planetary	Ecosystems, cities, nation-states
Geography of power	Concentrated in the hands of technical and political elites, with potential for either multilateral or unilateral state decision-making. 'Enlightened' states potentially making decisions on others' behalf.	Potentially more decentralised (e.g. at city/local government scales), but with potential for capture by economic elites (see Box 8.2)

(*Continued*)

Table 8.1 (Continued)

	Geo-management	*Resilience*
Political subjectivities, imagined or cultivated	Responsible geo-engineers, grateful publics	Resilient subjects, capable of self-reliance and community-mindedness in responding to disasters
Temporalities	Geoengineering deployment would require long-term (perhaps century scale) commitment to ongoing maintenance and governance (especially in the case of SAI); potential for climatic 'termination effects' if deployment stops	Emphasis on policies as provisional and partial solutions to problems, requiring constant evaluation and revision as situations change
Anthropocene narratives (see Chapter 1)	Aligns most closely with the 'Standard Narrative', and with elements of the 'Post-Natural Anthropocene' narrative which emphasise that humans have always transformed 'nature', and that the need now is to just do so better, on a grander scale	Aligns with elements of the 'Post-Natural' narrative, which emphasise complexity and interconnection in human/nonhuman relations. Also aligns with elements of the 'Eco-Catastrophist' narrative, which call for greater appreciation of the potential for socio-ecological systems to fundamentally break down. Many of the critiques of resilience come from an Eco-Marxist stance.

features epistemological standpoints, accounts of modernity, and implied geographies of decentralisation and even democratisation that could be nourished towards more transformative ends:

> As we search for other paths forward … we are each free to use the best lessons of resilience practice and discourse to challenge its worst aspects, to open up a much wider field of possibility and with a much broader set of actors. All of us, in all our various worlds, with our aptitudes, and skills are free to take back the powers of audacity, experimentation, and transformation, and put them toward other paths, alternative meanings for resilience, and more liveable ways of life. Up for grabs … is not only the survival of human life, but its very meaning. It's everyone's right to take up such a question.
>
> (Wakefield 2020, 57)

How various communities are taking up this question will be explored in the final chapters.

Conclusion

In Table 8.1, we summarise and compare some key features of the logics of geo-management and resilience. Both logics have emerged in direct response to the idea and material conditions of the Anthropocene. They address humanity's complex entanglement in fast-changing planetary conditions, and offer a new settlement of the human-nature distinction in response to its apparent breakdown. While geo-management invites humanity to embrace the power and responsibility of controlling key Earth System processes (e.g. the planetary radiation balance), resilience invites humanity to 'work with nature' in more situated, locally-attuned ways. Resilience discourse and policy is, however, self-consciously about engineering both nonhuman and human natures – it is about building resilience into material infrastructures, and about building resilience within communities. Geo-management might at face value be more concerned with technological engineering, but it too imagines certain political subjectivities – the responsible geo-engineer tinkering with the planet on humanity's behalf, and a grateful public (or subordinate states) who accept and defer to the need for long-term political commitment to radiation management. As such, both of these logics represent 'socio-technical imaginaries' – i.e. imaginations of what science and technology can do, but also serious fantasies about ideal forms of social order (Jasanoff and Kim 2015). They also, as we've suggested above, feature important real and imagined geographies. Both are discourses and practices with intellectual roots in the global North, particularly North America; resilience has been most visibly pursued in major cities, while geo-management practices aim to tinker with environmental processes on a planetary scale. Both logics construct imaginative geographies too – in the case of resilience, of places and systems enduring despite external 'shocks'; in geo-management's case of an entire planet amenable to control by an enlightened, geographically-situated few.

Taken together, we might interpret these two governing logics of the Anthropocene as themselves being assemblages. They are reasonably stable bundles of actors, institutions, technologies, ideas, and imaginaries, which cohere in certain places and times to transform – let's say – the material design of cities, or the goals and ambitions of climate change policy. But all assemblages are unpredictable. They can change in surprising ways, becoming more powerful or perhaps dispersing altogether. Far be it from us, then, to predict the fate of resilience and geo-management as we move deeper into the Anthropocene. Perhaps, as Baskin (2019) suggests, if the idea of the Anthropocene (and perhaps of an engineered 'good Anthropocene') gets hitched to the geo-management assemblage, geoengineering proposals might grow in influence. And perhaps further extreme weather events will lead more authorities to embrace the practice and goals of resilience planning. These are possibilities; the task for geographers is to track these assemblages (and others), to analyse how they manifest in particular contexts, how they remake urban and rural landscapes, and to examine their effects on the imagination, production, and experience of space and place.

In the next section we delve further into the speculative work of envisaging and making Anthropocene futures. In Chapter 9, we examine an emerging aesthetics of the Anthropocene across the sciences, the arts and popular culture, and question the role of representation in making geographical worlds. In Chapter 10, we examine the geographies of actually-existing attempts to build alternative lifeworlds in the shadow of the Anthropocene, and attempt to offer a cartography of hope that things could be otherwise.

Study questions

1. How might the deployment of geoengineering technologies transform global geopolitics?
2. To what extent can the popularity of the notion of 'resilience' in fields such as urban planning or disaster management be attributed to the rise of neoliberalism?
3. In light of the discussion in this chapter, does the Anthropocene represent the end of modernity, or its intensification?

Notes

1. Such questions have coalesced in a new field of inquiry into 'political geology'. See Bobbette and Donovan (2019).
2. We take the notion of 'geo-management' from (Swyngedouw and Ernstson 2018, 14), who memorably describe its pretensions to "manicure the dynamics of the earth system".
3. At the time of writing, neo-imperial jostling in the Arctic, driven in part by the US Trump administration's new geopolitical doctrine of great power 'spheres of influence' and resource-grabbing, appears to bear out Toal's earlier analysis (Foster 2025).
4. This is the stated logic of recent UK investments in geoengineering or 'climate cooling' research via the government's ARIA programme.
5. Consider the complex case of an ocean fertilization experiment part-financed by Indigenous residents of Haida Gwaii, Canada, in 2012 (Gannon and Hulme 2018)
6. Grove (2018) points out that Holling and other early resilience thinkers in the natural and social sciences were also heavily influenced by political scientist Herbert Simon and his work on human decision-making under conditions of uncertainty.

References

Anderson, B. 2015. "What Kind of Thing Is Resilience?" *Politics* 35 (1): 60–66.

Apostolopoulou, E., A. Chatzimentor, S. Maestre-Andrés, M. Requena-i-Mora, A. Pizarro, and D. Bormpoudakis. 2021. "Reviewing 15 Years of Research on Neoliberal Conservation: Towards a Decolonial, Interdisciplinary, Intersectional and Community-Engaged Research Agenda." *Geoforum* 124 (August): 236–56.

Baskin, J. 2019. *Geoengineering, the Anthropocene and the End of Nature*. Cham: Springer.

Becker, S. 2012. "The "Great Game": The History of an Evocative Phrase." *Asian Affairs* 43 (1): 61–80.

Bhowmick, M., S. Kanta Mishra, B. Kravitz, S. Sahany, and P. Salunke. 2021. "Response of the Indian Summer Monsoon to Global Warming, Solar Geoengineering and Its Termination." *Scientific Reports* 11 (1): 9791.

Biermann, F. 2021. "The Future of "Environmental" Policy in the Anthropocene: Time for a Paradigm Shift." *Environmental Politics* 30 (1–2): 61–80. https://doi.org/10.1080/09644016.2020.1846958

Biermann, F., and A. Gupta. 2024. "A Paradigm Shift? African Countries Call for the Non-Use of Solar Geoengineering at UN Environment Assembly." *PLOS Climate* 3 (5): e0000413.

Biermann, F., and I. Möller. 2019. "Rich Man's Solution? Climate Engineering Discourses and the Marginalization of the Global South." *International Environmental Agreements: Politics, Law and Economics* 19 (2): 151–67.

Biermann, F., J. Oomen, A. Gupta, et al. 2022. "Solar Geoengineering: The Case for An International Non-Use Agreement." *WIREs Climate Change* 13 (3): e754.

Bobbette, A. 2023. *The Pulse of the Earth: Political Geology in Java*. Durham: Duke University Press.

Bobbette, A., and A. Donovan, eds. 2019. *Political Geology: Active Stratigraphies and the Making of Life*. Cham: Palgrave Macmillan.

Borie, M., M. Pelling, G. Ziervogel, and K. Hyams. 2019. "Mapping Narratives of Urban Resilience in the Global South." *Global Environmental Change* 54 (January): 203–13.

Bronen, R., and P. Cochran. 2021. "Decolonize Climate Adaptation Research." *Science* 372 (6548): 1245–45.

Buck, H. J. 2015. "On the Possibilities of a Charming Anthropocene." *Annals of the Association of American Geographers* 105 (2): 369–77.

Buck, H. J. 2019. *After Geoengineering*. London: Verso.

Bulkeley, H. 2021. "Climate Changed Urban Futures: Environmental Politics in the Anthropocene City." *Environmental Politics* 30 (1–2): 266–84.

Cairns, R., and A. Stirling. 2014. "Maintaining Planetary Systems" or "Concentrating Global Power?" High Stakes in Contending Framings of Climate Geoengineering." *Global Environmental Change* 28 (1): 25–38.

Carey, M. 2011. "Inventing Caribbean Climates: How Science, Medicine, and Tourism Changed Tropical Weather from Deadly to Healthy." *Osiris* 26 (26): 129–41.

Castree, N. 2010. "Neoliberalism and the Biophysical Environment 1: What "Neoliberalism" Is, and What Difference Nature Makes to It." *Geography Compass* 4 (12): 1725–33.

Castree, N., R. Bellamy, and S. Osaka. 2021. "The Future of Global Environmental Assessments: Making a Case for Fundamental Change." *The Anthropocene Review* 8 (1): 56–82.

Castree, N., R. Kitchin, and A. Rogers. 2013. *A Dictionary of Human Geography*. Oxford: OUP Oxford.

Clark, N. 2011. "Inhuman Nature: Sociable Life on a Dynamic Planet." In *Inhuman Nature: Sociable Life on a Dynamic Planet*. London: SAGE.

Clark, N. 2013. "Geoengineering and Geologic Politics." *Environment and Planning A* 45: 2825–32.

Clark, N., and B. Szerszynski. 2020. *Planetary Social Thought: The Anthropocene Challenge to the Social Sciences*. Cambridge: Wiley.

Clark, N., and B. Szerszynski 2021. "Planetary Multiplicity, Earthly Multitudes: Interscalar Practices for a Volatile Planet." In *Narratives of Scale in the Anthropocene*. London: Routledge.

Clark, N., and K. Yusoff. 2017. "Geosocial Formations and the Anthropocene." *Theory, Culture & Society* 34 (2–3): 3–23.

Collomb, J. D. 2024. "A Paradox of the Anthropocene: The Radicalization of Techno-Scientific Modernity and the Future of Solar Geoengineering." In *Environmental Reflections on the Anthropocene*. London: Routledge.

Crutzen, P. J. 2002. "Geology of Mankind." *Nature* 415 (January): 23.

Crutzen, P. J. 2006. "Albedo Enhancement by Stratospheric Sulfur Injections: A Contribution to Resolve a Policy Dilemma?" *Climatic Change* 77 (3–4): 211–19.

Dalby, S. 2015. "Geoengineering: The Next Era of Geopolitics?" *Geography Compass* 9 (4): 190–201.

Dalby, S. 2020a. *Anthropocene Geopolitics: Globalization, Security, Sustainability*. Ottawa: University of Ottawa Press.

Dalby, S. 2020b. "Resilient Earth: Gaia, Geopolitics and the Anthropocene." In *Resilience in the Anthropocene*. London: Routledge.

Dalby, S. 2022. *Rethinking Environmental Security*. Cheltenham: Edward Elgar Publishing.

Davies, J. S., and T. Arrieta. 2024. "The Limits of "Resilience": Relationalities, Contradictions, and Re-Appropriations." *WIREs Climate Change* 15 (6): e911.

Derickson, K. D. 2018. "Urban Geography III: Anthropocene Urbanism." *Progress in Human Geography* 42 (3): 425–35.

Dittmer, J. 2014. "Geopolitical Assemblages and Complexity." *Progress in Human Geography* 38 (3): 385–401.

Dunleavy, H. 2021. "An Indigenous Group's Objection to Geoengineering Spurs a Debate About Social Justice in Climate Science." *Inside Climate News*. https://insideclimatenews.org/news/07072021/sami-sweden-objection-geoengineering-justice-climate-science/

Fan, W., J. J. McGuire, C. D. de Groot-Hedlin, M. A. H. Hedlin, S. Coats, and J. W. Fiedler. 2019. "Stormquakes." *Geophysical Research Letters* 46 (22): 12909–18.

Farbotko, C., E. Stratford, and H. Lazrus. 2016. "Climate Migrants and New Identities? The Geopolitics of Embracing or Rejecting Mobility." *Social & Cultural Geography* 17 (4): 533–52.

Farquharson, J. I., and F. Amelung. 2020. "Extreme Rainfall Triggered the 2018 Rift Eruption at Kīlauea Volcano." *Nature* 580 (7804): 491–95.

Fish, A. 2023. "Reforesting Native America With Drones: Rooting Carbon With Arborescent Governmentality and Decolonial Geoengineering." *Theory, Culture & Society* 40 (7–8): 157–77.

Fleming, J. R. 2012. "Will Geoengineering Bring Security and Peace? What Does History Tell Us?" *Sicherheit Und Frieden (S+F)/Security and Peace* 30 (4): 200–4.

Foster, J. B. 2025. "The MAGA Ideology and the Trump Regime." *Monthly Review* 77 (1): 1–24.

Foulger, G. R., M. P. Wilson, J. G. Gluyas, B. R. Julian, and R. J. Davies. 2018. "Global Review of Human-Induced Earthquakes." *Earth-Science Reviews* 178 (March): 438–514.

Furuhata, Yuriko. 2022. *Climatic Media: Transpacific Experiments in Atmospheric Control*. Durham NC: Duke University Press.

Gannon, K. E., and M. Hulme. 2018. "Geoengineering at the "Edge of the World": Exploring Perceptions of Ocean Fertilization Through the Haida Salmon Restoration Corporation." *Geo: Geography and Environment* 5 (1): e00054.

Group, ETC. 2017. *The Big Bad Fix: The Case Against Climate Geoengineering*. Montréal: ETC Group.

Grove, K. 2018. *Resilience*. London: Routledge.

Grove, K., G. Reid, S. Molinari, et al. 2024. "Absurd Geographies of Resilience and Justice." *Climate and Development* 16 (9): 739–50.

Gunderson, L. H., and C. S. Holling, eds. 2002. *Panarchy: Understanding Transformations in Human and Natural Systems*. Washington: Island Press.

Hamilton, C. 2014. "Geoengineering and the Politics of Science." *Bulletin of the Atomic Scientists* 70 (3): 17–26.

Heyen, D., J. Horton, and J. Moreno-Cruz. 2019. "Strategic Implications of Counter-Geoengineering: Clash or Cooperation?" *Journal of Environmental Economics and Management* 95 (May): 153–77.

Hickel, J., and A. Slamersak. 2022. "Existing Climate Mitigation Scenarios Perpetuate Colonial Inequalities." *The Lancet Planetary Health* 6 (7): e628–31.

Holling, C. S. 1973. "Resilience and Stability of Ecological Systems." *Annual Review of Ecology and Systematics* 4: 1–23.

Horton, J., and D. Keith. 2016. "Solar Geoengineering and Obligations to the Global Poor." In *Climate Justice and Geoengineering: Ethics and Policy in the Atmospheric Anthropocene*, edited by Christopher J. Preston, 79–92. Lanham: Rowman & Littlefield.

Horton, J. B., J. L. Reynolds, H. J. Buck, et al. 2018. "Solar Geoengineering and Democracy." *Global Environmental Politics* 18 (3): 5–24.

Hourdequin, M. 2020. "Climate Change, Climate Engineering, and the 'Global Poor': What Does Justice Require?" In *The Ethics of "Geoengineering" the Global Climate*. London: Routledge.

Hulme, M. 2014. *Can Science Fix Climate Change: A Case Against Climate Engineering*. Cambridge: Polity.

Jasanoff, S., and S. H. Kim. 2015. *Dreamscapes of Modernity: Sociotechnical Imaginaries and the Fabrication of Power*. Edited by Sheila Jasanoff and Sang-Hyun Kim. Chicago: University of Chicago Press.

Kaika, M. 2017. "'Don't Call Me Resilient Again!': The New Urban Agenda as Immunology … or … What Happens When Communities Refuse to Be Vaccinated With 'Smart Cities' and Indicators." *Environment and Urbanization* 29 (1): 89–102.

Kearns, G. 2025. "Critical Geopolitics 2.0: Geopolitics in the Web of Life." *Geography Compass* 19 (5): e70029.

Keith, D. 2013. *A Case for Climate Engineering*. Cambridge: MIT Press.

Kornbech, N., O. Corry, and D. McLaren. 2024. "Securing the 'Great White Shield?' Climate Change, Arctic Security and the Geopolitics of Solar Geoengineering." *Cooperation and Conflict* 60 (2): 286–307. https://doi.org/10.1177/00108367241269629.

Liu, C., A. T. Linde, and I. S. Sacks. 2009. "Slow Earthquakes Triggered by Typhoons." *Nature* 459 (7248): 833–36.

Livingstone, D. N. 2024. *The Empire of Climate: A History of an Idea*. Princeton: Princeton University Press.

Lorenzoni, I., S. A. Day, M. Mahony, T. J. Tolhurst, and R. H. Bark. 2024. "Innovation in Coastal Governance: Management and Expectations of the UK's First Sandscaping Scheme." *Regional Environmental Change* 24 (3): 101.

Mackinder, H. J. 1904. "The Geographical Pivot of History." *The Geographical Journal* 23 (4): 421–37.

MacKinnon, D., and K. Driscoll Derickson. 2013. "From Resilience to Resourcefulness: A Critique of Resilience Policy and Activism." *Progress in Human Geography* 37 (2): 253–70.

Matthews, A. J., J. Barclay, and J. E. Johnstone. 2009. "The Fast Response of Volcano-Seismic Activity to Intense Precipitation: Triggering of Primary Volcanic Activity by Rainfall at Soufrière Hills Volcano, Montserrat." *Journal of Volcanology and Geothermal Research* 184 (3): 405–15.

McConnell, F., and J. Dittmer. 2018. "Liminality and the Diplomacy of the British Overseas Territories: An Assemblage Approach." *Environment and Planning D: Society and Space* 36 (1): 139–58.

McGowran, P., and A. Donovan. 2021. "Assemblage Theory and Disaster Risk Management." *Progress in Human Geography* 45 (6): 1601–24.

McGuire, B. 2024. *Geophysical and Climate Hazards: A Very Short Introduction*. Oxford: Oxford University Press.

Moore, J. W. 2025. *Capitalism in the Web of Life: Ecology and the Accumulation of Capital*. 2nd ed. London: Verso Books.

Morrissey, W. 2024. "Avoiding Atmospheric Anarchy: Geoengineering as a Source of Interstate Tension." *Environment and Security* 2 (2): 291–315.

Naef, P. 2022. "Resistances in the 'Resilient City': Rise and Fall of a Disputed Concept in New Orleans and Medellin." *Political Geography* 96 (June): 102603.

O'Callaghan, J. 2025. "Controversial Geoengineering Projects to Test Earth-Cooling Tech Funded by UK Agency." *Nature* 641 (8063): 567–68.

Oksanen, A. A. 2023. "Dimming the Midnight Sun? Implications of the Sámi Council's Intervention Against the SCoPEx Project." *Frontiers in Climate* 5: 994193.

Parenti, C. 2020. "A Left Defense of Carbon Dioxide Removal: The State Must Be Forced to Deploy Civilization-Saving Technology." In *Has It Come to This?* edited by J. P. Sapinkski, H. J. Buck, and A. Malm, 130–42. New Brunswick: Rutgers University Press.

Peet, R. 1985. "The Social Origins of Environmental Determinism." *Annals of the Association of American Geographers* 75 (3): 309–33.

Pelling, M., and K. Dill. 2010. "Disaster Politics: Tipping Points for Change in the Adaptation of Sociopolitical Regimes." *Progress in Human Geography* 34 (1): 21–37.

Pickering, A. 2025. *Acting With the World: Agency in the Anthropocene*. Durham: Duke University Press.

Radcliffe, S. A., E. E. Watson, I. Simmons, F. Fernández-Armesto, and A. Sluyter. 2010. "Environmentalist Thinking and/in Geography." *Progress in Human Geography* 34 (1): 98–116.

Resilient Cities Network (2026) "New Orleans". https://resilientcitiesnetwork.org/new-orleans/. Last accessed 30 June 2026.

Ricke, K., J. S. Wan, M. Saenger, and N. J. Lutsko. 2023. "Hydrological Consequences of Solar Geoengineering." *Annual Review of Earth and Planetary Sciences* 51: 447–70.

Rivera, D. 2022. "Disaster Colonialism: A Commentary on Disasters Beyond Singular Events to Structural Violence." *International Journal of Urban and Regional Research* 46 (1): 126–35.

Schubert, J. 2021. *Engineering the Climate: Science, Politics, and Visions of Control*. Manchester: Mattering Press.

Siegert, M., et al. 2025. "Safeguarding the Polar Regions from Dangerous Geoengineering: A Critical Assessment of Proposed Concepts and Future Prospects." *Frontiers in Science* 3 (September). https://doi.org/10.3389/fsci.2025.1527393

Simon, S., and S. Randalls. 2016. "Geography, Ontological Politics and the Resilient Future." *Dialogues in Human Geography* 6 (1): 3–18.

Sovacool, B. K., C. Baum, and S. Low. 2023. "The Next Climate War? Statecraft, Security, and Weaponization in the Geopolitics of a Low-Carbon Future." *Energy Strategy Reviews* 45 (January): 101031.

Sovacool, B. K., C. M. Baum, R. Cantoni, and S. Low. 2024. "Actors, Legitimacy, and Governance Challenges Facing Negative Emissions and Solar Geoengineering Technologies." *Environmental Politics* 33 (2): 340–65.

Sovacool, B. K., C. M. Baum, and S. Low. 2022. "Climate Protection or Privilege? A Whole Systems Justice Milieu of Twenty Negative Emissions and Solar Geoengineering Technologies." *Political Geography* 97 (August): 102702.

Steffen, W., K. Richardson, J. Rockström, et al. 2020. "The Emergence and Evolution of Earth System Science." *Nature Reviews Earth & Environment* 1 (1): 1.

Surprise, K. 2019. "Stratospheric Imperialism: Liberalism, (Eco)Modernization, and Ideologies of Solar Geoengineering Research." *Environment and Planning E: Nature and Space* 2 (3): 444–64.

Swyngedouw, E., and H. Ernstson. 2018. "Interrupting the Anthropo-obScene: Immuno-Biopolitics and Depoliticizing Ontologies in the Anthropocene." *Theory, Culture & Society* 35 (6): 3–30.

Szerszynski, B., M. Kearnes, P. Macnaghten, R. Owen, and J. Stilgoe. 2013. "Why Solar Radiation Management Geoengineering and Democracy Won't Mix." *Environment and Planning A* 45 (12): 2809–16.

Thrift, N. 1999. "The Place of Complexity." *Theory, Culture & Society* 16 (3): 31–69.

Toal, G. 2024. *Oceans Rise Empires Fall: Why Geopolitics Hastens Climate Catastrophe*. Oxford: Oxford University Press.

Victor, D. G. 2011. *Global Warming Gridlock: Creating More Effective Strategies for Protecting the Planet*. Cambridge: Cambridge University Press.

Wakefield, S. 2020. *Anthropocene Back Loop: Experimentation in Unsafe Operating Space*. 1st ed. London: Open Humanities Press.

Wakefield, S. 2025. *Miami in the Anthropocene: Rising Seas and Urban Resilience*. Minneapolis: University of Minnesota Press.

Wakefield, S., K. Grove, and D. Chandler. 2020. "Introduction: The Power of Life." In *Resilience in the Anthropocene*. London: Routledge.

Walker, J., and M. Cooper. 2018. "Resilience." In *Companion to Environmental Studies*, edited by Noel Castree, Mike Hulme, and James D. Proctor, 90–94. Abingdon: Routledge.

Westgate, J. 2020. "Anthropocene Dwelling: Lessons from Post-Disaster Christchurch." *New Zealand Geographer* 76 (1): 26–38.

Whyte, K. P. 2018. "Indigeneity in Geoengineering Discourses: Some Considerations." *Ethics, Policy & Environment* 21 (3): 289–307.

Woods, M. 2017. "'Stop Calling Me Resilient': The Wrong Way to Save Louisiana." *Edge Effects*, May 9.

Yusoff, K. 2013. "The Geoengine: Geoengineering and the Geopolitics of Planetary Modification." *Environment and Planning A* 45 (12): 2799–808.

Zee, J. C. 2020. "Machine Sky: Social and Terrestrial Engineering in a Chinese Weather System." *American Anthropologist* 122 (1): 9–20.

Part IV

Geo-futures

9 Anthropocene Aesthetics

Introduction

Bright rectangles of azure blue, lime green, and aquamarine converge gently as they stretch towards the horizon. There, a seemingly barren mountain range bears flecks of snow on rocky peaks. In the middle distance, a flat, brown, sandy plane, upon which the neon rectangles appear to have been superimposed, perhaps through digital trickery. The dusty blue of the sky, and the grey-brown of the plane and the mountains, contrast sharply with the blocks of glowing colour, which look more like an array of paint samples than a feature of the landscape. The eye searches for scale, eventually finding it in one of the roadways that weave around the colourful blocks and the smaller areas of churned earth, which start to come into focus as some kind of open-cast mine. With no human figures visible in the scene, you put yourself in a vehicle traversing one of the tracks, and suddenly the enormity of the scene is thrust upon you. But it takes the sparse caption to give any kind of context. What you're looking at is a lithium mine in Chile's Atacama Desert, photographed in 2017 by Edward Burtynsky for his series *Anthropocene*, an exhibition of enormous photographs of enormous landscapes, which aimed to capture how our 'planetary system is affected by a magnitude of force as powerful as any naturally occurring global catastrophe, but one caused solely by the activity of a single species: us' (Figure 9.1; Burtynsky 2018).

How the rapidly changing world of the Anthropocene is represented is crucial for how societies make sense of environmental challenges, what they envision the future to look like, and therefore how to act in the present to try to make more desirable futures into a reality. Imagery like Burtynsky's, which suggestively combines the shininess of human creations with the environmental spoliation and toxicity that underpins so many such creations, has formed a core part of artistic responses to the heralding of the Anthropocene. In this chapter we begin this final section's work of exploring how futures are envisioned, represented and made, by here exploring an emerging *aesthetics* of the Anthropocene, and how artists of various kinds have responded to the provocations of this new age. To do so, we build on the work of cultural

DOI: 10.4324/9781003162469-13

Figure 9.1 Lithium Mines #1 by Edward Burtynsky. (© Edward Burtynsky, courtesy Nicholas Metivier Gallery, Toronto.)

geographers (and geographically minded cultural critics) who are interested in the relationships between the material world, cultural representations of that world, and the emotions, affects, and politics involved in imagining and actualising futures.

Cultural geographers and like-minded scholars offer 'a humanities-based approach to the concept of the Anthropocene [which] calls attention to the ways in which stories are told and to how crises are narrated or visualized' (DeLoughrey 2019, 8). These processes of narration and visualisation are crucial, we'll contend, to the politics of the Anthropocene – to how its geographies of causation and impact are understood and acted upon, its inequalities reckoned with, its futures re-imagined. As such, 'Anthropocene aesthetics' are a crucial part of the 'thicker' sense of the Anthropocene's geographies we advocated for in Chapter 2, and in this chapter, we join Bergmann and Briwa (2021, 889) in exploring the role of visual imagery and other artforms 'in shaping geographical imaginations of the Anthropocene'. While we can't claim to offer any kind of comprehensive view of the 'aesthetics of the Anthropocene', nor of geographers' manifold engagements with them, we nonetheless seek to offer an account of how the provocations of the Anthropocene are prompting new engagements with questions of scale, landscape, space and time in the field of creative arts, and how this in turn is filtering into new forms of both critical and creative practice among communities of geographers.

An aesthetic event

For art historian Salar Mameni, the onset of the Anthropocene is not just a geological, political, and discursive event, but an *aesthetic* one too:

> The Anthropocene is a work of art. Pyrotechnics paint the skies with pigments. Atomic lights tattoo bodies with charred lines. Clouds cry acid onto crabs and crawlers of the seas. Petroleum swirls in oceanic currents… who envisioned this damaged planet…. What does it mean to be beholders of the Anthropocene, to sense its sights and sounds, to vibrate with its tremors, to feel our way through its toxins, digest its chemical residues, and inhale its pollutants?
>
> (Mameni 2023, 51–52)

The intellectual project of studying aesthetic experience, 'or the attitudes, emotions, and perceptions wrought by art' and other human means of representing the external world, extends back at least to Ancient Greek philosophy (Bennett 2021, 922). Within the discipline of Geography, the study of aesthetics has been an important way in which cultural geographers have reconstructed how communities in different times and places have valued things like 'nature' differently. How nature is represented (in things such as paintings, poems, or even scientific models) can reveal a lot about societal values, preferences, and norms (Castree 2005). Such questions are consequential for the politics of contemporary environmental problems. Understanding how different ways of representing phenomena like climate change invoke different emotional responses in audiences has become a key research interest in disciplines like psychology and environmental sociology (Caracciolo 2021), with implications for how institutions might go about designing communication strategies to reach different communities (Hawkins and Kanngieser 2017; Zhang et al. 2018).

A growing number of scholars have offered descriptions and critiques of an emerging 'Anthropocene aesthetics'. Some of these critiques are informed by Marxist and decolonial arguments against the 'standard narrative' of the Anthropocene, and challenge dominant representations of the Anthropocene in both art and science, which reproduce the idea of an undifferentiated humanity whose actions can be guided by enlightened experts (i.e. the 'us' in the quote from Edward Burtynsky above). For example, TJ Demos argues that the dominant mode of scientific 'Anthropocene visualization' has relied on cutting-edge technologies like satellite remote sensing, producing images which claim to visualise the previously invisible dynamics of this new age (see also Engelmann 2023). Powerful and impactful though they may be, and reflective of impressive advances in scientific visualisation techniques, Demos argues that viewers are given little scope to interpret these images on their own terms, and that they are 'carefully edited in order to show generally positive examples of modern development' (Demos 2017, 16–17). As such, alternative interpretations offered by, for example, eco-Marxists or decolonial scholars about the Anthropocene as an age of rampant and destructive colonial capitalism are literally pushed out of the frame.

Other scholars of visual culture have critiqued notions of an undifferentiated nature in an emerging Anthropocene aesthetics, and foregrounded the work of artists who have responded to the idea of the Anthropocene with creations which challenge conventional categories of animal, vegetable, and mineral (e.g. Figure 9.2; Fowkes and Fowkes 2022). In so doing, artists and creatives have built on and extended the arguments of 'post-nature' and 'more-than-human' thinkers about the Anthropocene being a time of ever-deepening entanglement and hybridisation between species and other worldly entities.

Recognition of this entanglement and hybridisation often extends to calls to replace logics of control and domination with those of care, attentiveness, and hospitality to both human and nonhuman others (Ballard 2021). Consider, for example, the work of performance artist Evangelia Basdekis who, in her work *The Silence of the Monkeys* (2015), planted seeds of various nourishing plants in her own hair, transforming her body and its functions into a host for the growing plants. She replaced her social obligations with her new responsibilities to the seedlings who'd found refuge on her head, and in so doing posed questions about the meaning and boundaries of 'nature', about care for nonhuman others, and about our responsibilities as individuals and collectives towards radically different but co-dependent others.

Basdekis's exploration and harnessing of the agency and vitality of plants within a piece of performance art speaks to a range of theoretical concerns which have been influential within Geography: Jane Bennett's calls to rethink the boundaries between the human and nonhuman (Bennett 2016); Elizabeth Povinelli's related excavation of the 'geontologies' which regulate political distinctions between life and non-life (Povinelli 2016); and Elizabeth Grosz's emphasis on art as a thoroughly material, sensual practice, through which life experiments with matter in order to bring new futures in being (Grosz 2008). It is with such provocations in mind, along with the geoscientific revelations of the Anthropocene, that this geological event has been described as an aesthetic event as well:

> Our sensorial and perceptive systems are being refashioned at rates that we can barely keep up with, as the world around us changes so rapidly. We daily experience what used to be a sublime moment; anthropogenic mountains are now so ordinary that we don't even notice them.
>
> (Davis and Turpin 2015, 11)

Have we become desensitised by what is happening? Have the modes of Anthropocene visualisation critiqued by Demos (2017) begun to function as what Nicholas Mirzoeff (2014, 220) has called an 'imperial aesthetics', by

Figure 9.2 Diana Lelonek, *PET Environment*, from the series Centre for Living Things, 2017, photography of the found object. (Reproduced courtesy of the artist.)

which this fast-changing world 'comes to seem natural, right, then beautiful'? How can art and creative practice cut through this naturalisation and even beautification of the Anthropocene? Can the arts help to defamiliarise what has perhaps come to seem 'ordinary', and thus open new possibilities for political action? These are questions that environmental and cultural geographers have long been interested in, and the 'aesthetic event' of the Anthropocene has prompted renewed thinking about how 'world-making happens through the speculative, imaginative and engaging forms of politics propagated by creative practices' – both those of artists, and of geographers themselves (Hawkins 2019, 976).

Anthropocene ruins

For geographer Mia M. Bennett (2021), an emerging aesthetics of the Anthropocene has an unhealthy fixation on human ruins. The visual language of climate change in particular has long traded on ideas of desertified landscapes (or even planets – see Smith 2021) and inundated or ruined cities. In 1986, the German current affairs magazine *Der Spiegel* famously adorned its front cover with an image of Köln Cathedral standing in several metres of flood water. The image played on a historical lineage of images of war-ravaged cities, and connected the building's status as an icon of pre-unification Germany to a new threat to European political order – climate change, rising seas, and extreme weather – as they became new objects of political debate. Cities like London and Paris have played host to numerous exhibitions that imagine postbellum life, often with techniques such as digital photomontage used to manipulate familiar cityscapes into eerie worlds where buildings are crumbling, nature is taking over in the form of creeping vines and swamp flora, and largely isolated individuals carve out new modes of inhabiting the city (Mahony 2016, 10).

Imagining ruined cities in this way has a long lineage, which pre-dates concerns over carbon dioxide emissions and rising seas. Eighteenth-century Romanticism in Europe (see **Chapter 4**) was gripped by a 'cult of ruins', which offered a kind of 'delightful terror' as cities were re-appropriated by ecological and even geological forces – rampant vegetation, water, sand, volcanic debris, and more (Bennett 2021). For earlier Renaissance artists, ruins of ancient cities provided a means to explore and celebrate civilisational triumph, collapse, and memory. In the eighteenth and nineteenth centuries, imperial centres like London began to be imagined as ruins – not to celebrate or perhaps urge their destruction, but rather to imagine them being viewed from a distant future, from where their civilisational achievements could be afforded the same status as those of ancient Athens or Rome. At the same time, European tourists were fascinated by ruins in places like India, motivating colonial 'salvage missions' to save disappearing cultures from their less enlightened (to European eyes) modern-day stewards (Stoler 2008). The 'cult of ruins' helped motivate and rationalise colonial expansion, at the same time as it fantasised about future imperial decay.

This Romantic ruin aesthetic, which continues to shape how climate futures are imagined and visualised, emphasises 'the built environment lost to nature' (Bennett 2021, 923). Nonhuman nature is not necessarily presented as itself being in a state of ruination. But as we move deeper into the Anthropocene, artists and designers are increasingly grappling with both human and natural ruins. Images of burning rainforests and melting ice caps belong, Bennett argues, to the same genre of ruin aesthetics as counterpart images of decaying cities and infrastructure. Other geographical shifts have occurred too – in literature as well as the visual arts, ruination has been upscaled: 'ruins are no longer just buildings but rather whole regions, ecosystems', and even the whole planet (Bennett 2021). Bennett compares the examples of Weisman's bestselling book *The World Without Us* from 2007, which 'portrayed a planet from which humans had suddenly disappeared', with Wallace-Wells' *The Uninhabitable Earth* from 2019, 'a lyrically grim picture of a planet rendered imminently unliveable'. Between these two publications an important aesthetic shift occurred, which reflected broader developments in scientific understanding of the state of the planet:

> The crux of the matter has become not what will happen to all of the things humans have built on top of the planet but rather how humans will cope with an unrecognizable Earth whose ecosystem we have eviscerated.
> (Bennett 2021, 923)

In addition to this conceptual and spatial shift, Bennett identifies a temporal one too – a sense of accelerating ruination as things like the loss of summer Arctic sea ice appear to be happening on faster timescales than scientists and observers initially imagined. 'Large-scale human activity has hastened the normally creeping pace of geologic time', meaning that while Romantic and modern conceptions 'assumed that ruins would endure on the landscape in perpetuity', in the Anthropocene 'the landscape itself is in peril, and the Arctic might be one of the first ruins with an end date' (Bennett 2021, 923).

The picturesque and the sublime

Bennett draws on theories of aesthetics developed by the Romantics to critique the ways in which these processes of ruination are depicted. The *picturesque* 'is a style or affect that captures and curates sights worth seeing' (Bennett 2021, 923). Bennett suggests that the rise of social media and widespread image-sharing has led to a resurgence of picturesque aesthetics in the pursuit of visual pleasure. A new kind of 'tourist gaze' (Urry and Larsen 2011) has emerged, as 'Anthropocene-era tourists…scout out places with exotic, photogenic elements…perhaps those that future generations might not be able to capture' (Bennett 2021, 924). The Arctic region increasingly attracts such scouts, with Iceland, for example experiencing a recent tourism boom driven in part, it seems, by a Justin Bieber music video which featured scenes of the singer gazing out across icy vistas from mossy promontories in the fashion of many a Romantic landscape painting. Icebergs and polar bears likewise become objects of desire for 'ruin tourism', and objects of mourning and elegy for artists. But Bennett contends that the picturesque presentation of these disappearing icons of the Arctic 'evokes longing and mourning but rarely anger'. Such images often appear as part of a discourse which holds that the Arctic is 'already lost', and that nothing can be done to change the future, either by people living in the rest of the world or by residents of the Arctic themselves, who are all too often absent from such depictions of their homeland (Bennett 2021, 924).

If the picturesque can invoke an unhelpful fatalism about future geographies, Bennett suggests that, by contrast, the sublime can lay the aesthetic groundwork for much more radical and hubristic interventions in ecological and planetary systems, such as those we explored in Chapter 8. Conventionally, the sublime has concerned the representation and appreciation of the enormity and infinity of nature, and how it transcends human comprehension, or the inherent dynamism of nonhuman systems and how they transcend any efforts of human mastery and control (Yusoff 2009). In Romantic art, the sublime was represented through daunting vistas, terrifying storms, and inhuman geological forces and timescales. For theorists like Edmund Burke, 'while beauty could be found in small objects, things that are smooth, delicate, elegant, graceful and clean, fair, and mild in colour, the sublime exhibits vastness, privation, difficulty, infinity, magnitude, and magnificence' (Peeples 2011, 379)

In the nineteenth century, a distinct aesthetics of the 'Arctic sublime' emerged, which conveyed a cultural fascination with the vastness of the region's awe-inspiring icescapes, and which celebrated both the Arctic's enigmatic and seemingly unconquerable landscapes as well as the bravery and determination of those explorers who sought to conquer them in the name of national prestige (Loomis 2023). In the Anthropocene Arctic, the sublime lives on through images of another object of nineteenth-century fascination and awe – the ocean (Carroll 2015). Bennett points to planetary-scale maps of the consequences of a complete loss of polar ice – images which depict a radically transformed world but which, by placing such transformations in the far future (say, 5,000 years hence), engender a sublime sense of the grandeur of planetary processes while diminishing any sense of urgency or agency to act to forestall such a future. In the Arctic itself, the transition of ice to ocean was captured by a viral photograph made by a climatologist of sled dogs dashing across a frozen inlet, which was topped by new meltwater, giving the impression that the dogs were running on water. 'The striding sled dogs appear able to reach the shore; perhaps humans – or, more specifically – the scientists and their team of local hunters – will still triumph over the rising seas with their perseverance, knowledge, and technology'. As such, Bennett wonders whether the 'supernatural' scene might contribute to the idea 'that only *deus ex machina* solutions such as geoengineering' (see Chapter 8), mobilising the 'technological sublime' (Nye 1994), 'offer an escape from a diluvial future'(Bennett 2021, 925).

Technological sublime, toxic sublime

The technological sublime offers a sense of wonder at human accomplishments, rather than at the grandeur and dynamism of nature (Nye 1994). It is the aesthetics of the rocket ship, the jet plane, of globe-straddling infrastructures and information systems. It is an aesthetic and an affect that is prominent in discussions of the 'technosphere', which revel in the sense that humanity has created a set of interlinked technologies and infrastructures that seem to have a life of their own and which, like Romantic nature, elude any attempts at control and mastery, let alone democratic consent or accountability (Donges et al. 2017; Haff 2014). The technosphere has a dark side, of course (Peterson and Zahara 2016; UNESCO 2018). The counterpart to the technosphere's new version of the technological sublime has been described as the 'toxic sublime'. For environmental communication scholar Jennifer Peeples, the toxic sublime arises 'from recognizing the toxicity of a place, object or situation, while simultaneously appreciating its mystery, magnificence and ability to inspire awe' (Peeples 2011, 375). It is therefore:

> the counterpart and required "other" to the technological sublime. It shares with the technological sublime a marvel at human accomplishments. Instead of staring up as a rocket soars into the sky, the viewer stares a mile straight down into an open-pit mine - both constructed by humans, both amazing feats of technology capable of conjuring feelings of insignificance and awe. But the toxic sublime acts to counter that marvel with alarm for the immensity of destruction one witnesses...the horror of the toxic sublime calls to question the personal, social and environmental ethics that allows these places of contamination to exist.
>
> (Peeples 2011, 380)

Peeples is referring to imagery such as Edward Burtynsky's. Through a series of high-profile exhibitions and accompanying films (including one titled *Anthropocene: The Human Epoch,* co-produced with Jennifer Baichwal and Nicholas de Pencier to accompany the Anthropocene exhibition with which we began this chapter), Burtynsky has depicted the enormity of human transformations of planetary landscapes and the consequences of globalised industrial production. The *Anthropocene* series combines and juxtaposes characteristically wide-angled images of rapidly growing cities in India, rubbish dumps the size of small towns in Kenya, vast landscapes remade by agriculture in Spain and California, clear-cut forests in Malaysia and Canada, city-sized petrochemical complexes in Texas and Nigeria, quarries and open-cast mines in Italy, Germany, Russia, Mexico, and more. The visually-arresting images often feature a view from above, with photographs taken from an aeroplane or a promontory to create a sense of 'scalar derangement' (Woods 2021) as the eye struggles to make sense of the strange shapes and colours it encounters and searches for clues (a tiny human figure, a vehicle) as to the spatial scale of the landscape (Grotkopp 2023).

Burtynsky has encountered some sharp scholarly and journalistic criticism of his work from those who see his imagery as 'beautifying' the kinds of sacrifice zones we encountered in Chapter 6, and 'aestheticizing' the inherent violence of toxic contamination. A review in the UK newspaper *The Guardian* of *Anthropocene: The Human Epoch* lamented how the 'colossal eco doc prettifies disaster', with contemplation of humanity's destructive force reduced to a 'mere exercise in aesthetics, without any social or political context' (Le 2021). We would counter that there is no such thing as 'mere aesthetics', cleaved off from society and politics; on the contrary, aesthetic theorists will tell you that aesthetics are central to processes of collective meaning-making, communication, and the formation of political and ethical stances on the world. The French philosopher Jacques Rancière famously argued that aesthetics are inseparable from politics, as aesthetics determine what he called 'the distribution of the sensible' – i.e. a society's sense of the boundaries of itself, who or what is 'inside' and 'outside', and who or what is to be considered subject to its political and ethical accounting (see Tolia-Kelly 2019).

So, nothing is ever 'merely aesthetic'; rather, aesthetics are central to determining the kinds of things societies decide to care about – like disappearing rainforests, melting ice, or marginalised 'others'; they are thus central to making futures. But Le's criticisms of the apparent exclusion of 'social or political context' from Burtynsky's imagery are worthy of reflection. That exclusion is often a very deliberate compositional choice: the uncanny and otherworldly shapes and forms of the landscapes fill the frame, often without a horizon to provide any orientation. The viewer is gradually able to perceive the enormity of the infrastructures and wastelands that are the central objects of the photographs, but rarely the broader geographies in which they sit. People, for example, are few and far between. Like in earlier exhibitions of Burtynsky's work, with a few exceptions, there are no victims present:

> The suffering of the people who labor or live near these dangerous landscapes, who have had their health affected by them, who die early deaths, do not appear in these photographs.
>
> (Peeples 2011, 385)

The emphasis of the exhibition is on how humanity – 'us' – is transforming landscapes and ecosystems on vast scales (Milliken 2023). The narrative doesn't seemingly encompass how those changes negatively affect different communities in wildly unequal ways, as geographers and environmental justice scholars have spent decades pointing out (see Chapter 6), and as artists like Ignacio Acosta have strikingly documented.[1]

For some critics, Burtynsky's narrative is a problem. It risks depoliticising environmental change, silencing its frontline victims, and perhaps denying the agency of individuals or communities with more power and resources to change things. Burtynsky's images tell a super-human story of vast infrastructures which seem to elude any attempt at control or regulation, and which render the individual viewer insignificant, causing perhaps 'bewilderment and inertia at the thought of rectifying a problem as massive as one sees' (Peeples 2011, 383).

Nonetheless, Peeples offers another reading of Burtynsky's de-peopled landscapes:

> The purposeful omission of a victim leaves open the space for viewers to question whether they themselves might be the casualties of the toxins, since no other victims are present. In addition, excluding people from the image may allow for a more compelling sublime response. The viewer does not compete with a subject for a connection to the sublime object, nor does the viewer question the subject's response to the object. Instead, the viewer is left "alone" to confront the object and compare it to one's self, the final stage of the sublime response.
>
> (Peeples 2011, 385)

Peeples is suggesting that Burtynsky is very deliberately mobilising the aesthetic tradition of the sublime to force a confrontation, or a reckoning, between the individual viewer of the artwork and the objects depicted. While encounters with the sublimity of nature via Romantic art were intended to prompt moral introspection on the part of the viewer, Burtynsky's work perhaps invites a personal reckoning with our own connections to what we see, either as a possible victim of the toxic geographies themselves or as part of the machinery that produces them. For Peeples, Burtynsky's images demand:

> a confrontation with our consumptive habits, what we buy, where we buy it, what organizations and industries we directly or indirectly support, and how those choices are influential in creating the sites we see…In measuring oneself against the natural sublime, for example the towering walls of El Capitan, a person does not question what she has done to cause such a place to exist. In contrast, a comparison of the self to the toxic sublime raises questions of complicity, producing an internal reckoning (at least initially) as one measures one's life choices against the sites of destruction.
>
> (Peeples 2011, 387–8)

The toxic sublime then is, perhaps, a useful exercise in expanding the 'distribution of the sensible', in forcing a confrontation between the kinds of people who visit such exhibitions in wealthy, high-consumption parts of the world, and the distant, toxic geographies of the globe-spanning supply chains in which they are embedded, but of which they might be quite ignorant (Parsons 2023). Environmental justice scholar Sarah Jacquette Ray argues that

'Burtynksy's works make visible the externalised costs of global consumerism to very specific bodies and very specific environments, and draw audiences materially into relation with those costs' (Ray 2016, 203), and thus embody an aesthetic form that is vital to the task of achieving environmental justice. They enable new forms of *recognition* of unequal environmental harms, and therefore support efforts at ensuring *distributional justice* (see Chapter 6).

Senegal-based photographer Fabrice Monteiro has, like Burtynsky, come to prominence through works that take viewers on a tour of global toxic geographies. Yet Monteiro's work does not share the realism of Burtynsky and others labouring in the tradition of documentary photography. His series *The Prophecy* – widely viewable online – features scenes like plastic waste dumps, oil spills, and forest fires in Senegal, and was later expanded to take in similar sites around the world.[2] Yet rather than the wide-angled awe of Burtynsky's images, Monteiro centres a series of 'mythical creatures rising out of some dystopian environmental context'. These figures, which Monteiro has described as *djinn*, spirits, or genies, are positioned as emissaries of the Earth, sent to deliver prophetic warnings to humans (Hoffman 2022, 51). Monteiro collaborated with Senegalese fashion designer Doulsy to create fantastical costumes for the figures, assembling together plastic bags, fishing nets, cables, timber, and discarded metals. The figures rise out of and loom over the scenes, offering an uncanny beauty amid the landscapes of environmental despoliation. Far from the emerging conventions of the toxic sublime, Monteiro and Doulsy set out to make 'something that looked like a fashion editorial' while at the same time drawing attention to sites of environmental destruction (Monteiro and Siddons 2021, np.).

In contrast to what Bennett calls the 'tantalizingly picturesque' – landscapes arranged or arrayed for different forms of visual consumption – or the 'dauntingly sublime' – landscapes which inspire fear or even terror (Bennett 2021, 927) – images of beauty amid the Anthropocene can inspire 'devotion and affection' (Bennett 2021, 924). As such, they might more readily 'transform aesthetics into action' (Bennett 2021, 922). Monteiro's work has strong activist intentions: he makes art to change hearts and minds, and to draw attention to ignored or invisible problems. Nonetheless, while the mythic figures may serve as an 'indictment' of things like waste dumping and toxic colonialism, the images also convey 'creation and genesis':

> Whether one reads the emerging figures here as occult spirits or as science-fiction cyborgs, they are not simply victims of their debased environments. They are the products of it. They have found a way to inhabit their dystopic world…they invent new bodies and new relationships with the earth.
>
> (Hoffman 2022, 52)

Monteiro has described how he aimed to address a shortfall in attention to local cultures in conversations about environmental protection in places like

Senegal. 'I wanted to connect environmental issues with the cultural interests of the population', he told a UK newspaper, and as such developed an interest in local animist ontologies, which hold that 'objects and the natural world are imbued with spirits' (Monteiro and Siddons 2021). The spirits he populated his images with thus 'warn humankind about its neglect and destruction of the environment' (Monteiro and Siddons 2021), but they also seem to herald something else too:

> They express the latent possibility in any ruin space that the world might be imagined differently, that ameliorating a world of waste means not simply condemning it but employing it as part of the process of crafting new, locally meaningful visions of the future.
>
> (Hoffman 2022, 54)

So while Monteiro's images use strange new forms of beauty – rather than the sublime – to explore the local manifestation and iniquities of toxic global geographies (see also Lewis Hood 2023), they also prompt new questions about what can be made out of the debris of neo-colonialism and industrial modernity. As such, they join together history and the future, the global and the local. They function as a form of what David Matless might call the 'Anthroposcenic'.

Despatches from the future: folding space and time

Landscape and the Anthroposcenic

Across a series of articles, British cultural geographer David Matless has developed the notion of the 'Anthroposcenic' to capture how changing landscapes, as represented in paintings, photographs, literature, and film, become 'emblematic of environmental transformation' (Matless 2016, 118), and filter the global transformations described by geoscientists through the lenses of local cultures and processes of meaning-making. Landscape has long been a key interest of cultural geographers, who have studied how changing representations of landscape across time reveal a lot not just about societies' shifting aesthetic sensibilities, but also shifting attitudes towards 'nature', rural life, and social class (Cosgrove and Daniels 1988; Matless 2014). More recently, Matless has turned this critical eye towards the work of visual artists who have played with long-standing traditions of picturesque landscape painting, remaking the genre for an Anthropocene present, where landscapes may exist 'in a far-from natural, highly-altered state' (Tooth, quoted in Matless 2018a, 7).

Take Julian Perry's collection *An Extraordinary Prospect: The Coastal Erosion Paintings*, which depict scenes from the east coast of England marked by the steady attrition of sea level rise and crumbling cliffs. Human dwellings, 'notably caravans and bungalows, hover in mid-air, still grounded on grass and topsoil, but the ground below removed' (Matless 2016, 119). The paintings 'may stand as a warning of what happens to things on unprotected soft cliffs' (Matless 2018a, np.), and more metaphorically speak to the inherent instability of coastal landscapes – an instability that threatens to intensify as a result of anthropogenic climate change. 'The ground is disappearing under our feet', the paintings seem to say, as the apparent environmental stability of the Holocene gives way to the roiling turbulence of the Anthropocene (see Figure 9.3). Yet the paintings also play on longstanding concerns of landscape aesthetics. As Matless's work on the cultural geographies of the Norfolk Broads has shown, caravans and holiday bungalows have long annoyed those who wish to preserve idyllic rural landscapes from the 'vulgar' encroachments of mass tourism (Matless 2014). In literally elevating the humble caravan, Perry gives it a new value and attention: 'The Caravan Holiday was good, while it lasted' the paintings seem to say (Matless 2018a, np), as the viewer is invited to look back at a cultural artefact from an imagined future of vanished ground.

This is one of the key features of Anthroposcenic art, and arguably of Anthropocene discourse more broadly: its curious admixture of past and future.

Figure 9.3 *Fanfare 34* by Julian Perry, 2010. (Reproduced with kind permission of the artist.)

Excavating the Anthroposcenic imagination, in both past and present artistic practices, involves exploring how stories of the past are connected to expectations of the future. Cultural geographer Caitlin DeSilvey offers the notion of 'anticipatory history' as a way of telling new stories about landscape; stories which emphasise change and flux and which get away from notions of landscapes as fixed, unchanging sources of aesthetic appreciation and cultural meaning (DeSilvey 2012). Re-telling landscape in this way might, DeSilvey suggests, help us to better prepare for changes to come, such as the disappearance of much-loved coastal sites. The notion of 'anticipatory history' 'seems apt for the Anthropocene, itself an anticipatory designation, the only geological epoch in part prospectively defined, its rock record nascent' (Matless 2017, 371).

This 'intersection of the imagined future and the imagined past' (DeSilvey 2012, 35) shapes the Anthroposcenic imagination, and the aesthetic forms which artists reach for to express something about landscapes which might already be disappearing under our feet, or which are given new forms of meaning through the anticipation of future loss or change. See, for example, the work by artist Hannah Chalew, which grapples with the intersections of plantation and petrochemical economies in remaking the landscapes of the US South in the past and present, and which aims to draw attention to ongoing activist struggles to detoxify these landscapes for the future. David Matless traces related themes through past artistic and scientific engagements with coastal change, noting the long history of grappling with shifting coastlines, with slippery human-nature boundaries, with alternating presumptions of humility and control, and with possibilities of radical transformation (Matless 2018b, 2019). The coastline of the East Anglia region of the UK figures large in these studies. Oscillating between coastal marshes and soft, crumbly cliffs, this is a coastline whose changeability and vulnerability have long fascinated scientists and artists, as well as prompting complex debates over adaptation and the future of coastal communities in the Anthropocene (Arnall 2023).

Geographical thinkers have identified other 'Anthroposcenes', and employed creative methods to represent them (Griffiths et al. 2017; Hoskins and Whitehead 2015; see also Sajewska and Schade 2025). For example, Edgar Gómez Cruz offers a visual essay on the 'pervasive presence of human traces' using discarded adhesive bandages (or sticking plasters). Using techniques of urban street photography, he suggests that even a banal, everyday item, when discarded, can serve as a reminder of wider processes of pollution and destruction in the Anthropocene, and of a fundamental tension 'between human power and their own fragility' (Gómez Cruz 2021, 193). Gómez Cruz observes that so much Anthropocene art concerns humanity's impacts on the world at scales which transcend everyday experience – think of Burtynsky's imagery, for example. As a response, Gómez Cruz plays with scale, suggesting ways in which 'we can evoke the impacts of our cultural practices on the environment by focusing on the pervasive presence of small and quotidian objects discarded on the streets' (Gómez Cruz 2021).

The project *Quaternary Drift* similarly explored Anthropocene scalar derangement in urban settings, albeit this time in a fashion which emphasised unexpected encounters with nonhuman others, including 'the bones beneath the streets' which speak of cities' entanglements with deep time (Overend, Lorimer, and Schreve 2020). A collaboration between a performance artist, a human geographer and a physical geographer, the project developed the 'Situationist Drift', an early twentieth-century avant-garde walking practice, as a means of experimentally staging encounters 'with deep time and its nonhuman inhabitants' (Overend, Lorimer, and Schreve 2020, 454), as well as the 'feral animal worlds that are coming into existence' in Anthropocene cities (Overend, Lorimer, and Schreve 2020, 460). Playful methods of staging and representing these urban encounters – including 'pictures of tiny clay elephants on city streets, laminated woolly mammoths…geographers wearing hippo masks' in Trafalgar Square point to new ways in which geographers can convey the complexities of both deep, more-than-human histories and possible worlds-to-come (Hawkins 2021, 1713).

Speculative aesthetics

Exploring similar themes of the entanglement of pasts and futures, albeit on rather shorter timescales, geographers Nicolas Bergmann and Robert Briwa examine the work of artist and environmental campaigner Hannah Rothstein, whose series *National Parks 2050*[3] reimagined a series of historical posters promoting US National Parks as sites to encounter and enjoy natural 'wilderness'. By adapting the visual language of the originals, but projecting them forwards into a dystopian vision of depleted landscapes, widespread pollution, and extinction, Rothstein aimed to 're-instill horror and despair into national park wilderness'. Rothstein's images attempt to:

> make perceptible the invisibility of climate change, [to] challenge the myth of national parks as an unimpaired resource for the enjoyment of future generations, and [to] complicate contemporary romantic understandings of wilderness through depicting a possible future characterized by the toxic sublime.
>
> (Bergmann and Briwa 2021, 897)

Whereas Peebles defines the *toxic sublime* in opposition to the *technological sublime*, Bergmann and Briwa see Rothstein's work as a manifestation of the toxic sublime, which works by re-appropriating and re-mixing an earlier, Romantic visual language of the natural sublime. The original national parks posters featured vast mountain ranges, deep forests, and bellowing geysers, and had all the ingredients of earlier Romantic theorisations of sublime aesthetics. In Rothstein's re-workings, the toxic industries that have devastated these landscapes aren't visible; just the 'natural ruins' that are left over (Bennett 2021). So while earlier Romantic peddlers of the sublime sought to inspire awe in the face of nature, and occasionally some degree of terror through their depictions of the magnitude of earthly forces and of geological time, Rothstein's images offer a different flavour of terror, in the shape of 'starving grizzlies, burned-out redwoods, and melted permafrost' (Bergmann and Briwa 2021, 891). Like the images examined by Bennett (of the Arctic) and Matless (of disappearing coastlines), they offer a sense of accelerated geological time, with anticipated ruination dragged into the present to confront an audience who might still have time to help enact different futures.

While Bennett (2021) critiques sublime images of the impacts of climate change for creating a sense of distance between the topic and its audience, and perhaps feeding a sense of disempowerment, we might wonder too whether Rothstein's emphasis on 'horror and despair' (Bergmann and Briwa 2021, 897) might not also invite a nihilistic fatalism among those who encounter her undeniably striking images of decimated American landscapes (O'Neill 2025). Like much of the imagery discussed so far, and in consort with so many imaginative constructions of 'wilderness' (Cronon 1996; see Chapter 7), Rothstein's images are largely devoid of people. But they nonetheless belong to a wider genre of imagery which attempts to visualise future Anthropocene worlds; a genre within which we can also locate many more peopled landscapes.

In 2010, the Museum of London (and later the UK's National Theatre) hosted a high-profile exhibition entitled *Postcards from the Future*, a collection of images depicting a future London under climate change.[4] Huge, backlit photomontages hovered moodily in a darkened exhibition space, depicting a future London ravaged by heat, water, ice, new industries, and – controversially – more people. Mahony (2016, 11) observed that 'the ravaging was not exactly apocalyptic': buildings were not overrun with tropical vegetation as in the literary dystopia J.G. Ballard's *The Drowned World*, but stood proud and intact amid the waters of a swollen River Thames or amid the sand of a newly desertified Horse Guards Parade. The scenes seemed designed to appeal to an aesthetic sense of the *uncanny*: the 'propensity of the familiar to become defamiliarized and derealized' (Vasudevan 2015, 152). The *Postcards* presented a range of possible urban and climatic futures, rather than a single, internally coherent one. Many different 'Anthropocene narratives' are present. Alongside scenes of catastrophic inundation and perhaps collapse are scenes of ecological modernisation – of responses to environmental change which mitigate and adapt to its effects through the deployment of sophisticated technologies while not fundamentally challenging the social relations of capitalist modes of production. 'Flag poles on the Mall are replaced by Union Jack-bearing wind turbines. Kew Gardens, a centre of botanical and ecological research, now hosts a nuclear power station, its domed form towering over suburban southwest London. Water turbines float next to the Thames Barrier, offering a visual marriage of mitigation and adaptation strategies' (Mahony 2016, 11)

An image entitled *London as Venice* belongs to a longstanding genre of imagery of urban inundation which stretches back farther than the famous 1980s magazine image of a flooded Köln (see above). This time, however, the imagined flooding around the Houses of Parliament is carefully calibrated with scientific projections of a worst-case scenario for global sea level rise. The imagery of an inundated Westminster is echoed by a depiction of an inundated Buckingham Palace. Again, familiar street patterns are made strange by the montaged encroachment of new forms. Here though, it is not water, but 'shanty houses', stretching away across Green Park and the grounds of the palace. The artists used images of 90 Kenyan shanty dwellings to digitally create this sea, the caption claims, of 20 million individual houses filling this once (relatively) green corner of west London. The slum has come to be the tokenistic image of the non-Western city, contrasted to the ideals of urban design and rational planning (Davis 2006; Scott 1998). In images like *Buckingham Palace Shanty,* the slum is 'transported from the cities of the global South to the very heart of Great Britain's centre of power, wealth and privilege' (Mahony 2016, 12). As in other Anthroposcenic imagery, both space and time are folded, 'collapsing the distance between present and future, near and far, self and other' (Baldwin 2022, 112).

In *Trafalgar Square Shanty,* the artists similarly appropriate the imperial iconography of Trafalgar Square to further emphasise the potential for urban transformation. Images of 'street life in Kenya and covered souks in Morocco' are thrown against Nelson's Column, a piece of military memorialisation which here is pushed into the background. Climate change-induced migration, viewers are told, has seen London's population swell such that its open spaces are given over to new kinds of settlement, radically distinct from what would be allowed under current regulations and norms. In another image – *The Gherkin* – viewers are offered a close-up of Sir Norman Foster's 30 St Mary Axe, one of London's most well-known skyscrapers in the heart of the financial district. 'Using images of tenement blocks in Sao Paulo and Hong Kong, the artists recreate the so-called 'Gherkin' as a home to impoverished families improvising accommodation behind the crumbling facade of this former beacon of international finance' (Mahony 2016, 12). Behind grubby glass hang curtains and drying clothes, 'the signs and minutiae of life' individualising 'the otherwise uniform setting' (Graves and Madoc-Jones 2010). Again, a theme of *over*population is woven through this re-purposing of urban space.

The exhibition received mixed reviews – endorsement by many environmentally-concerned celebrities, and criticism from experts in climate change communication and migration politics, who condemned the images as lazily reproducing 'cheap stereotypes' about migration and vulnerable communities (Mahony 2016, 13). In working with the idea of a 'postcard' as a 'travelling landscape-object' (della Dora 2009) designed to capture the essence of a place (at least according to the 'tourist gaze' – Urry and Larsen 2011), the *Postcards* aimed to disrupt present-day viewers' own senses of London and its cultural geographies. 'The technique of photomontage allowed the transposition of particular assemblages of climate and people, with 'old' London still very much recognisable beneath the layers of posited transformation' (Mahony 2016, 13). We can say that the images were *dialectical* in intention, aiming to disrupt familiar visual continuums, and combine the seemingly incompatible, 'to reveal a society poised on an otherwise hidden threshold of change and a cessation of the normal' (Mahony 2016).[5] The images aimed to draw attention to this threshold, but in so doing positioned the 'normal' as a coherent whole, familiar and comfortable, 'tensed on the verge of transformation' (Baldwin 2016, 84). Rather than examining the dialectical and 'emancipatory possibilities that come with living on the threshold of a dramatically altered world' (Baldwin 2016, 87), the images are culturally and morally reactionary, offering the present as a bedrock of cultural order and urban stability in advance of a threshold of 'change, disorder, heterogeneity and alterity' (Mahony 2016, 13).

One image from the series – latterly removed from the project website – of monkeys on the balustrade of St Paul's Cathedral was captioned 'a new breed of tropical immigrants reminiscing about equatorial days' (quoted in Mahony 2016, 15). For political geographer Andrew Baldwin, images like these form part of an emerging 'racial futurism': a form of discourse and politics 'which stakes a claim on the future by constructing the racial other as that which could materialise if efforts are not taken in the present to preempt that future' (Baldwin 2022, 1). While Rothstein's images of ruined national parks, for example, try to urge their viewers to act in the present to stop such a future becoming a reality, images like the *Postcards* – and the wider aesthetics of race and climate change of which they form a part – offer a picture of a future wherein 'global racial order' (Baldwin 2022) has been upset by climate change and other environmental transformations. Such images urge action in the present to combat climate change not, so the implication goes, to ensure environmental justice for all, or the protection of nonhuman natures, but to preserve and protect that global racial order itself (Gergan, Smith, and Vasudevan 2020). As such, this uncanny aesthetic of the *yet-to-come* works by positioning migration as itself a *risk*. Vulnerable communities whose livelihoods and lifeworlds might be threatened by a changing climate are positioned as *themselves* embodying the threat that climate change poses to wealthy, liberal, capitalist societies in the global North (Baldwin, Fröhlich, and Rothe 2019; Turner and Bailey 2022).

Disruptive pasts and futures

The *Postcards from the Future* use techniques of digital montage to create their uncanny visions of situated Anthropocene futures. Montage – the arrangement and juxtaposition of imagery from different sources within one frame – has become a popular means of visualising environmental futures, and for deliberately disrupting conventional means of representation (Berger 2013). For example, scholar of contemporary Indigenous North American art Jessica L. Horton examines how Diné/Bilagaana artist Will Wilson mobilises photomontage techniques to connect histories of Indigenous dispossession to toxic futures in the southwestern United States. Wilson's series *Auto Immune Response* (*AIR*) connects issues around climate change with the conquest of Dinétah, the traditional homeland of the Diné or Navajo people (Figure 9.4). Across the twentieth century, these lands were frequently designated as 'wastes' by state authorities, enabling their transformation into 'sacrifice zones' (see Chapter 6) for toxic uranium mining and nuclear weapons testing.

Wilson's montages trace the story of a Diné man moving through these toxic landscapes, with 'the sublime Western landscape tradition' interrupted by 'clues to invisible poisons':

> As we follow the sightline of the Indigenous man in profile across the horizon of *AIR #1*, we encounter fence posts, power lines, an atrophied atom bomb, and his faint body double wearing a gas mask. These ghosts hail from a multilayered history of land uses and abuses that collectively create an environment inhospitable to life.... The *AIR* series is ostensibly set in a postapocalyptic future, not unlike the predicted horrors of climate change. Yet citations of multiple past events and processes suggest that catastrophe has already happened, again and again, at Dinétah.
>
> (Horton 2017, 65)

Figure 9.4 Auto immune response no. 2, 2005, Will Wilson. (Reproduced with permission from Portland Art Museum.)

Wilson's work reflects a broader critique of mainstream Anthropocene discourse, which frequently positions climate change and ecological destruction as a catastrophe that is either very recent or still to come (Curley and Smith 2024). For scholars of Indigenous histories in North America and beyond, socio-ecological upheaval and collapse has been a lived reality for hundreds of years, from the early stirrings of European colonialism right up to the present: 'indigenous peoples everywhere have been through repeated apocalypses', in the form of genocidal violence, deliberate cultural erasure, forced relocation, and the severing of ties to the nonhuman world (Whyte 2018, 238; see also Davis and Todd 2017; DeLoughrey 2019). While mainstream Anthropocene narratives position 'environmental change' as something relatively new and recent, scholars of Indigenous histories often frame the period of European colonialism in North America, for example, as a long story of oppression unfolding as a process of 'environmental change':

> Colonially-driven environmental change destroyed ecosystems on which Indigenous peoples relied, boxed Indigenous peoples into small reservations that were fractions of their original territories, or simply displaced Indigenous peoples from their homelands to new ecosystems.
>
> (Whyte 2017, 155)

For Potawatomi scholar Kyle Powys Whyte, these histories and ongoing lived experiences pose challenges to depictions of Indigenous communities that are often found in environmental discourse – for example, as communities that are 'uniquely' close to nature and which need protecting by well-meaning others from oncoming crises, or as storehouses of environmental wisdom that can be extracted and used by others. These depictions not only deny the agency of the kinds of Indigenous communities Whyte works with to fashion their own social, political, and environmental futures, but also erase the historical facts of ongoing socio-ecological disruption within Indigenous lifeworlds (Curley and Smith 2024).

This struggle between settler-colonial and Indigenous conceptions of 'environmental change' plays out in forms of artistic production. For example, Whyte compares the rapidly expanding literary genre of dystopian 'cli-fi' (*climate fiction*, as opposed to *science fiction*) with the forms of dystopia offered by Indigenous authors. Cli-fi futures have, until recently (see Cole 2022; Persinger, Vint, and Canavan 2024), tended to consist of 'dystopian times of rationing…major extinctions, social unrest, drastic measures, and defaced landscapes' – conditions which largely affect everyone in a given society, and which are coming down the track if we don't all get to grips with climate change soon (Whyte 2018, 226; Gergan, Smith, and Vasudevan 2020).

Indigenous North American narratives, by contrast, have often narrated 'a sense of ongoing crisis rather than an upcoming one' (Scott 2016, quoted in Whyte 2018, 227), and frequently aim to 'expose the inextricable links between the violences of settler colonialism and the violences of climate crisis in the Anthropocene era', past, present and future (Tillett 2023, 282). Key contrasts can be identified too in how these different literatures and genres explore the *agency* of different actors to push back against the destructive forces that are visited upon them. Two films help make this contrast clear. James Cameron's *Avatar* (2009) offers a retelling of the story of Pocahontas for the Anthropocene era, in which a future human society responds to ecological collapse on Earth by seeking resources from other worlds. One such world, the lush, forested Pandora, contains seemingly valuable minerals alongside the *Na'vi*, a race of 10-foot-tall humanoids who 'live in harmony with nature' (Wikipedia 2025). A recognisable story of environmental injustice against the *Na'vi* people ensues, as ecosystems are destroyed and they are forced to 'live under the dystopia of alien invasion from a more powerful military force'. Yet the hero-protagonist that emerges is not a *Na'vi* freedom fighter, but 'an alien, non-*Na'vi* white male who is able to pass for *Na'vi* and have a sexual relationship with a *Na'vi* gendered female character who becomes defined in terms of this romantic relationship' (Whyte 2018, 231). As in much Anthropocene discourse about Indigenous peoples, the *Na'vi* are presented as living admirable lives 'in harmony with nature', victimised by another society, but then reliant on more enlightened members of that society for their salvation.

Whyte contrasts *Avatar* with Cree/Métis filmmaker Danis Goulet's sci-fi short *Wakening* (2013). Again, an Indigenous community faces up to the horrors of an oppressive occupying force and accompanying eco-social crisis, albeit this time in near-future Canada. But in contrast to *Avatar*, the protagonists in *Wakening* emerge from the colonised community itself, and in turn from figures of Cree legend. *Weesageechak*, a trickster figure in Cree mythology, is portrayed in the film as a warrior woman who happens upon a theatre in which a few survivors plead for salvation from our other protagonist, *Weetigo*, a monster of Cree legend. Weesageechak suggests that 'the occupiers have tricked Weetigo into being so destructive, and that it is the occupiers who are more powerful, Weetigo now being forgotten' (Whyte 2018, 231). Weetigo eventually turns away from killing Weesagechak, and instead attacks two members of the occupying forces who have entered the theatre lobby. 'The film ends with both protagonists staring into each other with the noise of the occupiers in the background, as Weetigo disappears and Weesageechak stares into a brighter horizon with a wistful look' (Whyte 2018).

Drawing on analysis of the film by environmental humanities scholar Salma Monani (2016, 2024), Whyte emphasises how the protagonists of *Wakening* 'occupy social identities that are disrespected or villainized in Canadian or U.S. settler colonialism, whether owing to gender, Indigeneity or being nonhuman'. The female and nonhuman protagonists 'have to figure out how to relate to each other again to resist the genocide and environmental destruction of the occupiers who are the true force of destruction'. Both must recalibrate their relationships and responsibilities – Weesageechak to 'respect and confront Weetigo', and Weetigo to 'not to be fooled by the occupiers'. Weesagechak's wistful look towards a sunny horizon at the end of the film suggests a brighter future, brought about by the recognition of 'the importance of reestablishing a relationship of reciprocal responsibility between the two protagonists, and emphasising diverse gendered and nonhuman agencies' (Whyte 2018, 231).

Studies of Indigenous narratives of environmental change, settler colonialism, and dystopian futures have emphasised how they frequently refuse linear conceptions of time, which see a straight line between past, present, and future, with a future potentially marked by catastrophe (Curley and Smith 2024). Instead, cyclical or spiralling temporalities are prevalent, in which one considers oneself as 'living alongside future and past relatives simultaneously' (Whyte 2018, 229). Narratives may emphasise 'cyclicality, reversal, dream-like scenarios, simultaneity, counter-factuality, irregular rhythms… The spiralling narratives unfold through our interacting with, responding to, and reflecting on the actual or potential actions and viewpoints of our ancestors and descendants'. As such, these narratives frequently involve 'a certain form of philosophizing about what actions we or our communities ought to take to respond to the issues and problems that characterize our current situations' (Whyte 2018, 229). This is why Whyte titles his article 'Indigenous science (fiction) for the Anthropocene' – *science* to denote the forms of knowledge and expertise practiced by Indigenous communities in the stewardship of socio-ecological relations, and *(fiction)* to denote the power of things like counterfactual or speculative narratives and intergenerational dialogue in deliberating over the ethics and politics of certain actions in the present (LaDuke & Cowen 2025). The aesthetics of narrative and storytelling here are inseparable from considerations of how to live, both in the present and the future (see also Death 2025). This is a conjunction we will continue to explore in the next chapter.

Anthropocene allegories

In his 2016 book *The Great Derangement*, Indian novelist and critic Amitav Ghosh argues that the climate crisis is primarily a crisis of the imagination; a crisis of humanity's failure to grapple with the pace and scale of environmental change, and to imagine how the world could be organised differently – in ways which promote justice, sustainability, and respect for human and nonhuman others. He argues that dominant Western literary forms, such as the novel, are ill-equipped to deal with the challenges of the Anthropocene. Since the eighteenth century, the Western novel has offered a form of literary realism that focuses attention on the inner lives and interpersonal relationships of individuals, in a fashion that makes a large-scale phenomenon like climate change difficult to adequately represent. Ghosh claimed that as a result, no great 'climate change novel' has yet been written. Now, some students of early twentieth-century fiction might beg to differ, detecting in the modernism of writers like Virgina Woolf and James Joyce an attunement to contemporary environmental change and planetary processes past, present, and future that is surprisingly 'Anthropocene' (see Adkins 2022, 2026). Nonetheless, Ghosh describes how earlier, non-Western literary forms, such as the epic, provide a better model for storytelling in the Anthropocene. Ursula Heise (2019, 281) likewise contends that the genre of science fiction, with its narrative forms that bounce between spatial scales from the personal to the galactic, and which move through time with devices such as time-travel, represents a 'continuation of the epic tradition in the age of the novel'. As such, sci-fi is well-suited to narrating the 'scalar derangements' of the Anthropocene (Woods 2014) – something which seems to be borne out in the more recent explosion of 'cli-fi' (Persinger, Vint, and Canavan 2024).

Sci-fi, and other genres which draw on the tradition of the 'epic', often makes use of allegory – stories in which particular characters or things are symbolic of wider aspects of human experience. 'Allegory stages other worlds to draw parallels and disjunctions between the present and an often dystopic future' (DeLoughrey 2019, 5), in a fashion which is frequently pedagogical in intent and aims to inspire action. *Wakening* is an excellent example of this kind of allegorical narrative. Literature scholar Elizabeth DeLoughrey examines recent writing and visual arts from the Caribbean and the Pacific, arguing that many of the narratives and themes explored by artists such as Kathy Jetñil-Kijiner, Tony Capellán and Erna Brodber work as *allegories*, in which their postcolonial island settings allow the telling of stories that are both local and global at the same time, and which speak to a 'perceived disjunction between humans and the planet, between our "species" and a dynamic external "'nature"' (DeLoughrey 2019, 4). In a range of global literary traditions, islands have long served allegorically to 'telescope' the reader between 'local and global, island and Earth'; their relative boundedness as well as their permeability to new arrivals and rapid change make them useful literary devices for exploring relationships across both space and time (DeLoughrey 2019, 6).

DeLoughrey notes an upsurge in the use of allegorical forms, typified by works such as Erna Brodber's novel *The Rainmaker's Mistake* (2007), which DeLoughrey interprets as using Caribbean plantation histories to allegorise the wider colonial disjuncture between humans and the Earth, or Dominican visual artist Tony Capellán's striking installation *Mar Caribe* (1996; see DeLoughrey 2019, 105–11). Capellán's installation gathers together abandoned flip-flops, salvaged from Dominican beaches, and arranges them together to create an impression of a striking azure seascape, which seems to speak a sedate language of human leisure and mobility, of time spent relaxing on exotic beaches and enjoying tropical natures. Closer inspection, however, reveals that the straps of the flip-flops have been replaced with barbed wire, a striking juxtaposition of leisurely mobility and a violent technology of control and containment. The work reminds us that the oceans have long been a space not of unfettered flows but of bordering and containment, and of violence towards and abandonment of those lives considered as disposable as a pair of flip-flops. Capellan's piece therefore prompts a reckoning with both the material waste of globalisation, and the 'wasted lives' of the poor, the enslaved, and the refugee, who haunt so much Caribbean writing and art. An allegory of Anthropocene disposability.

DeLoughrey's overall argument is a very geographical one – she emphasises how literary allegory works to tell stories which straddle the spatial scales of human experience and planetary change, and urges renewed attention to artistic forms which are emerging outside of Western traditions. In so doing, she aims to 'parochialise' our understanding of the Anthropocene – to examine how the past, present, and future of environmental change is being both experienced and represented differently in different geographical and cultural contexts.[6] In so doing, she joins cultural critics like Mark Bould (2021) who have argued that Ghosh's lament for the lack of good literature addressing Anthropocene challenges like climate change relies on too literal an interpretation of what that would look like. Bould argues that an 'Anthropocene unconscious' has pervaded swathes of artistic production over the last few decades, as creatives have grappled with the implications of unruly natures, of human hubris coming home to roost, and of a sense of time running out. From Hollywood movie franchises like *The Fast and the Furious* and *Sharknado* to the more 'high-brow' works of Karl Ove Knausgaard and Thomas Hardy, Bould wonders whether 'all the stories we tell are stories about the Anthropocene' (Bould 2021, 17). He examines novels (including some of Ghosh's own) concerned with inundation by water or fire, with catastrophic land use changes, with the ecological violence of colonialism, with complex interspecies relationships and the fall-out of globalisation, and suggests that they can all be read as allegories of the Anthropocene, even if that wasn't necessarily their authors' intentions (see also Johns-Putra 2019; Pfeifer 2025).

Nonetheless, Bould contends that the stories that we tell ourselves matter. And many of the possible stories of the Anthropocene, Bould suggests, contain within them the seeds of new kinds of activism:

> If we start from the position that all cultural texts are about climate change – even if only in the most fleeting, evanescent way…then they are wide open. Not just as opportunities for dialogue, discussion and debate, but as adventure playgrounds, workshops, studios, festivals, carnivals, invitations to creative play, to thinking through, to action.
>
> (Bould 2021, 132–33)

These stories, these 'cultural texts', are places where we can begin to not only grasp the enormity of the implications of climate change, but also engage in creative, collaborative work of imagining how the world could be otherwise; tasks which cultural geographers in particular are embracing.

Creative geographies

From David Matless's analysis of depictions of the possible fates of East Anglian coastal scenes, to Kyle Powys Whyte's work on Indigenous North American narrative traditions, we can see how Anthropocene aesthetics invite reckonings with both pasts and futures. Work by cultural geographers has shown how grappling with the future of treasured landscapes amid the upheavals of the Anthropocene requires re-appraising landscape histories, and perhaps requires a new acceptance of transience, impermanence, and even decay and loss (DeSilvey 2017). Scholarship on Indigenous American histories of loss and upheaval also provides means of thinking in new ways about futures; Whyte's arguments similarly invite re-appraisal of histories of people and land, but foreground acts of violent dispossession and dislocation, and prompt imaginaries of how such injustices might be undone. Meanwhile, the imagery of Edward Burtynsky and colleagues aims to make connections across geographical space, between producer and consumer, shiny product and toxic landscape. Fabrice Monteiro's photographs aim for something similar, albeit in a fashion which responds more fully to local lifeways, and which hints at capacities to make new worlds amid the ruins and toxicity of the old.

If 'art is thought from the future', as the philosopher Timothy Morton (2016, 1) suggests, then perhaps it is a form of human expression which can help crack-open our imagination about worldly possibilities at the very moment that that future appears to become ever more foreclosed and predetermined, either by the supposedly iron laws of geophysics or of capitalism (Fisher 2009; Noxolo 2025). In this chapter, we've explored geographically-themed artworks and geographical analyses of such artworks which have helped tell new stories about Anthropocene pasts, presents, and futures. But while geographers can join Mark Bould and like-minded literary critics in excavating the hidden meanings of artworks, they are also increasingly working in collaboration with artists and creatives, bringing geographical knowledges and artistic practices together (Barry and Keane 2019). A 'creative turn' (Hawkins 2019) has brought geographers into closer working relationships with painters, poets, filmmakers, and others – as contributors to their work, and not just analysts of the finished products.

While the practice of geography has, for decades or perhaps even centuries, encompassed creative and artistic modes of working (in the form, for example, of field sketches, paintings, and creative prose and poetry), more recently these practices have been re-appraised as part of the discipline's methodological and conceptual toolkit, particularly in cultural geography. Journals like *cultural geographies* and *GeoHumanities* make space for things like photo-essays and poetry alongside conventional research papers, while geographers have taken on roles and identities as exhibition curators (Driver 2013; Kinkaid 2019), theatre producers (Richardson 2015), filmmakers (Gandy 2021), visual, sound and performance artists (Barry and Keane 2019;

Engelmann 2023; Kanngieser 2023; Planetary Portals 2026) photographers (Davies 2013) and poets (Cresswell 2013) in ways which are central (rather than supplemental to) their roles as researchers and educators.

The project *Oceanic Refractions* is an excellent example of this intersection of geographical research and creative practice. An immersive, multisensory installation, *Oceanic Refractions* was 'created to convey the testimonies of Oceanic communities who are bearing witness to ongoing and accelerating ecocide' (Hunchuck 2024, np.), combining video and sound works produced in collaboration with Pacific island scholars, artists and communities 'to showcase the nuanced and variegated ways that people understand, produce knowledge about, and collectively attend to their lived experiences of ecocide' in settings like Kiribati (Kanngieser 2023, 170). Curved ground-level and ceiling-hung screens immerse the viewer in oceanic sights, sounds, and smells, alongside field recordings and voices of community members. Rather than simply trying to invoke empathy across geographical distance, *Oceanic Refractions* 'moves towards immersion and resonance', and pays respect to Pacific onto-epistemologies:

> If you look at the environment, it's part of the land and everything that surrounds that. In fact, in Kiribati, if you look at the language very closely, the word of our people and the word for land is the same. We call it *te aba. ABA*. You can talk to *te aba and* you mean *land*. You can talk about *te aba* and you mean people. So, there's a very close connection between land and people.
>
> (Teweiariki Teaero, quoted in Kanngieser 2023, 171)

This is a philosophical framework which is 'almost impossible to reconcile with European ways of approaching environmental relations, which always inevitably centre the human even when attempting to dissolve this distinction' (Kanngieser 2023, 171). As such, *Oceanic Refractions*, in its appearances at major European art and media festivals, does not simply offer testimony of suffering, or invite sympathy for vulnerable, passive communities on the frontlines of Anthropocene ecocide (O'Neill 2025; Tuck 2009). Rather, it 'invites audiences into a comportment that may be more suited to approaching Pacific ways of interdependence' (Kanngieser 2023, 171). It is an invitation into other ways of thinking about human-environment relations (beyond such dualisms), to gaze at the painful past and present of colonial exploitation and environmental degradation, and to consider how different worlds might be built on different philosophical and political foundations. And in so doing, it shows how practices of geographical research (such as interviews and field recordings) can exist alongside creative practices of composing soundscapes, making films, and designing art installations, to generate works which challenge and push back against both the material violences and conceptual blind spots of the Anthropocene (see also Barry and Keane 2019).

Conclusion

In this chapter, we've offered an initial overview of how geographers and other scholars, artists, activists, and critics have engaged with emerging visual and literary languages of the Anthropocene. Throughout the chapter, we've emphasised how such engagements cannot be reduced to 'mere aesthetics' – instead, such acts of representation of the world are always also interventions in the world. Whether that be artists hoping to draw attention and inspire action, or scholars experimenting with new ways of collaborating with communities and creatively sharing their insights, these aesthetic interventions are changing the 'distribution of the sensible' in the Anthropocene. They redraw the maps of who or what is seen, heard, or felt, and by whom, and force new considerations of individual and collective relationships with proximate or distant others, with landscapes and nonhuman natures, and with Anthropocene pasts and futures. Decolonial scholarship concerned with exploring and elevating the narrative and aesthetic forms developed by members of Indigenous or colonised cultures is particularly central, we would suggest, to this new politics of Anthropocene aesthetics – to disrupting received notions of time, space, progress, and crisis, and offering alternative means of imagining and actualising just futures.

In the next and final chapter, we turn to that final point, about envisaging and building futures. The seeds of such futures not only lie in the stories we tell about the Anthropocene, but they also lie in the active experiments which scholars and activists are undertaking in fashioning alternative models of collective life on an increasingly troubled planet. So, onwards, to the future!

Study questions

1. How are notions of the 'sublime' being re-made by an emerging Anthropocene aesthetics?
2. How can we understand the relationship between aesthetics and politics in the context of environmental art?
3. Evaluate the role that geographers have played as both critics and (co-) creators of environmental art.

Notes

1. See e.g. his series *Copper Geographies* (2012–2016), which documents the spatial and ecological transformations wrought by the mining, production, use and disposal of copper (Acosta et al. 2019).
2. Monteiro reports a concern that if he focused only on Africa he would be "drawing attention to Africa for the wrong reasons. I was concerned it made the continent look uniquely polluted, as though this isn't a problem all over the world. The only reason Europe doesn't look like this is because it ships its waste out to us" (Monteiro and Siddons 2021, np.)

3 https://www.hrothstein.com/national-parks-2050
4 www.london-futures.com
5 On thresholds and liminality, see Chapter 8.
6 See also Fowkes and Fowkes (2025), who argue for the power of art history in understanding the history and politics of environmental change in places which are relatively under-studied in Western academia, like the former Soviet Union.

References

Acosta, I., L. Purbrick, and X. Ribas. 2019. "Trafficking the Earth: Documents on Nitrate, Copper and Capitalism." *Transformations* 19 (1): 21–21.

Adkins, P. 2022. *The Modernist Anthropocene: Nonhuman Life and Planetary Change in James Joyce, Virginia Woolf and Djuna Barnes*. Edinburgh: Edinburgh University Press.

Adkins, P. 2026. *Virginia Woolf and the Anthropocene*. Edinburgh: Edinburgh University Press.

Arnall, A. 2023. "Encountering the Anthropocene: Reconfiguring Human-Nature Relations on the North Norfolk Coast, UK." *Geoforum* 143 (July): 103768.

Baldwin, A. 2016. "Premediation and White Affect: Climate Change and Migration in Critical Perspective." *Transactions of the Institute of British Geographers* 41: 78–90.

Baldwin, A. 2022. *The Other of Climate Change: Racial Futurism, Migration, Humanism*. Lanham: Rowman & Littlefield.

Baldwin, A., C. Fröhlich, and D. Rothe. 2019. "From Climate Migration to Anthropocene Mobilities: Shifting the Debate." *Mobilities* 14 (3): 289–97.

Ballard, S. 2021. *Art and Nature in the Anthropocene: Planetary Aesthetics*. London: Routledge.

Barry, K., and J. Keane. 2019. *Creative Measures of the Anthropocene*. Singapore: Springer.

Bennett, J. 2016. *The Enchantment of Modern Life: Attachments, Crossings, and Ethics*. Princeton: Princeton University Press.

Bennett, M. M. 2021. "Ruins of the Anthropocene: The Aesthetics of Arctic Climate Change." *Annals of the American Association of Geographers* 111 (3): 921–31.

Berger, J. 2013. *Understanding a Photograph*. London: Penguin.

Bergmann, N. T., and R. M. Briwa. 2021. "Re-Envisioning the Toxic Sublime: National Park Wilderness Landscapes at the Anthropocene." *Annals of the American Association of Geographers* 111 (3): 889–99.

Bould, M. 2021. *The Anthropocene Unconscious: Climate Catastrophe Culture*. London: Verso Books.

Burtynsky, E. 2018. 'Photographs: Anthropocene'. www.edwardburtynsky.com.

Caracciolo, M. 2021. "Being Moved by Nature in the Anthropocene: On the Limits of the Ecological Sublime." *Emotion Review* 13 (4): 299–305.

Carroll, S. 2015. *An Empire of Air and Water: Uncolonizable Space in the British Imagination, 1750–1850*. Philadelphia: University of Pennsylvania Press.

Castree, N. 2005. *Nature*. London: Routledge.

Cole, M. B. 2022. "At the Heart of Human Politics": Agency and Responsibility in the Contemporary Climate Novel." *Environmental Politics* 31 (1): 132–51.

Cosgrove, D. E., and S. Daniels. 1988. *The Iconography of Landscape: Essays on the Symbolic Representation, Design and Use of Past Environments*. Edited by Denis E. Cosgrove and Stephen Daniels. Cambridge: Cambridge University Press.

Cresswell, T. 2013. *Soil*. London: Penned in the Margins.

Cronon, W. 1996. "The Trouble With Wilderness; or, Getting Back to the Wrong Nature." In *Uncommon Ground: Rethinking the Human Place in Nature*, edited by William Cronon. New York: Norton.

Curley, A., and S. Smith. 2024. "The Cene Scene: Who Gets to Theorize Global Time and How Do We Center Indigenous and Black Futurities?" *Environment and Planning E* 7 (1): 166–88.

Davies, T. 2013. "A Visual Geography of Chernobyl: Double Exposure." *International Labor and Working-Class History* 84: 116–39.

Davis, H., and Z. Todd. 2017. "On the Importance of a Date, or Decolonizing the Anthropocene." *Acme* 16 (4): 761–80.

Davis, H., and E. Turpin. 2015. "Art & Death: Lives Between the Fifth Assessment & the Sixth Extinction." In *Art in the Anthropocene: Encounters Among Aesthetics, Politics, Environments and Epistemologies*, edited by Heather Davis and Etienne Turpin. London: Open Humanites Press.

Davis, M. 2006. *Planet of Slums*. London: Verso.

Death, C. 2025. *African Climate Futures*. Oxford: Oxford University Press.

DeLoughrey, E. M. 2019. *Allegories of the Anthropocene*. Durham: Duke University Press.

Demos, T. J. 2017. *Against the Anthropocene: Visual Culture and Environment Today*. Berlin: Sternberg Press.

DeSilvey, C. 2012. "Making Sense of Transience: An Anticipatory History." *Cultural Geographies* 19 (1): 31–54.

DeSilvey, C. 2017. *Curated Decay: Heritage Beyond Saving*. Minneapolis: University of Minnesota Press.

Donges, J. F., W. Lucht, F. Müller-Hansen, and W. Steffen. 2017. "The Technosphere in Earth System Analysis: A Coevolutionary Perspective." *The Anthropocene Review* 4 (1): 23–33.

Dora, V. 2009. "Travelling Landscape-Objects." *Progress in Human Geography* 33 (3): 334–54.

Driver, F. 2013. "Hidden Histories Made Visible? Reflections on a Geographical Exhibition." *Transactions of the Institute of British Geographers* 38 (3): 420–35.

Engelmann, S. 2023. "Weathering Three Storms: Experiments in an Elemental Geohumanities." *GeoHumanities* 9 (2): 1–17.

Fisher, M. 2009. *Capitalist Realism: Is There No Alternative?* Ropley: John Hunt Publishing.

Fowkes, M., and R. Fowkes. 2022. *Art and Climate Change*. London: Thames & Hudson.

Fowkes, M., and R. Fowkes. 2025. "The Socialist Anthropocene in the Visual Arts." *ARTMargins* 14 (2): 3–10.

Gandy, M. 2021. "Film as Method in the Geohumanities." *GeoHumanities* 7 (2): 605–24.

Gergan, M., S. Smith, and P. Vasudevan. 2020. "Earth Beyond Repair: Race and Apocalypse in Collective Imagination." *Environment and Planning D: Society and Space* 8 (1): 91–110.

Gómez Cruz, E. 2021. "Everyday Anthropo-Scenes: A Visual Inventory of Human Traces." *Cultural Geographies* 28 (1): 193–201.

Graves, R., and D. Madoc-Jones. 2010. "Postcards from the Future." http://www.london-futures.com/. Accessed 14 May 2026.

Griffiths, H. M., G. Goodwin, T. Keevil, E. Salisbury, S. Tooth, and D. Roberts. 2017. "Searching for an Anthropo(s)Cene in the Uplands of Mid Wales." *GeoHumanities* 3 (2): 567–79.

Grosz, E. 2008. *Chaos, Territory, Art: Deleuze and the Framing of the Earth.* New York: Columbia University Press.

Grotkopp, M. 2023. 'Tipping the Scales. The Interfering Worlds of Anthropocene: The Human Epoch'. *Interfaces. Image Texte Language* (50): 1–23.

Haff, P. 2014. "Humans and Technology in the Anthropocene: Six Rules." *The Anthropocene Review* 1 (2): 126–36.

Hawkins, H. 2019. "Geography's Creative (Re)Turn: Toward a Critical Framework." *Progress in Human Geography* 43 (6): 963–84.

Hawkins, H. 2021. "Cultural Geography I: Mediums." *Progress in Human Geography* 45 (6): 1709–20.

Hawkins, H., and A. Kanngieser. 2017. "Artful Climate Change Communication: Overcoming Abstractions, Insensibilities, and Distances." *WIREs Climate Change* 8 (5): e472.

Heise, U. K. 2019. "Science Fiction and the Time Scales of the Anthropocene." *ELH* 86 (2): 275–304.

Hoffman, D. 2022. "The Pedagogy of Trash: Photography Environmental Activism, and African Dumpsites." In *New Mediums, Better Messages?: How Innovations in Translation, Engagement, and Advocacy Are Changing International Development.* 1st edn, edited by David Lewis, Dennis Rodgers, and Michael Woolcock. Oxford: Oxford University Press.

Horton, J. L. 2017. "Indigenous Artists Against the Anthropocene." *Art Journal* 76 (2): 48–69.

Hoskins, G., and M. Whitehead. 2015. *Mining Memories: Placing the Anthropocene.* Aberystwyth: Aberystwyth University.

Hunchuck, E. M. 2024. "Ways of Reading, Ways of Listening: An Editorial Introduction to Oceanic Refractions." *Oceanic Refractions.*

Johns-Putra, Adeline. 2019. *Climate Change and the Contemporary Novel.* Cambridge: Cambridge University Press.

Kanngieser, A. M. 2023. "Ethics and Ecocidal Listening: Oceanic Refractions as an Artistic Case Study." *Acadia* 52 (2): 169–72.

Kinkaid, E. 2019. "At the Limits of Critical Geography: Creative Interventions into the Exclusionary Spaces of U.S. Geography." *Gender, Place & Culture* 26 (12): 1784–811.

LaDuke, Winona, and Deborah Cowen. 2020. "Beyond Wiindigo Infrastructure." *South Atlantic Quarterly* 119 (2): 243–68.

Le, P. 2021. "Anthropocene: The Human Epoch Review – Colossal Eco Doc Prettifies Disaster'. Film." *The Guardian*, September 4. https://www.theguardian.com/film/2021/sep/01/anthropocene-the-human-epoch-review-jennifer-baichwal-nicholas-de-pencier-edward-burtynsky

Lewis Hood, K. 2023. "Residual Repertoire: Black Geo-Aesthetics after the Mine." *GeoHumanities* 9 (1): 45–63.

Loomis, C. C. 2023. "The Arctic Sublime." In *Nature and the Victorian Imagination*, edited by U. C. Knoepflmacher and G. B. Tennyson. Berkeley: University of California Press.

Mahony, M. 2016. "Picturing the Future-Conditional: Montage and the Global Geographies of Climate Change." *Geo: Geography and Environment* 3 (2): e00019.

Mameni, S. 2023. *Terracene: A Crude Aesthetics. ANIMA: Critical Race Studies Otherwise*. Durham: Duke University Press.

Matless, D. 2014. *In the Nature of Landscape: Cultural Geography on the Norfolk Broads*. Chichester: Wiley.

Matless, D. 2016. "Climate Change Stories and the Anthroposcenic." *Nature Climate Change* 6 (February): 118–19.

Matless, D. 2017. "The Anthroposcenic." *Transactions of the Institute of British Geographers* 42 (3): 363–76.

Matless, D. 2018a. "The Anthroposcenic: Landscape in the Anthroposcene." *British Art Studies* (10). https://britishartstudies.ac.uk/issues/10/landscape-anthroposcene

Matless, D. 2018b. "Next the Sea: Eccles and the Anthroposcenic." *Journal of Historical Geography* 62 71–84.

Matless, D. 2019. "Checking the Sea: Geographies of Authority on the East Norfolk Coast (1790–1932)." *Rural History* 30 (2): 215–40.

Milliken, C. 2023. "Documenting the Anthropocene: The Burtynsky Trilogy." *Canadian Journal of Film Studies* 32 (2): 87–108.

Mirzoeff, N. 2014. "Visualizing the Anthropocene." *Public Culture* 26: 213–32.

Monani, S. 2016. "Feeling and Healing Eco-Social Catastrophe: The 'Horrific' Slipstream of Danis Goulet's Wakening." *Paradoxa* 28: 192–213.

Monani, S. 2024. *Indigenous Ecocinema: Decolonizing Media Environments*. Morgantown: West Virginia University Press.

Monteiro, F., and E. Siddons. 2021. "Fabrice Monteiro's Best Photograph: A Spirit Emerges from a Rubbish Dump in Senegal'. Art and Design." *The Guardian*, November 24.

Morton, T. 2016. *Dark Ecology: For a Logic of Future Coexistence*. New York: Columbia University Press.

Noxolo, P. 2025. "Quantum Black Creative Geographies: Embodiment, Coherence and Transcendence in a Time of Climate Crisis†." *Singapore Journal of Tropical Geography* 46 (1): 6–17.

Nye, D. E. 1994. *American Technological Sublime*. Cambridge: MIT Press.

O'Neill, S. 2025. *The Visual Life of Climate Change*. Bristol: Bristol University Press.

Overend, D., J. Lorimer, and D. Schreve. 2020. "The Bones Beneath the Streets: Drifting Through London's Quaternary." *Cultural Geographies* 27 (3): 453–75.

Parsons, L. 2023. *Carbon Colonialism: How Rich Countries Export Climate Breakdown*. Manchester: Manchester University Press.

Peeples, J. 2011. "Toxic Sublime: Imaging Contaminated Landscapes." *Environmental Communication* 5 (4): 373–92.

Persinger, K., S. Vint, and G. Canavan. 2024. "Science Fiction in the Anthropocene." *Annual Review of Environment and Resources* 49: 51–71.

Peterson, J., and A. Zahara. 2016. "Anthropocene Adjustments: Discarding the Technosphere." *Discard Studies*. https://discardstudies.com/2016/05/26/anthropocene-adjustments-discarding-the-technosphere/

Pfeifer, Lena. 2025. Anthropocene Affordances: Scale, Narrative Form, and the Human in US-American Literature. Bielefeld: Transcript Verlag.

Planetary Portals. 2026. 'Planetary Portals'. https://unthinking.photography/projects/planetary-portals/. Accessed 26 Feb 2026.

Povinelli, E. A. 2016. *Geontologies: A Requiem to Late Liberalism*. Durham: Duke University Press.

Ray, S. J. 2016. "Environmental Justice, Vital Materiality, and the Toxic Sublime in Edward Burtynsky's Manufactured Landscapes." *GeoHumanities* 2 (1): 203–19.19.

Richardson, M. J. 2015. "Theatre as Safe Space? Performing Intergenerational Narratives With Men of Irish Descent." *Social & Cultural Geography* 16 (6): 615–33.

Sajewska, D., and J. Schade. 2025. "Transforming the Anthropo(s)Cene." *Insert* 8 (November). https://doi.org/10.5281/ZENODO.17771620

Scott, C. 2016. "(Indigenous) Place and Time As Formal Strategy: Healing Immanent Crisis in The Dystopias of Eden Robinson And Richard Van Camp." *Extrapolation* 57(1–2): 73–93.

Scott, J. C. 1998. *Seeing Like a State: How Certain Schemes to Improve the Human Condition Have Failed*. New Haven: Yale University Press.

Smith, T. B. M. 2021. "The Anthropocene in Frank Herbert's Dune Trilogy." *Foundation* 50 (140): 62–75.

Stoler, A. L. 2008. "Imperial Debris: Reflection on Ruins and Ruination." *Cultural Anthropology* 23 (2): 191–219.

Tillett, R. 2023. "Criminal Violences: The Continuum of Settler Colonialism and Climate Crisis in Recent Indigenous Fiction." In *The Routledge Handbook of Crime Fiction and Ecology*, edited by Nathan Ashman. London: Routledge.

Tolia-Kelly, D. P. 2019. "Rancière and the Re-Distribution of the Sensible: The Artist Rosanna Raymond, Dissensus and Postcolonial Sensibilities Within the Spaces of the Museum." *Progress in Human Geography* 43 (1): 123–40.

Tuck, E. 2009. "Suspending Damage: A Letter to Communities." *Harvard Educational Review* 79 (3): 409–27.

Turner, J., and D. Bailey. 2022. "Ecobordering: Casting Immigration Control as Environmental Protection." *Environmental Politics* 31 (1): 110–31.

UNESCO. 2018. *The Unbearable Burden of the Technosphere*. Paris: UNESCO.

Urry, J., and J. Larsen. 2011. *The Tourist Gaze 3.0*. London: SAGE.

Vasudevan, A. 2015. "The Artifice of Landscape: Photomontage in the Work of Beate Guetschow." In *Geographical Aesthetics: Imagining Space, Staging Encounters*, edited by Harriet Hawkins and Elizabeth Straughan. Farnham: Ashgate.

Whyte, K. 2017. "Indigenous Climate Change Studies: Indigenizing Futures, Decolonizing the Anthropocene." *English Language Notes* 55 (1–2): 153–62.

Whyte, K. P. 2018. "Indigenous Science (Fiction) for the Anthropocene: Ancestral Dystopias and Fantasies of Climate Change Crises." *Environment and Planning E: Nature and Space* 1 (1–2): 224–42.

Wikipedia. 2025. '*Avatar* (2009 Film)'. In *Wikipedia*. March 10.

Woods, D. 2014. "Scale Critique for the Anthropocene." *The Minnesota Review* 2014 (83): 133–42.

Woods, D. 2021. "Geomedia and Michael Madsen's Into Eternity." In *Narratives of Scale in the Anthropocene*. London: Routledge.

Yusoff, K. 2009. "Excess, Catastrophe, and Climate Change." *Environment and Planning D: Society and Space* 27 (6): 1010–29.

Zhang, B., S. van der Linden, M. Mildenberger, J. R. Marlon, P. D. Howe, and A. Leiserowitz. 2018. "Experimental Effects of Climate Messages Vary Geographically." *Nature Climate Change* 8: 370–4.

10 Placing the Future

Introduction

In a radio interview in 1993, the speculative fiction writer William Ford Gibson made a now oft-quoted remark: 'the future is already here – it's just not very evenly distributed' (The Economist 2001).

The remark was part of his reflections on the themes that have continued to animate his work (like his influential 'cyberpunk' novel *Neuromancer*): the transformation of culture and economies by information technologies, the nature of post-industrial societies, and the shadows and margins of urban life. The quote has been taken to refer to how futuristic, advanced technologies are always unevenly distributed, with inequalities of access leading some members of society to seemingly be living in a different time period than others – a phenomenon sometimes referred to as the 'digital divide' (Pick and Sarkar 2020). In the Anthropocene, if we take a more gloomy view of the environmental future, we might reflect instead on how certain predicted future impacts of processes like climate change are already causing problems in certain parts of the world, like sea-level rise in coastal Bangladesh or drought in the Horn of Africa. Such places perhaps give a glimpse into a future that awaits many others.

But we can also read the quote to imply something about how the future is necessarily *indeterminate* – manifold possible futures exist, and the seeds of them are scattered around in the present, waiting to germinate under the right conditions. This is the theme we want to develop in this final chapter. While some discourse about the Anthropocene and environmental change offers a sense that the future is already determined, and doesn't look particularly rosy, in this chapter, we engage with geographical research concerned with the hopeful *making* of different futures. So, this isn't about evaluating different scientific predictions of the future, or indeed about us making our own predictions! Rather, this chapter is concerned with how societies imagine and picture possible futures, how that imaginative process can be made more democratic, and about the communities, organisations, inventors, activists, and researchers that are actively trying to build alternative worlds in the present.

DOI: 10.4324/9781003162469-14

Whole books could be written on this topic of course, so we've had to be selective. We start by revisiting the Anthropocene narratives we identified in Chapter 1, to identify the kinds of Anthropocene futures they imply and imagine. We then move on to sample how communities in different parts of the world are actively seeking to bring some of those futures into being. We are particularly drawn to examples that have been engaged with by geographers, and to initiatives which self-consciously build on some of the historical and political insights about the nature of the Anthropocene that we've explored in previous chapters. In engaging with these efforts to build alternative futures, we're interested in the conduct of 'prefigurative politics' – 'popularly imagined as "being the change you wish to see"' (Jeffrey and Dyson 2021, 641) – and the material geographies which are brought into being in the present through the enactment of different visions of desirable futures. Along the way, we encounter geographical researchers engaging in different ways with these 'spaces of hope' (Anderson and Fenton 2008; Head 2016) – as observers, analysts, allies, and participants – thus pointing the way to new kinds of engaged geographical scholarship for the Anthropocene.

From narratives to futures

If we think right back to our argument in Chapter 1 about the existence of various ways of telling the story of the Anthropocene, we can see already that there are multiple visions of Anthropocene futures (or post-Anthropocene futures) abroad in the world. In that chapter, we identified five overarching narratives of the Anthropocene which each contain, implicitly or explicitly, certain moral messages or political calls-to-arms about how societies should respond to the challenges of the Anthropocene. They are visions of possible futures – some desirable, to be made real through political or cultural change – and others undesirable, the cost and consequence of humanity failing to act.

We've already discussed some of these imagined futures. For example, in Chapter 8, we explored how for some proponents of the 'standard narrative' of the Anthropocene, a commonly envisioned future is one in which humans embark on engineering planetary systems in order to secure the environmental conditions seemingly necessary for civilisational survival. This is a future where scientific and technical experts are elevated to positions of extraordinary power, tinkering with components of the Earth System on behalf of the rest of humanity. Of course, pressing questions remain over how such an undertaking would be governed, and whether in envisioning a geoengineered future, some envision (and perhaps desire) a future of autocratic decision-making where democracy is put on hold while global environmental problems are solved (Willis 2020)

Quite a dystopian future then – even enthusiastic proponents of geoengineering often talk about it as a 'least worst' option; a bitter pill for society to swallow in order to 'buy time' while it puts its house properly in order.

Many proponents of the standard narrative still baulk at geoengineering, and envisage instead a future where governments at all scales listen to what the physical sciences are telling them about global environmental change, and act with the required urgency and ambition – to decarbonise, to conserve nature, to stop perturbing biogeochemical flows through the Earth System, and so on. A similar vision of society heeding the calls of scientists and experts is offered by proponents of the 'post-natural Anthropocene' narrative, particularly those of an 'ecomodernist' persuasion. For them, a desirable and possible future is one in which humanity embraces its role as 'stewards' of the Earth System, and as a maker of environments. The call is to embrace human ingenuity and inventiveness, to not pull back from interfering in the non-human world, but simply to do it better. Ecomodernists envision a future where economic growth is 'decoupled' from environmental destruction, where liberal market economies are spread to all corners of the planet, where people are 'liberated' from the drudgery of subsistence agricultural labour by the deeper technologisation and intensification of food production, and where more land can be set aside for nonhuman nature (if not a pristine, primordial 'Nature') to thrive (Breakthrough Institute 2015)

This is another grandiose vision of the future, where most of the population is expected to consent to new forms of life and new forms of human-environment relationships. It's a vision that's been criticised for making a range of assumptions about the kinds of futures desired by people working in the agricultural sector, for example, while others have questioned whether market capitalism can ever be 'decoupled' from environmental damages (whether in the shape of greenhouse gas emissions, habitat destruction, species loss, etc.). Some critics 'think ecomodernism pushes one way of life – modern, urban, and based on higher material consumption – as the best and ignores ways of life that do not depend on industrial modernity' (Breewood and Garnett 2022, 14), such as those practiced by many Indigenous communities, or by groups who value spiritual wellbeing over material wealth. We would note that ecomodernism is a set of ideas with a particular geography, 'a particular and partial vision' of the world 'offered from a specific place and time' – significantly perhaps, 'from an institutional base in Silicon Valley, California' (Collard, Dempsey, and Sundberg 2015, 230). It is difficult not to read the projection into the future of a vision of an urban, high-tech, engineered world as a projected universalisation of the lifeworld enjoyed by some in a particular place and time, i.e. wealthy corners of early twenty-first century California (Utrata 2024).

Philosopher Clive Hamilton points out the contradictions, as he sees them, between ecomodernist thinking and the geography which apparently helped shape it. While many in California thrive in the kind of world envisaged for the rest of us in the *Ecomodernist Manifesto*, many others in the state have been reeling from a years-long mega-drought, brought about by reduced run-off from snow melt in the Sierra Nevada mountains, and regular spates of catastrophic wildfires. These are perhaps representative of wider impacts of climate change, which are starting to become visible worldwide, and a harbinger of a much more challenging world to come. Hamilton struggles to square these more troubling images of the future with the more sanguine, optimistic images of a 'Good Anthropocene' offered by the ecomodernists:

> I ask myself what kind of thinking clicks in when the ecomodernist reads these stories or sees the images on the television. How does one overlay images of mega-drought or Hurricane Sandy or the sinking Maldives with a narrative of a good Anthropocene? For me, this remains a mystery.
>
> (Hamilton 2016, 237)

This more pessimistic (some would say realistic) view of a future defined not by 'decoupling' the human pursuit of prosperity from environmental problems, but by repeated socioecological disasters, is one shared by proponents of what we called in Chapter 1 the eco-catastrophist narrative. In this case, however, when attempting to analyse the futures which are being imagined, it is a little harder to disentangle futures which are to be avoided and futures which are to be desired and built. Eco-catastrophists are rather more fatalistic about the trajectory of climate change, holding to scientific findings that further climatic change and sea level rise are already 'baked-in' to the system, and that even drastic action today to reduce emissions would still be followed by intensifying climatic disasters which will shake the very foundations of modern industrial society. For thinkers like Jem Bendell and Rupert Read (2021), the task for thought-leaders and social movements now is to help prepare communities for what is to come, to help them process their grief for a world slipping away and their fear for a world to come, and to start building modes of life that might have some resilience to much more challenging environmental conditions.

This combined emotional and social work has been labelled 'deep adaptation', a framework which encourages communities to think through what aspects of modern, industrial lifestyles are worth protecting and saving, and what aspects can be let go in favour of developing new modes of living with each other and with nonhuman natures. Adherents envision much more localised food systems, cooperative living arrangements, considerably reduced consumption of things like energy and water, and a whole new set of attitudes towards human and more-than-human community:

> [Deep Adaptation] seeks to develop so-called collapse-readiness (i.e., creating an equitable system for distribution of life essentials such as food, water, energy, and health care) and collapse-transcendence (fostering psycho-social-spiritual-cultural shifts to accept and live through collapse with some composure and stability).
>
> (Mark and Lewis 2020, np)

For Deep Adaptation practitioners and other adherents to the catastrophist narrative, these aren't futures to be necessarily desired. But they are futures within which explicit reconsideration can happen of how societies collectively define the 'good life', and in which a set of positive values (such as love, compassion, and generosity) can be nurtured, and around which new ways of living can be organised. So, while the overall vision is pessimistic (at least as far as the fate of industrial modernity is concerned), these narratives see an opportunity for communities to rebuild themselves in ways that might be beneficial for their spiritual, emotional, and cultural wellbeing, even in the face of challenging environmental conditions.

Adherents to the eco-Marxist narrative want to remake the world too, albeit along different lines. With *capital* being the major super-human actor driving the planet into the Anthropocene, the desirable solution and therefore desirable future is a post-capitalist world, where new forms of socioeconomic life are brought into being through collective action or even revolution. Visions of possible 'ecosocialist' futures are many and varied, but they largely 'share a commitment to struggling for a socioecological transition beyond capitalism by democratizing the means of production, subjecting markets to more ecologically rational planning, and subordinating private profit to social use-value and ecocentric production' (Albert 2023, 12). Forms of collective ownership and planning would direct production processes towards those things society really needs, while levels of energy and material consumption would be carefully managed to respect biophysical limits and narrow inequalities between the global north and south. Some envision the knitting-together of existing, local sustainability initiatives to build a worldwide movement for ecosocialism from the ground up (Kallis et al. 2020); others see the need for global revolution and forms of control and management that transcend the existing nation-state system (Magdoff and Williams 2017). Ecosocialists are also divided on whether the challenge is to convince people of the need for 'degrowth' and economic contraction, or whether the move, for example, from fossil fuels to solar energy instead promises a new age of abundance, if only the means of energy production can be wrested out of capitalist control and placed under some form of collective ownership (Huber 2022). But despite these differences, visions of ecosocialist futures are united in offering a vision of human beings as primarily 'ecological beings who have basic needs to reproduce their lives (food, energy, housing, health care, love, leisure)' – needs which can be provided by systems other than capitalism, and in a fashion which doesn't lead to ecological breakdown (Huber 2019, np.).

Adherents to the 'colonial Anthropocene' narrative share similar ambitions, albeit ones that centre dismantling and moving beyond colonial relations, rather than just capitalist ones. Building a decolonial future would mean undoing the relationships of colonial exploitation that undergird things like climate change (see Chapters 3 and 4), putting right the patterns of racialised environmental injustice we saw in Chapter 6, and fashioning a world of equity and justice (Sultana 2022). As we'll explore below, this is a prominent line of thought and action within Geography as it grapples, as a discipline, with the future of human and nonhuman life within what Stephanie Wakefield calls the 'Anthropocene back loop'.

Life in the back loop

The future, baseball star Yogi Berra once quipped, *ain't what it used to be.* Those words have taken on a new set of meanings in the Anthropocene, as citizens of modernity go about recalibrating their orientation to the future. One of the defining features of modernity was the optimism that the future would always be brighter, a progressive and inexorable improvement upon the present (Head 2016). However, if in modernist thought 'time was seen to flow from the present to the future, today we increasingly experience time coming toward us, from the future to the present' (Braun 2015, 239). It is for this reason that philosopher-anthropologist Bruno Latour described the times we are living through as 'apocalyptic' – '[n]ot in the sense of catastrophic (although it might be that also), but in the sense of the revelation of things that are coming *toward* us' (Latour 2015, 12). The heralding of the Anthropocene is this moment of *revelation* – an aesthetic, philosophical, and scientific revelation that further reveals not only 'the full horror of the past' but also 'the shape of things to come' for nature-society relations (Braun 2015, 239).

Yet the modernist drive to predict and perhaps control the future has not gone away. States, corporations, and international organisations have developed a whole suite of means for anticipating and therefore managing these problematic 'things to come'. From sophisticated computer simulations and risk models to emergency management protocols, the future is increasingly crunched as numbers and rehearsed as a reality to come, in a fashion that largely aims at ensuring existing political and economic orders can survive through whatever shocks or challenges that future brings (Anderson 2010). Yet for the resilience theorists we encountered in Chapter 8, the uncertainties of the present moment pose fundamental challenges to modernist attempts to predict, control, and manage the future. For many designers and planners, this has informed a move towards a new paradigm of resilience planning, geared around 'working with nature', adaptiveness, and experimentation (Pickering 2025). But for urban geographer Stephanie Wakefield, this version of resilience remains problematic, as again it merely seeks to make current systems and structures safe from disaster, rather than addressing the root causes of those potential disasters. Nonetheless, Wakefield also finds hope in other forms of resilience work, where 'a much wider field of possibility and … a much broader set of actors' are evident (Wakefield 2020, 57). This is a space and time of experimentation and transformation suited to what she calls the 'Anthropocene back loop'.

The notion of the 'back loop' comes from the resilience ecology of C.S. Holling and the idea of the 'adaptive cycle' through which all kinds of natural, human, and more-than-human systems pass: swamps, forests, cities, organisations, perhaps even civilisations. First comes a 'front loop' stage of rapid growth leading to a period of stability, conservation, and persistence, in which a few entities – species, people, organisations – come to dominate. But these periods of stability aren't permanent. Sudden or slow disturbances can push systems into a 'back loop' in which 'energies and elements previously captured in conservation phases are set free, unexpected new combinations emerge, and wild, exuberant experimentation becomes the modus operandi' (Wakefield 2017, np). The back loop is a time of radical transformation, in which 'structures come apart, leading to a period of destabilization, fragmentation, confusion, and release but also great potential for experimentation, reorganization, and transformation' (Wakefield 2020, 22). Holling himself wondered in 2004 whether global society was already experiencing back loop conditions (Holling 2004). Wakefield has since used his ideas to describe the evolution of the Anthropocene: the historical geographies we narrate in Chapters 3–5 can be read as the 'Anthropocene front loop', but you read this book now as a participant in the Anthropocene back loop.

Both Holling and Wakefield use the example of a forest after a fire. In the post-fire situation, much of the energies and nutrients which were previously organised around feeding the mature forest are 'scattered and released':

> Space is opened for new species to colonize the area. Pioneer species sprout from stumps of burned trees. Birds nest in their charred branches. Genetic mutations prove useful. Undergrowth is cleared, making way for the floor [to] receive sunlight. Ash settles in, returning previously locked-in nutrients to the soil. Surviving species are freed from long-standing relationships, available and open to new combinations, exploring the new zone using seeds in the soil, debris and existing vegetation — "biotic legacies" left behind by the disturbance and creating new combinations and feedbacks.
>
> (Wakefield 2020, 23)

The back loop, then, 'is a time of great possibility' (Wakefield 2020). A seed bank could be used to help re-establish the old forest, or all of these new relations and combinations could be allowed to do their own experimenting in building something fundamentally new. Wakefield sees evidence of the back loop not only in the many indicators of human impacts on the environment, which scientists have used as evidence of the Anthropocene. She sees back loop conditions too in the splintering of some of the central tenets of modern, liberal life and politics, which undergirded the Great Acceleration. The corrosion of the modernist promise of 'predict-and-control' government (see Chapter 8) is accompanied by declining levels of trust afforded to long-standing institutions of democratic government (most notably perhaps, at the time of writing, in the United States); stagnating wages and increasingly insecure work betray the post-war promises of industrial Fordism and undo the achievements of the labour movement; warfare and violence flare-up through new state and para-statal means; social identities increasingly swirl through technologies of interconnection while 'the crumbling of once-coherent narratives' see 'fractious discourses and new tribalisms of all stripes' abound (Wakefield 2020, 30). Since 2011 or so, the world has seen 'riots, insurrections, and revolutions from left to right that, to the front loop mind, may look insane, but are very real', and which add to the evidence that we perhaps not only inhabit the Anthropocene, but inhabit more precisely its back loop (Wakefield 2020, 31).

But Stephanie Wakefield is no catastrophist. She doesn't deny the *possibility* of catastrophe and societal collapse, but sees in the model of the adaptive cycle and its back loop the imminent possibilities of other, perhaps better worlds to come. She keeps her critical eye trained on those who would exploit back loop conditions to further concentrate power and privilege amongst a few. But she also draws our attention, through close empirical research, to 'a vast proliferation of experimentation in redefining what human life will be' (Wakefield 2020): to individuals, communities and organisations who are using the scattered materials and energies of the back loop to experiment with and build new forms of life. This is not simply about finding technical 'solutions' to the problems of the back loop – like pulling carbon out the atmosphere or switching to electric cars. Rather, Wakefield is interested in those who are embracing what it means to live in and through threshold times – these conditions of liminality in which all sorts of natural and human systems appear to be in flux, conditions which could be met with fear and trepidation but which can also be met with an ethos of experimentation, to find new modes of social life and order that might be better suited to an increasingly chaotic and unpredictable world.

To explore these tensions and possibilities, let's examine one expression of this emerging geography of Anthropocene anticipation – the space of the bunker and the figure of the 'prepper'.

Geographies of preparation

The backyard or underground bunker, with its shelves piled high with non-perishable food, medical supplies and perhaps even lethal weapons, has become a key source of personal security for those imagining and anticipating some kind of civilisational breakdown, often to be brought on by either an environmental or biological catastrophe – climate-induced agricultural failure perhaps, or a fast-spreading and lethal global pandemic – through anticipated civil conflict drawn along racial lines, or even divine apocalypse (Garrett 2020; Kirsch and Ray 2025). These kinds of spaces – where vulnerable bodies

Figure 10.1 Providing high-tech, luxury 'bunkers' for the super-rich, across the United States and in places deemed sufficiently 'isolated' like Alaska and New Zealand, has become big business for companies like SAFE. (Reproduced with the permission of SAFE – Strategically Armored & Fortified Environments.)

can be sealed off from the wider world and sustained for long periods, before an expected re-emergence into a world either re-established or fundamentally changed – are an increasingly key part of the anticipatory geographies of the Anthropocene (Figure 10.1). They have a particular prevalence in North America, where Cold War fears of nuclear conflict led many to invest in a protective and sustaining cocoon in which they could ride out the effects of radioactive fallout, and perhaps also any ensuing breakdown in social order. Repurposed amid new fears of environmental, biological, and social insecurity, Casey Ryan Kelly (2016) examines how these spaces – and the people that build them, show them off to others, and expect to inhabit them – function as an expression of an 'apocalyptic manhood'. When we look closely at how these spaces are represented in popular culture (like National Geographic's *Doomsday Preppers* show, for instance), we can detect a vision of masculinity which seeks to recover an imagined and idealised pre-industrial model whereby certain 'manly' skills – to hunt, to kill, to make, to protect – enable individual or familial self-reliance. In this vision, a moment of apocalypse represents something to be feared and avoided, but also an opportunity – perhaps even a welcome one – to re-assert a version of masculinity which some feel has been lost in increasingly 'feminised' post-industrial societies. In the bunker, a much-mythologised version of manhood can be recovered, and its value asserted anew (Kelly 2016).

Geographer and urban explorer Bradley Garrett has mapped this emerging geography of bunkers and 'prepping'. He finds both an 'architecture of dread' as well as speculative hopes that a bunker can function as an 'ark' in which to ride out catastrophe and await the dawning of a new, better world (Garrett 2021; Garrett and Klinke 2019). Stephanie Wakefield has also conducted extensive ethnographic fieldwork amongst 'prepper' communities, and has aimed to dislodge certain preconceptions:

> While popular media and academic literature often stereotype preppers as a mixed bag of white racist men or conspiracy-obsessed wingnuts, there is a much broader participation of races and genders and a deeper interest than personal self-preservation.
>
> (Wakefield 2020, 87)

Wakefield examines how a much wider cast of people have begun engaging in prepper culture, for example by attending classes or weekend retreats on disaster preparedness, 'bugging out' (leaving a place quickly, perhaps on foot, to escape danger), and survival skills. Popular depictions of prepper culture have presented it as primarily a rural phenomenon, existing in spaces whose isolation from urban centres enables the cultivation of further disconnection and cocoon-building. But geographical researchers have emphasised the urban geographies of prepping too, whether in the spread and normalisation of 'go-bags' in the urban or suburban homes of those threatened by wildfires in Australia, or in North American urban centres where catastrophic storms are increasingly prevalent.

Often this wider cast of preppers is motivated not by imaginations of catastrophes-to-come, but by experiences of recent disasters. For example, in post-Hurricane Katrina New Orleans, private military contractors were dispatched to rescue residents of wealthy neighbourhoods while poor people were left to die by the roadsides. Such events have prompted many to ask 'how not to be hostage to relief agencies…or governments that disdain whole populations?' (Wakefield 2020, 86). Wakefield reports how, for many diverse reasons, significant numbers of urban and rural Americans are 'taking up learning new skills':

> Attend survival skills classes in upstate New York or deep in Queens, and you will be sitting on wood logs sharing a fire with Dominican dads trying to help their kids learn about life; white and black suburban families with teenagers in hot pink bandanas; kids from Flushing leading field trips to the woods to help their fellow youth toughen up; all trying to skill up on basic life capacities they see themselves as lacking. From classes to gear, neighbors and individuals are organizing to help each other and themselves in the next storm.
>
> (Wakefield 2020, 86)

Participants in such initiatives inhabit a range of political standpoints, and transcend conventional images of prepper demographics and their 'apocalyptic manhood'. Instead, both Wakefield (2020) and Barker (2020) emphasise the increasingly communal aspect of emerging prepper cultures, and their orientation around practices of care for both the self and for others. They are informed by recent post-disaster experiences, where communities exhibited not the rugged, individualistic survivalism of apocalypse movies but rather an upwelling of voluntary mutual aid and cooperation, as folks banded together to develop new ways of provisioning food, water, energy, and care. As such, prepping becomes a form of social action independent of the state, and a practice of *deepening* existing social relations rather than preparing for their severance:

> Some preppers see learning to care for oneself and others as a basic trait of maturity and dignity. Some distrust or fear government … Others are curious, motivated by the joy of learning. Others are on a quest for something, whether dignity, autonomy, security, or a degree of knowledge about or connection to the conditions of their existence.
>
> (Wakefield 2020, 88)

Wakefield (2020) and Barker (2020) therefore seek to rehabilitate the 'prepper', from a figure of popular and academic ridicule to a figure who perhaps quite presciently discloses the fragilities of contemporary infrastructures and the myth of the universality of state protection, while fostering new kinds of community and new levels of what Barker calls 'crisis attunement'. In preparing for a rocky future in the back loop, such communities are not simply trying to find ways to hunker down and ride out a crisis, but rather are actively experimenting in new ways of organising autonomous communities and provisioning the essentials of life.

3-D printing a civilisation

In a similar vein, the Open Source Ecology (OSE) movement is experimenting in building new forms of autonomous, self-reliant communities that are fit for the Anthropocene back loop. But while prepping is often (although certainly not entirely) about 'awaiting an exceptional moment when everything would change', OSE is, for Wakefield, more thoroughly suited to the back loop, orientated as it is not just towards singular catastrophic events but towards leaving 'current modes of social and economic organization' behind entirely (Wakefield 2020, 92).

OSE is a project started by Polish-American physicist Marcin Jakubowski, whose main aim is to develop and make widely available a 'Global Village Construction Set' – 'a modular, DIY, low-cost, high-performance platform that allows for the easy fabrication of the 50 different Industrial Machines that it takes to build a small, sustainable civilization with modern comforts' (OSE 2025a). Blueprints for the modular, scalable machines – from brick makers

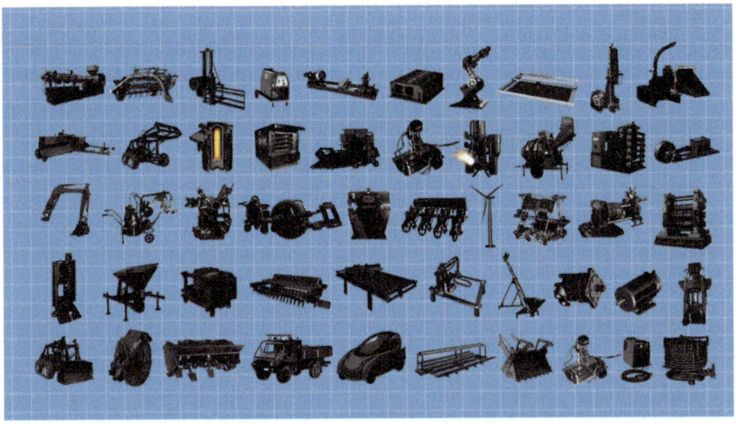

Figure 10.2 OSE's Global Village Construction Set – the machinery you need to build a new civilisation from scratch. (Open Source Ecology/Creative Commons Attribution-ShareAlike 4.0 International.)

and bread ovens to tractors, 3-D printers, wind turbines – are designed such that communities can engage in new forms of building and production that are largely autonomous and sustainable, and which can provision not just the essentials of life but allow people to live both in comfort and independently of capitalist means of production and consumption (Figure 10.2). OSE compares their designs to a giant LEGO set, with machines using interchangeable parts and adding up to construction systems that in turn can produce new machines. The goal is to redistribute technical know-how and 'raw productive power to the people', to enable communities to autonomously and locally solve the complex, 'wicked problems' of environmental and social change (OSE 2025b).

As with any vanguard idea for potentially world-changing technology, the OSE offers a vision that is not just technological but is social and political too. It proffers a wholesale example of a 'socio-technical imaginary', which are:

> collectively held, institutionally stabilized, and publicly performed visions of desirable futures, animated by shared understandings of forms of social life and social order attainable through, and supportive of, advances in science and technology.
>
> (Jasanoff 2015, 4)

In this case, the 'social order' being imagined is politically radical: 'Governments as we know them become obsolete', with the redistribution of economic and technological capacities creating complete autonomy from remote, centralized forms of authority like the nation-state. Instead, the OSE foresees 'an equal playing field of competent, well-organized, small-scale, decentralized republics', springing up 'after the borders of empires dissolve' (quoted in Wakefield 2020, 91). From a modest base in rural Missouri, USA, OSE is experimenting in new ways of buildings things *and* communities, for example by offering studentships and fellowships to young people who can learn the skills and concepts of the open-source technology movement while contributing to the material development of OSE's campus – itself a 'living model' of the kind community it hopes can be replicated elsewhere (OSE 2025c).

As such, the OSE movement practices and embodies a form of prefigurative politics which Stephanie Wakefield suggests is well-suited to the Anthropocene back-loop and the 'need and possibility of transforming the very modes of our lives here and now' (Wakefield 2020, 92). While the movement may yet be some way off remaking the political geography of nation-states and corporate control over technological innovation, it nonetheless offers a powerful vision – articulated not just in written manifestos but in existing material practices – of how things could be otherwise.

Reparation and abolition

Prepper culture and the OSE movement are, by definition, forward-looking, and see communities experimentally building new geographies of autonomous life amid a ruined or at least fundamentally transformed modernity. Other communities are grappling more overtly with the legacies and inheritances of past upheavals and violence. In this section, we move from preparation to reparation, examining the growing climate reparations movement alongside emerging geographies of 'abolition ecology' as another means of building new forms of life in the Anthropocene back loop.

Climate reparations

In climate change politics, the notion of 'reparation' has recently come to renewed prominence.[1] Bolstered by the increasing emphasis placed on experiences of, and responsibilities for, ongoing 'Loss and Damage' within international climate policy, the notion of climate reparations has come to mainstream prominence. While conventional approaches to climate change mitigation and adaptation aim to avoid or deal with *future* impacts of climate change, 'climate reparations address present damage' (Mohan 2025, 1). Climate reparations thus:

> represent a broader acknowledgment of historical inequities and ongoing harms stemming from the disproportionate contribution of industrialized nations to greenhouse gas emissions … Advocates for climate reparations

> suggest that nations with significant historical emissions should recognize their role in the current climate situation and provide assistance to those communities, frequently developing countries, [Smalls Island Developing States], and marginalized populations, who are experiencing the most severe consequences.
>
> (Mohan 2025, 2)

Climate reparations aim for a form of *reparatory justice*, concerned with writing historical wrongs, undoing the present-day legacies of historical iniquities, and fostering a future of ecological justice (Obeng-Odoom 2024). The climate reparations movement grew out of a wider movement for historical redress relating to slavery, Indigenous genocide, indentureship, and other forms of racialised exploitation (Beckles 2019; Táíwò 2022). Within international climate policy and the UN Framework Convention on Climate Change (UNFCCC), notions of reparatory justice have begun to inform policies around financial compensation for Loss and Damage, but reparation scholars have urged the international community to go further.

Loss and Damage policies are arguably motivated and informed by the kind of 'thin geography' we described in Chapter 2 – that is, descriptions of where the damages are taking place, in comparison to where the historical sources of greenhouse gas emissions have been concentrated. A 'thicker geography' might account for the research findings we introduced across Chapters 3 and 4, which showed how the geography of industrialisation was not independent of the geography of colonial exploitation. A world-systems or world-ecology analysis shows how plantation slavery in the 'New World' was a crucial condition of possibility of, for example, a profitable coal industry in Great Britain (Moore 2015; Yusoff 2024), and that the legacies of colonialism and the plantation system continue to produce vulnerabilities to further forms of socio-ecological harm (Luke and Heynen 2020). Therefore:

> reparative justice is forward-looking and seeks to transform ongoing structures of oppression by translating into housing, energy, land, and water policy the debts that are owed to those disproportionately Black and Brown communities whose ancestors made this accumulation of energy and energy-powered infrastructure possible in the first place.
>
> (Fitz-Henry and Klein 2024, 3)

Such a conceptualisation, based on a reading of the 'thick geographies' of colonialism and the reproduction of inequality and vulnerability in the present, lead Fitz-Henry and Klein to some more radical policy proposals for ensuring reparative justice, which go far beyond those of Loss and Damage mechanisms in the UNFCCC. They draw on the work of philosopher Olúfẹmi Táíwò (2022), who argues that reparations should be about much more than repair and redress; rather, they should be a form of what he calls 'worldmaking'.

Rather than just seeking justice for a specific, historically-wronged group, the worldmaking project of reparative justice seeks to transform and decolonise the systems that created those inequalities in the first place, and in so doing actively build 'the just world to come' by redesigning the systems that distribute rights, advantages, harms, and resources (Táíwò 2022, 84, 74).

Táíwò's project of reparative justice is a self-consciously ambitious one, concerned with global systems and generational change. Fitz-Henry and Klein (2024), writing in the journal *Political Geography*, show how the kind of reparative principles espoused by Táíwò are being put to work in initiatives such as Bolivia's Cochabamba Declaration and the Movement for Black Lives (M4BL) in the United States.

The Cochabamba Declaration (or *The Universal Declaration of the Rights of Mother Earth*) was a proposal put forward by a consortium of civil society groups led by Indigenous activists from South America and beyond following a conference in Bolivia in 2010. The declaration includes a range of reparative demands concerning payment of 'climate debts', resistance to ongoing forms of colonialism and imperialism, the establishment of an International Climate and Environmental Justice Tribunal, and calls for countries in the Global North 'to do the work of much more robustly 'decolonizing the atmosphere'' (Fitz-Henry and Klein 2024, 4). This is about much more than financial compensation, but rather about restoring integrity to *Mother Earth* and all its beings. Fitz-Henry and Klein argue that the declaration is 'one of the most important global articulations of the interlocking forms of restitution – or reparation – required for any significant structural transformation of a still violently colonial and racist world-system' (Fitz-Henry and Klein 2024, 4).

The M4BL is most widely known for being at the forefront of global demands for the end of anti-Black police violence and for the abolition of carceral responses to social and racial injustice. Nonetheless, the M4BL has also developed 'one of the most ambitious climate agendas anywhere in the world', called the Red, Black and Green New Deal (RBGND; Fitz-Henry and Klein 2024, 4). The 9-point programme centres abolitionist ideas and reparative principles in its calls for global climate justice. It calls for financial divestment from corporate agents of securitisation and extractive capitalism, for investment in the health and well-being of Black and Indigenous communities, the return of land to those communities, and the design of new financial mechanisms to help them navigate both the transition to new forms of energy production and consumption as well as the challenges of adapting to a changing environment.

Reparative justice principles are prominent, and like the Cochabamba declaration, the RBGND (and the campaigns that have followed in its wake) offer a 'thick geography' of Anthropocene (in)justice. Fitz-Henry and Klein (2024) point out how both initiatives utilize more 'expansive temporalities' than climate and environmental justice movements have operated with to date. While environmental justice campaigns have been driven by readings of an unequal geographical distribution of risks and harms in the present (see Chapter 6), these initiatives both 'repeatedly invoke the histories of more than 400 years of slavery, genocide, colonization and racial discrimination as part of their forward-looking proposals for transformation'. For example, when it comes to re-thinking land tenure practices and ownership, the RBGND emphasises the long history of Black stewardship and care for land in North America and beyond, which isn't mirrored in the geography of present-day land ownership. The Cochabamba Declaration makes similar points about Indigenous-led land stewardship practices and how they've been steadily undermined by centuries of colonial and neo-colonial practices (Fitz-Henry and Klein 2024, 6).

Alongside these expanded temporalities, both initiatives base their demands on understandings of the thoroughly 'relational geographies' of environmental change. For Fitz-Henry and Klein, this means that they both go beyond the 'methodological nationalism' that has shaped analyses of, and policy responses to, environmental injustice to date. Rather than accounting for distributions of risks and harms *within* nation-states, or just comparing the aggregated vulnerabilities and responsibilities *of* nation-states, the two initiatives think about justice across scales. Emphasis is placed on how relations and processes of coloniality reach across borders, and how they result in differential impacts at regional and local scales that are often invisible to national and international policymakers. For example, while the RBGND is very clear about the historical responsibilities of the United States as a country, it stresses the differential experiences and fortunes of Black communities both within the United States and beyond. The authors 'repeatedly trace the linkages to communities in the global Black diaspora, highlighting connections between US energy consumption and the 'continuing harms' inflicted on Black communities throughout the Global South' (Fitz-Henry and Klein 2024, 7).

These 'thicker' understandings of the geography of the Anthropocene and environmental injustice can inspire radically different policy proposals to those normally considered by international policymakers; from land redistribution through to the establishment of new global institutions for adjudicating on the realities of environmental injustice. Fitz-Henry and Klein propose an important role for geographers in all this:

> our hope is that…geographers will pursue more detailed quantitative and qualitative studies of how and where these reparative demands are gaining traction, how and where they are being resisted or reconfigured (and by whom), and how and where they are bumping up against more technocratic approaches or being transformed by other social and political struggles.
>
> (Fitz-Henry and Klein 2024, 8)

They call for detailed research on how existing environmental initiatives around the world, whether activist-, community- or government-led, are already channelling reparative justice principles, and thus prefiguring, across different spatial scales, the kind of global worldbuilding that Táíwò (2022) advocates for. In the next section, we show how some geographers have begun pursuing this task under the guise of 'abolition ecology'.

Abolition ecology

This notion has been developed by geographers concerned that the field of Political Ecology, with its interest in how social inequalities are reproduced through relationships between people and their environments, often in the context of agrarian communities, has paid insufficient attention to the production and reproduction of racial inequalities through the workings of racial capitalism (Heynen and Ybarra 2021). Adapting the notion of 'abolition geography' from scholar-activist Ruth Wilson Gilmore, Nik Heynen and others have proposed 'abolition ecology' as a concept with which to both study and advocate for practices of world-making and place-making among Black and Indigenous communities, which seek to transcend racialised processes of environmental appropriation and destruction. It builds from 'the political ecological imperatives of access to fresh air, clean water, sufficient land, amelioration of toxic chemicals, and beyond', but seeks to not 'define communities by the shared violence they suffer – racism, capitalism, and settler colonialism', but rather according to 'the places they have made for themselves' (Heynen and Ybarra 2021, 23). Abolition thinking draws inspiration from the scholarship and activism of the Black Radical Tradition (see Pulido and De Lara 2018), with its roots in anti-colonial and anti-slavery movements and its ongoing commitment to dismantling oppressive institutions that continue to produce racial inequalities (Gilmore 2007). But it is a tradition of thought that is not solely about taking apart powerful structures; rather, 'abolition is a theory of change…It's about making things' (Gilmore, quoted in Heynen and Ybarra 2021, 26). For Gilmore, liberation from oppressive power structures is achieved through place-making, through building new geographies in which radically different forms of life can be fashioned and practiced.

Abolition ecology, then, is about cultivating human-environment relationships which reject the imperatives of racial capitalism, which stand outside of dominant modes of production and exchange, and which seek to prefiguratively experiment with new ways of organising life, land, and labour. It is a response to an Anthropocene conceived as fundamentally colonial in its origins, and to the 'racialised ways that nature has always been unevenly socially produced through relations of empire, settler colonialism and racial capitalism' (Heynen and Ybarra 2021, 27).

Much theorising of abolition ecology has been driven by work on the geographies of the US South. States like Mississippi, once the cradle of the plantation-slavery nexus, continue to see vast inequalities of wealth and power, often structured along racial lines. Despite the material wealth of the

Mississippi Delta, Black populations continue to endure grinding poverty. While sitting atop the richest agricultural land in the country, rural Mississippians face profound food insecurity and high rates of ill-health. Geographers like Clyde Woods have documented how the promises of post-emancipation economic development were broken, and how racialised inequalities of wealth and power have been maintained through the actions of economic and political elites (Woods 2017). In the face of the continuing power of what Woods called the 'Plantation Bloc' – the coalition of interests that included plantation owners, White politicians, financial and development institutions and law enforcement (often with significant overlaps) – Black communities have devised new modes of resistance which have sought to either overthrow oppressive institutions or, echoing earlier geographies of marronage (see Chapter 3), to craft new modes of life outside of their reach (Freshour and Williams 2023).

For example, in the 1960s, civil rights activist Fannie Lou Hamer responded to impoverishment and starvation of Black agrarian communities by setting up the Freedom Farm Cooperative, which 'rejected capitalist logics of property ownership and agricultural production, substituting sustainability, cooperative economics, and Black self-sufficiency for the pursuit of profit' (Ludwig 2021, 6). At the time, most of the Black community of Sunflower County, Mississippi worked in agriculture, but fewer than 0.2% owned land. Mechanisation was forcing agricultural labourers off the land, poverty was deepening, and infant mortality and diet-related illnesses were the highest in the country. The county was 'the epicenter of food insecurity, hunger, racism, and poverty in the U.S.' (Smith 2019, 36). The Cooperative sought to address food insecurity within the Black community, as well as to provide essential social services, education opportunities, and housing, which the state refused. The initiative expanded and 'created an alternative food system' which met the needs of impoverished families while also allowing local communities to put their deep agricultural knowledge to work in the process of cultivation (Smith 2019, 37). In the mid-1970s, the Cooperative began to struggle to support its social programmes with its agricultural revenues, which were hampered by repeated floods and droughts, and the death of Hamer in 1977 saw some of its most important backers drift away. But while the initiative was relatively short-lived, it continues to inspire and influence the fashioning of new Black agrarian and urban geographies (see McCutcheon 2019), and experiments in post-capitalist communities and economies.

Within such spaces, geographers are developing new ways of working with, alongside, and for communities who are at the frontlines of environmental, racial, and social injustice (Derickson 2022). For example, Nik Heynen has developed a years-long collaboration with residents of Sapelo Island on the coast of Georgia, USA. The island is home to the largest population of Gullah Geechee People in the United States – descendants of people taken from the rice-growing regions of West Africa and enslaved on the rice, indigo, and cotton plantations of the lower Atlantic coast. These coastal plantations saw the emergence of a distinctive culture, language, foodways, and music, which community organisations are now fighting to preserve. Heynen's partnership with the Sapelo Island community has involved supporting efforts to resist further land appropriation for tourism developments, to ensure equity in planning for sea level rise, and to re-establish agricultural traditions in a way that can support the community's economic development. Heynen writes with Maurice Bailey – whose ancestry on Sapelo reaches back to some of the earliest slaves brought to the island – to describe a practice of 'liberation farming' which 'combines contemporary agricultural practice with age-old struggles for freedom' (Bailey and Heynen 2020, np). Together, they have sought to reclaim the African heritage of sugar production, identifying the enslaved peoples whose agricultural expertise was used by plantation owners to increase their profits, and reestablishing heritage crop varieties like Purple Ribbon Sugarcane, Geechee Red Peas, and sour oranges. The aim is both to generate revenue to support the community and to 'lock Geechee culture into the landscape for generations to come' (Bailey and Heynen 2020). The project shows how geographical scholarship on questions of race and environmental pasts and futures need not stop at collating, analysing and publishing data, but can also involve close, enduring collaboration with communities, mud on the boots, and long days out working together under a hot sun.

The most radical city on the planet

A few hundred miles east of Sapelo Island, in another heartland of the plantocracy, lies the city of Jackson, Mississippi. In 2017, civil rights activist Chokwe Antwar Lumumba was elected mayor by a landslide, having run on the promise of making Jackson 'the most radical city on the planet'. Lumumba was supported by Cooperation Jackson, an eco-socialist cooperative movement aiming to cement a 'just transition' in the city. The organisation's work is deeply rooted in the historical geographies and racial politics of the region, and takes direct inspiration from initiatives like the Freedom Farm Cooperative, as well as from peasant and Indigenous movements from across the Americas. Jason Ludwig describes the 'critical historical consciousness' which drives Cooperation Jackson's work, and which 'interprets the overlapping political, economic, and environmental crises' of the Mississippi region – a regional back loop, perhaps – through the histories and geographies of racial violence and oppression, and the legacies of slavery and plantation agriculture (Ludwig 2021, 7).

With an anti-capitalist praxis 'that centers environmental sustainability, cooperative economics and working class power' (Ludwig 2021), the organisation aims to tackle head-on the consequences of changing global and regional economic geographies, and the local socioecological legacies of the plantation system. While the state saw economic growth during the nineteenth century through global trade in 'cash crop' agricultural commodities, the declining significance of such commodities, and the broken promises of post-emancipation reconstruction, has left the region – and particularly many of its Black populations – facing a lack of employment opportunities and food deprivation (Akuno and Nangwaya 2017). Furthermore, as we saw in Chapter 6, the imprinting of the petrochemical industry on former plantation landscapes has led to the contamination of land, air, and water, with disproportionate impacts on already-marginalised communities.

As such, Cooperation Jackson 'situates itself within the historical trajectory of resistance to the intertwined structures of racial violence and environmental exploitation that have produced Black disposability and degraded ecosystems' (Ludwig 2021, 7–8). At the centre of the vision are models of 'Sustainable Communities' and 'Community Production'. The organisation seeks to 'create a Just Transition for Jackson from extractive energy and industries'. Projects include developing part of West Jackson into an eco-village with affordable, rent-stabilised and energy-efficient housing, while a Fabrication Laboratory ('Fab Lab') aims, like the OSE movement, to democratise forms of sustainable production and technical innovation in the service of developing a city-wide economy that actively builds community (rather than corporate) wealth (Cooperation Jackson 2025; Figure 10.3).

While rooted in the struggles of Black and other marginalised communities in urban Mississippi, Cooperation Jackson has begun to inspire and inform related projects in other geographical contexts. Cooperation Hull in the UK deliberately namechecks the Jackson movement, and has taken inspiration from its emphasis on economic democracy and the 'right of all people to

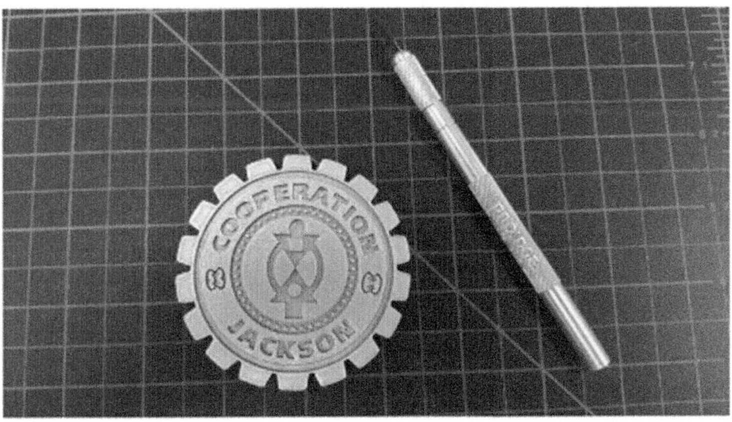

Figure 10.3 A keyring 3-D printed in co-operation Jackson's maker space. (Copyright: Cooperation Jackson, reproduced under Creative Commons Non-commercial 4.0 International.)

shape the economic conditions that influence their lives'. The Hull movement has emphasised cooperative working arrangements as a way of fostering a more sustainable, democratic local economy, supporting initiatives like the Hull Delivery Coop and The People's Land Cooperative, while Cooperation Jackson's emphasis on People's Assemblies to shape city financial decisions has inspired Cooperation Hull's efforts to establish a city-wide assembly in order to democratise civic politics and help create an urban economy that works for both people and planet (Cooperation Hull 2025). In these two cities on either side of the Atlantic, we can therefore see how community-led initiatives to transform urban life are responsive to local histories, material conditions, and cultural values, but also tied together by globally circulating ideas and processes of learning and adaptation. As we explore more below, this is an important aspect for thinking about how 'local' initiatives like these can become more 'global'.

Experimental spaces and the geography of hope

All of the above examples of communities attempting to build radically new modes of living can be thought of as experiments in building alternative futures. When we talk about 'experiments' in this context, we don't mean it in the sense of the laboratory tinkering and hypothesis testing of positivist science, but rather in terms of an open, ongoing, always unfinished process of collaboratively trying out different solutions to different problems (Sabel and Victor 2022). They may be very practical initiatives, 'born out of necessity', and not necessarily starting 'from a particular radical politics'. But, as various observers have noted, through their experimental processes of 'becoming and making', a new kind of radical politics can emerge, often as participants see first-hand the intransigence of the state as a potential ally, or the forces of commercial interests ranged against efforts to, for example, remake local food systems (Pickerill 2021, 250).

There is a *hopefulness* in these spaces of experimentation – a belief that the world can be made better, both locally and perhaps even globally, 'that stands in contrast to gloomy prognostications of catastrophic climate change and earth systems pushed past tipping points' (Braun 2015, 241). As encounters between people, practices, ideas, and materials, grounded often in particular places and shaped by both local and global histories, they represent attempts to prefigure a different future – 'to act in the present how they intend the future to be' (Pickerill 2021, 250). In this vein, geographer Leslie Head explicitly describes *hope* as a 'process that creates possibility and potential', and 'a starting point for how we might imagine worlds otherwise' (Head 2016, 74).

Geographers have been busy in recent years tracing the origins, history, struggles, and victories of many such movements for change. For example, Patrick Bresnihan and Naomi Millner document 60 years of feminist, anticolonial, Indigenous, workers' and agrarian movements which have coalesced around questions of land, labour and social reproduction, making important 'ecological' gains even though often operating outside of the framework of Western environmentalism (Bresnihan and Millner 2023; see also Undisciplined Environments Collective, 2024). Another important strand of geographical work has concerned urban environmental 'experiments': new ways of dealing with things like climate change in cities through collaborations between public and private actors to intervene in cities' material infrastructures and make them more suited to the challenges of either reducing greenhouse gas emissions or of a changeable, perhaps more extreme local climate (Bulkeley and Castán Broto 2013; Edwards and Bulkeley 2018). Meanwhile, geographers and environmental social scientists have been busy 'mapping' the diverse ways publics in countries like the UK are engaged with sustainability transitions – whether through participation in protests, community energy groups, deliberative processes like Citizens' Assemblies, or through engaging with simple behavioural changes (Chilvers et al. 2021; Eden 2016). The aim is to convince governments that their role is not simply to educate or persuade the public about the need to adopt new technologies or consumption habits, but rather to listen to, learn from, and help foster the system-level transformations that are already taking place (Pallett, Chilvers, and Hargreaves 2019).

Some Anthropocene thinkers have gathered together these spaces of hopeful experimentation and future-making under the banner of 'Seeds of Good Anthropocenes'. While the notion of the 'Good Anthropocene' has often been associated with techno-optimism and market fundamentalism of the 'ecomodernist' movement (Hamilton 2016; Salleh 2025), the 'Seeds' project takes a more open and potentially radical stance towards curating together examples of community and organisational experiments in building just and sustainable futures. From blockchain technologies to community gardening projects, the Seeds project offers a database of 'existing innovations that aim to address social-ecological challenges', but which 'are not yet mainstream' (Pivo 2025, np). Drawn in large part from the Global South, they are 'small-scale experimental projects that embody new ways of thinking and acting, often at the margins of current systems and worldviews' (Preiser, Hichert, and Biggs 2024, 5):

> They can be social initiatives, new technologies, economic tools, or social-ecological projects, or organisations, movements or new ways of acting that appear to be contributing to the creation of a future that is just, prosperous, and sustainable.
>
> (Pivo 2025, np)

The emphasis is on initiatives which facilitate socio-ecological transformation not through government or corporate policies, but through collective, democratic effort organized through localized networks and 'reflecting the values and aspirations of local actors' (Preiser, Hichert, and Biggs 2024, 5). The Seeds project therefore recognises something fundamental about the prefigurative geographies of the Anthropocene: that the world is complex and heterogeneous, and that Anthropocene futures are unlikely to be successfully imposed 'from above' and in a uniform fashion; instead, multiple futures will be made in multiple places, products of local values, material conditions, histories, and aspirations (LaDuke and Cowen 2020; Priyadharshini 2021).

This stands in contrast to how Anthropocene futures have tended to be imagined. Recall our discussion above about how the different dominant narratives of the Anthropocene contained different imaginations of what the future could or should look like. Most of them offered fairly undifferentiated, global stories. And that's how a lot of environmental scenario-planning has worked to date. For example, the Intergovernmental Panel on Climate Change (IPCC) publishes scenarios of global socioeconomic changes to predict possible future greenhouse gas emission pathways and the impacts of international policy agreements. Climate change futures, then, are commonly talked about in terms of aggregate metrics of global emissions, global temperature, global impacts, and global policy (Hulme 2010; Stoddard, Anderson, and Capstick 2021). However, this emphasis on global policies, helped along by 'well-established global models', has meant that 'these futures have emphasized global and regional processes and commonalities at the expense of national and local dissimilarities, questions and challenges' (Bennett et al. 2021, 173). The global perspective assumes – perhaps not entirely unreasonably – 'that it will be possible to have globally coordinated and orchestrated actions that propel the Earth System toward one path or another or that global goal setting will suffice to produce change' (Bennett et al. 2021). But geographers and other social scientists whose instincts draw them to local particulars rather than global abstractions will tell you that things are rarely that simple (e.g. Death 2025).

Current geopolitical trends would seem to bear that out – just at the moment when the 'poly-crises' of the Anthropocene seem to call for a coordinated, cooperative global response, a politics of national isolationism is on the rise, with large, economically influential states pulling back from institutions of international collaboration (Wainwright and Mann 2018). But even without these 'macro-level' trends, many sustainability scholars would join geographers in emphasising that the world is inherently 'complex and contradictory'. Echoing anthropologist Anna Tsing's emphasis on the spatial heterogeneity and 'patchiness' of Anthropocene landscapes (Tsing, Mathews, and Bubandt 2019), and geographer Leslie Head's emphasis on the Anthropocene as a set of geographically differentiated conditions (Head 2016), Elena M. Bennett and her sustainability science colleagues describe 'a patchwork of local and regional contexts' which together shape the ways in which 'global drivers' produce vastly different outcomes in different places (Bennett et al. 2021, 174).

The Seeds database helps bear this out – it offers a literal map of local communities and initiatives struggling with local material conditions, historical inheritances, structural inequalities, and communal hopes to build better socio-ecological futures; phenomena which aren't very easy to model at a global level! But the Seeds project goes beyond the empirical work of collecting and mapping examples of emerging Anthropocene futures; it also aims to use these seeds to develop new ways for other communities to envision their own futures. Members of the Seeds project have used collections of existing seeds as starting points for collaborative visioning and scenario-planning exercises. Starting with a set of seeds that vary in their 'worldviews, problematisation of the current state of the world, and approaches to a better future', participants are invited to imagine how the world might look if the seeds became fully mature and mainstream, how they might interact, and how they might reshape both local and global socio-ecological relations (Bennett et al. 2021, 174). This then informs the development of a positive vision for a chosen future and pathways for its realisation, including analysis of how currently marginal practices might grow, how currently powerful institutions might decline, and the various 'enabling conditions required for a successful transition' (Preiser, Hichert, and Biggs 2024, 8). The Seeds methodology has so far been used to develop positive visions for Anthropocene futures, often with a regional focus (e.g. Hamann, Biggs, and Pereira 2020), and has begun to influence the future-visioning practices of international science-policy organisations like the Intergovernmental Science-Policy Platform on Biodiversity and Ecosystem Services, dislodging mainstream approaches to global scenario-planning (with their 'thin geographies') and emphasising 'more of the diversity and complexity of the world' (Bennett et al. 2021, 175) – what we've been calling here 'thick geographies'.

The seeds, then, are doubly generative: they describe the many ways in which communities are already experimenting in building different futures – experiments with their own potential to grow, scale-up and transform; and they offer a starting point for other communities to begin the complicated work of democratically envisioning and thereafter building alternative futures for themselves.

Sceptical readers might, at this point, question whether this emphasis on small-scale, community-led initiatives to, for example, remake local food systems or preserve local pockets of biodiversity, is really up to the task of dealing with the monumental global challenges of global climate change, ecosystem collapse, perturbed biogeochemical cycles, and so on. And in the face of, for example, an oil and gas industry seemingly doing whatever it can to maintain its profit margins by pumping evermore planet-heating gases into the atmosphere, or in the face of some growing government intransigence towards the policies seemingly required to meet 'net zero' emissions, can a network of urban gardens, community currencies, and the like really be the answer?

Community-led sustainability initiatives regularly face the charge of insignificance, with critics arguing that they can never truly 'scale-up' to achieve emissions reductions or biodiversity savings that can make more than a scratch on the surface of those graphs that detail the global trajectory of Anthropocene environmental devastation (see Chapter 5). But such criticisms tend to come from a place of *solutionism* – the idea that environmental problems are easy to define and relatively straightforward to solve with a single fix – and often a technological one (Head 2016). Proponents of such a view tend to point towards humanity's relative success in fixing the hole in the ozone layer over Antarctica – scientists identified the problem, the international community came together to regulate the gaseous emissions (chlorofluorocarbons) that were causing the problem, and engineers came up with alternative technologies, which enabled things like refrigerators to continue doing their jobs without contributing further to the problem. As a result, the ozone layer seems to be steadily recovering (Wang, Solomon, and Santer 2025). Yet the manifold other environmental challenges of the Anthropocene are arguably much more complex, and far less amenable to relatively simple policy and technological 'fixes'. So while the geoengineers of Chapter 8 might push for a global technological fix to global climate change, for example, others advocate a more bottom-up approach to building environmentally and socially just futures (Horváth and Lovasz 2025).

Decades of research on such bottom-up initiatives have shown how, despite their experimental nature and necessary embeddedness in local contexts, change can nonetheless be infectious (Seyfang and Haxeltine 2012). Bennett and colleagues describe four ways in which change at a local level can combine to produce global impacts:

1 *Aggregation.* Many local initiatives, added together, produce outcomes of global significance. For example, just as global food demand is a function of aggregate regional food demand, many local food initiatives can add up to a significant change in the functioning of the global food system;
2 *Compensation.* Less welcome perhaps! In this case, changes in one place are offset by changes in another. For example, slowing deforestation in one place might not change the global demand for wood, in which case deforestation could accelerate elsewhere;
3 *Learning.* Action in one place can enable action to take place elsewhere. For example, we could apply this both to how innovations in solar panel manufacturing have pulled down production costs globally, or to how initiatives like Cooperation Jackson have informed work in other places, like Cooperation Hull;
4 *Contagion.* Action in one place can spread to a whole range of other places – think, for example, of how Greta Thunberg's school strike sparked similar movements around the world (Priyadharshini 2021), or how the Black Lives Matter movement emerged out of the particular racial politics of the US justice system but quickly became an international movement for racial justice (Bennett et al. 2021).

So, while global modellers and scenario planners might worry about *aggregation*, and whether local initiatives will 'add up' to significant-enough change, human geographers and others have become more interested in variants of *learning* and *contagion* as ways of conceptualising the power and agency of both individual actors and seemingly 'small', local experiments in transforming human-environment relations. For example, University of Oslo human geographer Karen O'Brien has argued, in her book *You Matter More Than You Think*, that such experiments have much more agency and power than we might recognise. Building on insights into our fundamental entanglement with one another and with the nonhuman world, as well as into the workings of quantum mechanics and social change, O'Brien argues that change commonly happens through unpredictable 'ripple effects' that can tie places together in a cascade of transformation (O'Brien 2021). We exist not in mechanistic systems, O'Brien contends, in which things interact with each other in a linear, predictable way. Instead, we exist in complex relations where small things can have big effects, like the metaphorical butterfly flapping its wings on one side of the planet and setting off a hurricane on the other.

It is with these kinds of ideas in mind that Lesley Head urges us to embrace not a straightforward *optimism* about Anthropocene futures, but rather to practice *hope*. Building on a strong Antipodean tradition of feminist geographical work with communities building alternative economic futures (Gibson, Rose, and Fincher 2015), Head links optimism to the thoroughly modernist idea that the future is always better than the present, with 'progress' driven by the inevitability of technological advancement and economic growth. An *optimistic* stance towards Anthropocene futures might therefore cling to the idea that some technology will eventually ride to the rescue, or that some heroic figure – an inventor, industrialist, politician, activist – will suddenly appear on the scene to bring all our collective problems under control. We said we wouldn't make predictions in this chapter, but allow us one: that moment of heroic rescue almost certainly won't happen. Instead, Head describes hope as both a disposition and a practice – it is about working with the understanding that things *could* be otherwise, 'that the future has the capacity to be changed…and remains open in the face of what might appear to be foregone conclusions' (Bulkeley et al. 2018, 6). Head advocates not for blind optimism in a likely saviour, but for a 'gritty, keep-going kind of hope' (Head 2016, 11); a hope which is not simply about the world as we wish it to be, but about savouring, challenging and re-making the world as we find it in the present. Part of that engagement, Head makes plain, will involve grief – grief for what is being lost: loved places, loved ones, cherished species – both through the steady attrition of, for example, biodiversity loss, or through the trauma of extreme events. 'Grief will be our companion', Head writes, as we figure out what it means to live in the Anthropocene. If hope is practiced through efforts to disturb and remake existing orders, then forms of melancholy and grief will never be far away; nor will occasional disappointments and frequent frustrations. But in the grit and persistence of community organisers and activists, urban gardeners, farmers' cooperatives and more, Head detects not just 'scalable' ways to transform systems of production and consumption, but also valuable lessons about the difficult emotional work and the dispositions to change, uncertainty and instability that will be the hallmark of citizens not of modernity, but of the Anthropocene (Head 2016).

Conclusion

We remarked in the **Prologue** to this book that one of our aims was, paraphrasing Raymond Williams, to not just 'make environmental despair more convincing', but to 'make hope possible'. If we follow the lead of geographers like Lesley Head, then we can see how the possibility of hope isn't about having a rosy, optimistic view of the future, or having trust that some kind of saviour will come along to shorten the duration of the Anthropocene. Rather, hope is made possible by the realisation that, however gloomy some predictions of the future might be, nothing is a foregone conclusion, and other worlds are possible. As environmental geographer Jenny Pickerill puts it, 'despite being in the midst of multiple global crises, there is reason to be hopeful'. She points to the many, many examples of the kind of transformative grassroots action which we have sampled in this chapter, finding in them processes of change that are 'generative, inclusive, radical, political, creative, and making a difference to people's everyday lives'. Propelled by an experimental spirit and a faith in the power of collective action, they demonstrate:

> that whatever histories, state relations, reactionary far right politics, and problematic privileges complicate community efforts, many possibilities still thrive. Even in the darkest of times, such as the Covid pandemic, the togetherness of community responses means we should remain full of hope that, at whatever scale and wherever they happen, community initiatives can help facilitate on-going socio-ecological transformations to create a more equitable, ecologically sensitive, and just world.
>
> (Pickerill 2021, 251)

Focusing on the importance of these experimental spaces and community initiatives should not detract, of course, from the importance of state and corporate actors discharging their own responsibilities to build the kind of world Pickerill describes. But in putting these kinds of practices front-and-centre in this final chapter, our aim has been to:

1. Show how the 'seeds' of alternative futures are already here – unevenly distributed geographically, of course, as they are necessarily local phenomena responding to local conditions, needs, and aspirations. Nonetheless, they are 'seeds' which are ready to germinate and shape new ecologies of transformation, through processes of aggregation, learning, and contagion;

2 Show how different readings of the Anthropocene as a set of differentiated material conditions inform community responses to the challenge of building new lifeworlds. The future is never just built out of the materials to hand in the present, but is also a product of forms of reckoning with historical processes, injustices, and legacies. As such, the historical geography of the 'Anthropocene front loop' (Wakefield 2020) is not just a subject of intellectual curiosity. Instead, it is central to how scholars and activists alike conceptualise what alternative futures could and should look like;
3 Emphasise the multiple ways that researchers and students in Geography can engage with these worldmaking projects. Some are good at informing and critiquing government policies, others at working in the fields with communities trying to build new agricultural systems (Owens 2005). Others might be good at staying in the library or the lab and shedding new intellectual light on the nature of Anthropocene challenges. It takes all sorts! But for geographers in the Anthropocene, there are a range of choices available and multiple ways – some of them perhaps quite unexpected – to contribute to the collective project of building a more socially and ecologically just world.

What kind of geographer one chooses to be, and how we each respond to the material and intellectual challenges of the Anthropocene, will be a product of our own values, skillsets, politics, and ethics, as well as our own reading of the scientific evidence for the physical changes that are going on around us. It will be a product of which 'narrative' of the Anthropocene we find the most convincing, and what visions of possible local and global futures we find most compelling. It is a potentially daunting but also exciting time to be a geographer. At this moment in history, people the world over are grappling with problems that have long been a core part of the discipline's intellectual project: human-environment relations, changing landscapes and ecosystems, global interconnections and feedbacks, multi-scalar causes and impacts, and so on. So, whether you find yourself drawn intellectually to global biogeochemical cycles or geopolitics, or to local landscape transformations or community sustainability initiatives, consider what kind of world your knowledge and intellectual labour can help to build, for and with whom, and then get to work. To paraphrase Karen O'Brien (2021): as a geographer, you matter more than you might think.

Study questions

1 Critically examine how community-level sustainability initiatives can grow and spread through processes of learning and contagion.
2 In light of the geographies explored in this and preceding chapters, evaluate W.F. Gibson's statement that 'the future is already here – it's just not very evenly distributed'.

3 *Researcher, policy advisor, activist* – what kind of geographer are you, or would you like to become? Evaluate the challenges and possibilities of such a role with reference to examples.

Note

1 Not to be confused with notions of 'climate repair', which are technological in orientation and often include the kind of geoengineering proposals discussed in Chapter 8. See Buck (2019).

References

Akuno, K., and A. Nangwaya, eds. 2017. *Jackson Rising: The Struggle for Economic Democracy and Black Self-Determination in Jackson, Mississippi*. Montreal: Daraja Press.

Albert, M. J. 2023. "Ecosocialism for Realists: Transitions, Trade-Offs, and Authoritarian Dangers." *Capitalism Nature Socialism* 34 (1): 11–30.

Anderson, B. 2010. "Preemption, Precaution, Preparedness: Anticipatory Action and Future Geographies." *Progress in Human Geography* 34 (6): 777–98.

Anderson, B., and J. Fenton. 2008. "Editorial Introduction: Spaces of Hope." *Space and Culture* 11 (2): 76–80.

Bailey, M., and N. Heynen. 2020. "Sweet (and Sticky) Redemption." *Scalawag*. https://scalawagmagazine.org/2020/09/gullah-geechee-crops-agriculture/

Barker, K. 2020. "How to Survive the End of the Future: Preppers, Pathology, and the Everyday Crisis of Insecurity." *Transactions of the Institute of British Geographers* 45 (2): 483–96.

Beckles, H. 2019. "The Reparation Movement: Greatest Political Tide of the Twenty-First Century." *Social and Economic Studies* 68 (3/4): 11–30.

Bendell, J., and R. Read, eds. 2021. *Deep Adaptation: Navigating the Realities of Climate Chaos*. Chichester: Wiley.

Bennett, E. M., R. Biggs, G. D. Peterson, and L. J. Gordon. 2021. "Patchwork Earth: Navigating Pathways to Just, Thriving, and Sustainable Futures." *One Earth* 4 (2): 172–76.

Braun, B. 2015. "Futures: Imagining Socioecological Transformation: An Introduction." *Annals of the Association of American Geographers* 105 (2): 239–43.

Breakthrough Institute. 2015. *An Ecomodernist Manifesto*. Oakland: Breakthrough Institute.

Breewood, H., & Garnett, T. (2022). *What is ecomodernism? TABLE Explainer Series*. TABLE, University of Oxford, Swedish University of Agricultural Sciences and Wageningen University & Research.

Bresnihan, P., and N. Millner. 2023. *All We Want Is the Earth: Land, Labour and Movements Beyond Environmentalism*. Bristol: Policy Press.

Buck, H. J. 2019. *After Geoengineering*. London: Verso.

Bulkeley, H., and V. Castán Broto. 2013. "Government by Experiment? Global Cities and the Governing of Climate Change." *Transactions of the Institute of British Geographers* 38 (3): 361–75.

Bulkeley, H., G. Drew, R. Hobbs, and L. Head. 2018. "Conversations With Lesley Head About Hope and Grief in the Anthropocene: Reconceptualising Human-Nature Relations." *Geographical Research* 56(3): 325–335.

Chilvers, J., R. Bellamy, H. Pallett, and T. Hargreaves. 2021. "A Systemic Approach to Mapping Participation With Low-Carbon Energy Transitions." *Nature Energy* 6 (3): 250–59.

Collard, R. C., J. Dempsey, and J. Sundberg. 2015. "The Moderns' Amnesia in Two Registers." *Environmental Humanities* 7 (1): 227–32.

Cooperation Hull. 2025. "What Can Hull Learn from 'the Most Radical City on Earth?'" https://www.cooperationhull.co.uk/blog/what-can-hull-learn-from-the-most-radical-city-in-the-world

Cooperation Jackson. 2025. *Community Production Initiative*. Cooperation Jackson. https://cooperationjackson.org/the-community-production-initiative/

Death, C. 2025. *African Climate Futures*. Oxford: Oxford University Press.

Derickson, K. D. 2022. "Disrupting Displacements: Making Knowledges for Futures Otherwise in Gullah/Geechee Nation." *Annals of the American Association of Geographers* 112 (3): 838–46.

Eden, S. 2016. *Environmental Publics*. London: Taylor and Francis.

Edwards, G. A. S., and H. Bulkeley. 2018. "Heterotopia and the Urban Politics of Climate Change Experimentation." *Environment and Planning D: Society and Space* 36 (2): 350–69.

Priyadharshini, E. 2021. Pedagogies for the Post-Anthropocene: Lessons from Apocalypse, Revolution & Utopia, 14. Cham: Springer.

Fitz-Henry, E., and E. Klein. 2024. "From Just Transitions to Reparative Transformations." *Political Geography* 108 (January): 103004.

Freshour, C., and B. Williams. 2023. "Toward "Total Freedom": Black Ecologies of Land, Labor, and Livelihoods in the Mississippi Delta." *Annals of the American Association of Geographers* 113 (7): 1563–72.

Garrett, B. 2020. *Bunker: Building for the End Times*. London: Penguin.

Garrett, B. 2021. "Doomsday Preppers and the Architecture of Dread." *Geoforum* 127 (December): 401–11.

Garrett, B., and I. Klinke. 2019. "Opening the Bunker: Function, Materiality, Temporality." *Environment and Planning C: Politics and Space* 37 (6): 1063–81.

Gibson, K., D. B. Rose, and R. Fincher. 2015. *Manifesto for Living in the Anthropocene*. Brooklyn: Punctum.

Gilmore, R. W. 2007. *Golden Gulag: Prisons, Surplus, Crisis, and Opposition in Globalizing California*. Berkeley: University of California Press.

Hamann, M., R. Biggs, L. Pereira, et al. 2020. "Scenarios of Good Anthropocenes in Southern Africa." *Futures* 123: 102526.

Hamilton, C. 2016. "The Theodicy of the "Good Anthropocene." *Environmental Humanities* 7 (1): 233–38.

Head, L. 2016. *Hope and Grief in the Anthropocene : Re-Conceptualising Human-Nature Relations*. London: Routledge.

Heynen, N., and M. Ybarra. 2021. "On Abolition Ecologies and Making 'Freedom as a Place'." *Antipode* 53 (1): 21–35.

Holling, C. S. 2004. "From Complex Regions to Complex Worlds." *Ecology and Society* 9 (1): 11.

Horváth, M., and A. Lovasz. 2025. "Against Eco-Authoritarianism and Ecomodernism: Towards a Critique of "Planetary" Governmentality and Fantasies of Steering." *Environmental Values* 34 (2): 144–66.

Huber, M. 2019. "Ecosocialism: Dystopian and Scientific." *Socialist Forum*. Winter 2019. https://sfarchive.dsausa.org/issues/winter-2019/ecosocialism-dystopian-and-scientific/

Huber, M. T. 2022. *Climate Change as Class War: Building Socialism on a Warming Planet*. London: Verso Books.

Hulme, M. 2010. "Problems With Making and Governing Global Kinds of Knowledge." *Global Environmental Change* 20 (4): 558–64.

Jasanoff, S. 2015. "Imagined and Invented Worlds." In *Dreamscapes of Modernity: Sociotechnical Imaginaries and the Fabrication of Power*, edited by Sheila Jasanoff and Sang-Hyun Kim. Chicago: University of Chicago Press.

Jeffrey, C., and J. Dyson. 2021. "Geographies of the Future: Prefigurative Politics." *Progress in Human Geography* 45 (4): 641–58.

Kallis, G., S. Paulson, G. D'Alisa, and F. Demaria. 2020. *The Case for Degrowth*. Cambridge: Polity Press.

Kelly, C. R. 2016. "The Man-Pocalpyse: Doomsday Preppers and the Rituals of Apocalyptic Manhood." *Text and Performance Quarterly* 36 (2–3): 95–114.

Kirsch, R. E., and E. Ray. 2025. *Be Prepared: Doomsday Prepping in the United States*. New York: Columbia University Press.

LaDuke, W., and D. Cowen. 2020. "Beyond Wiindigo Infrastructure." *South Atlantic Quarterly* 119 (2): 243–68.

Latour, B. 2015. "Telling Friends from Foes in the Time of the Anthropocene." In *The Anthropocene and the Global Environmental Crisis: Rethinking Modernity in a New Epoch*, edited by Hamilton Clive, Bonneuil Christophe, and Francois Gemenne. London: Routledge.

Ludwig, J. 2021. "The Anthropocene Blues: Notes from Mississippi." *The Anthropocene Review* 8 (2): 213–28.

Luke, N., and N. Heynen. 2020. "Community Solar as Energy Reparations: Abolishing Petro-Racial Capitalism in New Orleans." *American Quarterly* 72 (3): 603–25.

Magdoff, F., and C. Williams. 2017. *Creating an Ecological Society: Toward a Revolutionary Transformation*. New York: NYU Press.

Mark, B., and J. Lewis. 2020. "Group Interventions for Climate Change Distress." *Psychiatric Times*. Dec 10 2020. https://www.psychiatrictimes.com/view/group-interventions-climate-change-distress

McCutcheon, P. 2019. "Fannie Lou Hamer's Freedom Farms and Black Agrarian Geographies." *Antipode* 51 (1): 207–24.

Mohan, P. S. 2025. "Climate Reparations for a Just Response to Climate Change: A Review of Historical Responsibility and Future Implications." *WIREs Climate Change* 16 (2): e70007.

Moore, J. W. 2015. *Capitalism in the Web of Life: Ecology and the Accumulation of Capital*. London: Verso.

O'Brien, K. 2021. *You Matter More Than You Think: Quantum Social Change for a Thriving World*. Oslo: Change Press.

Obeng-Odoom, F. 2024. "Reparations." *The Review of Black Political Economy* 51 (3): 458–78.

OSE. 2025a. "Machines: Global Village Construction Set." *Open Source Ecology*. https://www.opensourceecology.org/

OSE. 2025b. "About." *Open Source Ecology*. https://www.opensourceecology.org/
OSE. 2025c. "The Bigger Picture of OSE." OSE Apprenticeship. *Open Source Ecology*. https://www.opensourceecology.org/
Owens, S. 2005. "Making a Difference? Some Perspectives on Environmental Research and Policy." *Transactions of the Institute of British Geographers* 30 (3): 287–92.
Pallett, H., J. Chilvers, and T. Hargreaves. 2019. "Mapping Participation: A Systematic Analysis of Diverse Public Participation in the UK Energy System." *Environment and Planning E: Nature and Space* 2 (3): 590–616.
Pick, J. B., and A. Sarkar. 2020. "Geographies of Global Digital Divides." In *Geographies of the Internet*, edited by Warf, Barney. London: Routledge.
Pickerill, J. 2021. "Hopefulness for Transformative Grassroots Change." *Environmental Policy and Governance* 31 (3): 249–51.
Pickering, A. 2025. *Acting With the World: Agency in the Anthropocene*. Durham: Duke University Press.
Pivo, D. 2025. "What Are Seeds." *Seeds of Good Anthropocenes*. https://goodanthropocenes.net/
Preiser, R., T. Hichert, R. Biggs, et al. 2024. "Transformative Foresight for Diverse Futures: The Seeds of Good Anthropocenes Initiative." *Development Policy Review* 42 (S1): e12791.
Priyadharshini, Esther. 2021. Pedagogies for the Post-Anthropocene: Lessons from Apocalypse, Revolution & Utopia. Singapore: Springer.
Pulido, L., and J. De Lara. 2018. "Reimagining 'Justice' in Environmental Justice: Radical Ecologies, Decolonial Thought, and the Black Radical Tradition." *Environment and Planning E: Nature and Space* 1 (1–2): 76–98.
Sabel, C F., and D. G. Victor. 2022. *Fixing the Climate: Strategies for an Uncertain World*. Princeton: Princeton University Press.
Salleh, A. 2025. *DeColonize EcoModernism!* London: Bloomsbury Publishing.
Seyfang, G., and A. Haxeltine. 2012. "Growing Grassroots Innovations: Exploring the Role of Community-Based Initiatives in Governing Sustainable Energy Transitions." *Environment and Planning C: Government and Policy* 30 (3): 381–400.
Smith, B. J. 2019. "Building Emancipatory Food Power: Freedom Farms, Rocky Acres, and the Struggle for Food Justice." *Journal of Agriculture, Food Systems, and Community Development* 8 (4): 4.
Stoddard, I., K. Anderson, S. Capstick, et al. 2021. "Three Decades of Climate Mitigation: Why Haven't We Bent the Global Emissions Curve?" *Annual Review of Environment and Resources* 46 (1): 653–89.
Sultana, F. 2022. "The Unbearable Heaviness of Climate Coloniality." *Political Geography* 99: 102638.
Táíwò, O. O. 2022. *Reconsidering Reparations*. Oxford: Oxford University Press.
The Economist. 2001. "Broadband Blues." *The Economist*, July 21.
Tsing, A. L., A. S. Mathews, and N. Bubandt. 2019. "Patchy Anthropocene: Landscape Structure, Multispecies History, and the Retooling of Anthropology: An Introduction to Supplement 20." *Current Anthropology* 60 (S20): S186–97.
Undisciplined Environments Collective. 2024. *Insurgent Ecologies: Between Environmental Struggles and Postcapitalist Transformations*. Halifax: Fernwood Publishing.
Utrata, A. 2024. "Engineering Territory: Space and Colonies in Silicon Valley." *American Political Science Review* 118 (3): 1097–109.
Wainwright, J., and G. Mann. 2018. *Climate Leviathan: A Political Theory of Our Planetary Future*. London: Verso.

Wakefield, S. 2017. "Field Notes from the Anthropocene: Living in the Back Loop." *The Brooklyn Rail*, June 1.

Wakefield, S. 2020. *Anthropocene Back Loop: Experimentation in Unsafe Operating Space*. 1st ed. London: Open Humanities Press.

Wang, P., S. Solomon, B. D. Santer, et al. 2025. "Fingerprinting the Recovery of Antarctic Ozone." *Nature* 639 (8055): 646–51.

Willis, R. 2020. *Too Hot to Handle? The Democratic Challenge of Climate Change*. Bristol: Bristol University Press.

Woods, C. 2017. *Development Arrested: The Blues and Plantation Power in the Mississippi Delta*. London: Verso.

Yusoff, K. 2024. *Geologic Life: Inhuman Intimacies and the Geophysics of Race*. Durham: Duke University Press.

Epilogue

What Future for the Earth?

This book has ranged widely. We've sought to be relatively comprehensive and largely dispassionate. We've surveyed different narratives about the Anthropocene, and attended to different aspects of this multifaceted 'condition' – a condition that, as climate change illustrates, inescapably envelopes all present and future people. In this epilogue, we take the liberty of sharing with readers our own sense of where the world is heading. Which arguments about the future of the Earth will shape what sorts of actions, and with what knock-effects? While we obviously lack a crystal ball, we want to sketch some sort of answer about how the 'tight dialectic' between knowledge and practice will 'co-produce' our shared future in the decades ahead.

You may well disagree with this analysis, diagnostically and/or politically. We hope it will at least give readers grounds for reflection and debate! Throughout this book we've emphasised how knowledge is never separate from action, and how our shared understandings of how the world *is* are always bound-up with our sense of how the world *ought to be*. So it only seems right that we give a sense of where our reading of the evidence and arguments about the 'Anthropocene' leads us.

To do this, we want to employ a distinction between the 'realm of necessity' and the 'realm of contingency'. As social scientists have long argued, the former encompasses those aspects of life that are, in essence, fixed and unchangeable. The latter, by contrast, encompasses that which can be altered if some combination of effort, purpose and good luck is achieved. As the social anthropologist Ernest Gellner once sagely observed, 'a great deal can happen without being necessary' (1988, 3). The way people delvier life's necessities can vary a great deal, while non-necessary practices can be created and eliminated with sufficient will.

The realm of necessity

Let's begin, though, with what *is* necessary. Clearly, some basic *biological* requirements apply to all humans (e.g. the need for clean water). Life can't proceed unless these are satisfied. But some *social* arrangements are, in effect,

DOI: 10.4324/9781003162469-15

also necessary aspects of twenty-first-century life for the majority of over 8 billion human beings. They include the following:

- *Capitalism*: Now truly global in scale, this mode of production is so embedded in the life of nations that it's virtually impossible to foresee a set of coordinated political revolutions toppling this system. Oppositional movements, at the time of writing, are too fragmented and lack the resources to mount a serious challenge to capitalism. The only radical political options seem to be to withdraw from the system or partially de-couple from it (see Chapter 10).
- *Nation state sovereignty*: While many countries remain internally unstable, such as Somalia, most of the world's territory is divided into self-governing nations led by elected or non-elected governments. It is difficult to imagine the governance of human affairs moving beyond this so-called 'Westphalian' foundation. For instance, a 'world government' is extremely unlikely to emerge any time soon. Far more likely is a new alliance of powerful countries in the interests of coordinated and effective management of the Earth's environment and other things – or perhaps multiple alliances and coalitions, reflecting an increasingly multipolar geopolitics (Toal 2024). The United Nations might continue to be the face of this 'asymmetrically distributed' global governance.
- *Minority political rule of majorities*: Because populations numbers are large in most countries – notably China and India – political decision-making is necessarily concentrated in relatively few hands in national government. This seems unlikely to change any time soon. It means that the future of the Earth will largely be determined by a mixture of autocratic and elected leaders spread across secular and religious polities. They will use the power of the national state – a very *considerable* power in states such as the USA, the EU, Russia, and China – to enact certain measures in response to Earth System change. The perceived legitimacy of their actions will depend on the quality of their relationship with the citizens whose lives they govern.
- *Advanced technology*: Advanced technologies suffuse our world ever more. They are highly complex and require a myriad of materials and components for their construction and maintenance. Such technologies will surely be seen as indispensable aids to improving human welfare, including by mitigating and adapting to various environmental problems in a warming world. Such technologies will be a key means by which capitalists continue to make money in competition with each other. Most countries will want access to these technologies, and many governments will fund research and development to stay ahead and avoid undue reliance on overseas suppliers.
- *An international division of labour*: All modern modes of production require a social and spatial 'division of labour' i.e. different workers and

places supplying materials and parts to other locations in various extended 'commodity chains'. In a future where capitalism and advanced technology are seemingly here-to-stay, so too is an international division of labour. Such a division implies the need for trade, multilateral cooperation and relative geopolitical stability. It pushes against nations isolating themselves in order to 'go it alone'.

- *A massive expansion in 'zero carbon' energy*: To avoid the planet becoming too hot for human life in many parts of the world, it's now necessary to keep fossil fuels in the ground. This implies the need to massively upscale renewable energy supplies, which will be a mix of solar, hydro, tidal, wind, geothermal and nuclear power. Moreover, these renewable systems will need to supply *more* energy, not less, than current fossil fuel powered energy systems. This is because of poverty: the UN Sustainable Development Goals rightly target the eradication of chronic poverty, which at least a fifth of all humans suffer from. Reliable and affordable energy supply is key to raising living standards.
- *A massive expansion in CCS, even SRM*: As GHG emissions grow, despite 30 years of attempts to reduce them, the countries of the world are fast running out of time to keep the atmosphere sufficiently cool. This now all but guarantees that wide-scale terrestrial and aquatic carbon capture and storage (CCS) will be required, and even that risky solar radiation management (SRM) might be attempted (see Chapter 8). These measures will have significant implications for land use, biodiversity, ice cover and so on, just as climate change itself does. They suggest that careful stewardship or even muscular management of the biosphere, hydrosphere, atmosphere and cryosphere will be necessary in order to regulate the effects of such technological interventions.
- *The use of new biophysical materials on an enormous scale*: Notwithstanding any future success in 'decoupling' energy supply systems from entropy production, the troubling fact remains that *creating* those systems, and *maintaining* them, will involve a significant *additional* appropriation of biophysical resources. Because a 'resource' is a social category, new materials will become – are already becoming – a focus of significant interest as research and innovation proceed. The uneven geographies of these resources will produce new patterns of economic wealth creation, new trade relations, new injustices, and new attempts to ensure access for those with the technological know-how to design different parts of renewable energy systems (across generation, transmission and storage). In the meantime, lives powered by fossil fuel combustion will be remain a feature of our world during the 'green transition'.
- *A step-change in materials recycling and re-use*: While product recycling is an all-too-familiar aspect of modern life, in order to bring entropy nearer to zero, a simply enormous increase in effective materials re-use will be imperative, along with a huge leap in efficiency (i.e. far less waste per unit

of material re-used). Something close to circular economies will need to be created worldwide, based on reliable renewable energy supplies – these economies will largely adapt and reconfigure materials that have already been in use for years or decades. Among these materials will be hundreds of millions of vehicles and industrial engines retired because they're powered by petrol or diesel fuel. As for food, this 'material' will need to be produced in ways that are much less disruptive to land- and water-based ecosystems. Clearly, a 'post-material', 'weightless' and 'post-extractive' mode of living would require a very different global geography of production, investment, employment and trade.

The realm of contingency

So much for things that, at some point, seem destined to occur as combined human actions push the planet further out of its Holocene envelope. But within what we might call these social 'boundary conditions' there's some room for manoeuvre. Indeed, the precise actions that will be taken in order to realise the above-listed necessities will make a simply *enormous* difference to the quality of future human lives and the 'thick geographies' that will shape them.

Here we need to declare our own political aspirations, before speculating on whether they have a reasonable chance of being realised in the twenty-first century and beyond. Like many people, we aspire to a world where:

- **Universal human needs are recognised and satisfied.** These include biological needs (e.g. adequate food) but also social needs such as a proper education.
- **Universal rights are recognised and upheld.** These include rights to choose one's government, to worship (or not), to be free from violence, and so on.
- **People's freedoms are maximised, within collectively chosen constraints.** Freedom of thought and action is rightly prized the world over, yet is currently denied to millions of people. But freedom must be enjoyed within limits to avoid one group's freedoms impeding other peoples'.
- **Considerable care for the non-human world becomes customary**: One doesn't have to be 'ecocentric' politically and morally to understand the folly of continuing to treat the Earth as a giant zone of resource supply and waste disposal. It is in humanity's self-interest to accord the non-human world greater moral importance in our lives.

A world in which these four things were achieved would be a more just world where stark social inequality is eliminated, where solidarity among otherwise different societies was strong, where democratic government was the norm, where socio-cultural difference still flourished (though largely without enmity and wars), where far less biophysical damage was inflicted, and where liveable environments were a priority worldwide. In aggregate, the capacities of more people would flourish. Of course, this is all very idealistic. It's a vision of what we might call a 'green democratic internationalist socialist cosmopolitanism' (G-DISC, for short). It would need to exist within a capitalist, national state framework but, even without being revolutionary, it would pose a serious challenge to current arrangements in most democratic and autocratic capitalist states.

To get a handle on just how much of a challenge, consider the following non-exhaustive inventory of the global state-of-play:

- Geopolitically, American hegemony on the world stage is waning. While it still possesses enormous power, it's now heavily indebted and has lost out economically to China, in particular, since 2000. China's government has used the country's economic success to project its power politically, using such things as the Belt-and-Road and Digital Silk Road initiatives. Then there's Russia which, economically, isn't that strong but has diverted economic resources into the military and intelligence services to a degree that it's a serious geopolitical player in Europe and the Middle East. It also possesses a large nuclear weapons arsenal. The point is that these three countries are not currently aligned in terms of political ambitions and are competing. The European Union sides with America, but the recent actions of President Donald Trump have made EU leaders rethink strategy regarding Russia, China and other large powers (such as India and Brazil). A new 'multi-polar' world is taking shape, geopolitically.
- Economically, despite Trump's 'trade war' of 2025, the world economy remains relatively open. Capital flows in search of opportunities to profit. For instance, India's economy is developing rapidly, for good reason: much of its enormous population exists on very low incomes and lacks access to quality public services. This must be addressed. But these capital flows serve to perpetuate, not equalise, uneven geographical development on the world stage. For instance, some oil rich Middle Eastern states are now creating complex modern societies, while parts of the otherwise wealthy United States see unemployment and poverty rise. This changing spatial unevenness in economic development fuels the changing geopolitics mentioned above.
- Socially, income inequality is acute worldwide. Despite major improvements in China and elsewhere in recent years, there's plentiful evidence that, across most societies, the relative income share enjoyed by the top 1% of income earners has increased prodigiously. And it's not only incomes. Worldwide, a shocking number of people lack reliable access to clean water, food, vaccines, medicines, school education and so on. Famines, civil wars and inter-state wars (e.g. Israel's scorched earth approach to Gaza at the time of writing) plague the twenty-first-century world. The

desire for social stability and the basics of a normal life are felt acutely by countless people.

- Culturally, we live in 'a world of difference'. For instance, many people's identities are deeply bound-up with their faith (such as as Catholics, Jews, Hindus, and Muslims). Meanwhile, an estimated 360 million people identify as 'first peoples', whose predecessors were assailed by various European colonists from the 1600s onwards. Cultural differences can and should be celebrated. But they can also form the basis for rivalries and prejudices, impeding the pursuit of shared interests through empathy and mutual understanding. We are seeing this in the United States and parts of Western Europe at the time of writing, with anti-immigrant sentiment rising. Furthermore, the *MeToo* and *Black Lives Matter* movements are reminders that sexism and racism remain rife in many places.

We could go on. But we've said enough to remind readers that humanity today is both deeply interconnected yet, in many ways, sharply divided. It's against this background that responses to the looming challenges of the Anthropocene will be formulated and actioned. These challenges can't *override* the things listed above. Instead, the responses must work to *dovetail with* various other existing worldwide challenges, such as poverty reduction and mutual respect for differences of faith. Ideally, they need to be aspirational and even inspirational (Hajer and Oomen 2025). Clearly, this is no mean feat. What we're calling a G-DISC political programme would aim for such alignment by balancing a commonality of human interest against positive differences in ways of living within, or perhaps delinked from, the frameworks of capitalism and national political sovereignty. It would venture some substantive sense of 'the common good' and be an inclusive narrative. This programme would combine elements of the 'eco-Marxist' and 'decolonial' narratives considered earlier in this book, but also some elements of the 'post-natural' narrative. It would seek to head-off a catastrophic future for as many as possible. And in contrast to the 'standard narrative', it would have less of a fixation on science and technology alone as saviours of humankind, and a far stronger emphasis on justice and on pursuit of the good.

The realm of politics: what's most likely to transpire?

In closing, let's briefly consider what chance a G-DISC programme has of gaining traction in any country of influence. This brings us back to politics, which we've discussed many times in this book in different contexts. The most important political actors in the world are national states. Uniquely, compared to all others who influence big-P and small-p politics, national governments possess a special sort of social power: namely, the power to make and enact laws and binding rules, backed in most cases by the police and the military (as need arises). This power was exercised for all the world to see during the

COVID19 pandemic, a state of emergency that saw extraordinary measures taken to prevent mass mortality. This power would be core to the 'activist state' that would be needed to deliver the G-DISC programme we favour. But this power is best exercised in a democratic political environment where there is feedback between rulers and the ruled.

This immediately suggests something that political strategists and tacticians have recognised for over two centuries. Namely, that to make significant changes to a society one needs to occupy the cockpit of national government. This is very difficult in many autocratic societies, where opposition organisations are suppressed and free elections prohibited. It is quite difficult in erstwhile democratic states where monied interests can use political lobbying and the power of the news media to keep certain political ideas 'off the agenda'. Indeed, some critical social scientists believe, understandably, that we now live in a 'post-political' age where alternative ideas – especially progressive ones like our G-DISC programme – are written-off as 'unrealistic' and 'extreme'. In the meantime, a 'nativist' populism has risen-up in the US, the UK, Sweden, Holland, France, Italy, Hungary and elsewhere. An internal fracturing of many capitalist, now highly multi-cultural, democracies is occurring against the background of external tensions with China, Russia and various Muslim states seen as potentially hostile to 'Western values'. It's no wonder that many political commentators see this is a dangerous moment in world history, often comparing it with the lead-up to the second world war that broke out in 1939.

And yet: anthropogenic climate change, and the other direct and indirect changes to the Earth System recounted earlier in this book, stand as a *shared challenge across all existing societies*. Various forms of cooperative action will need to occur and, as we've said, we'd very much hope to see a G-DISC world created as the basis for this multi-lateral endeavour. The only place, realistically, a progressive political shift could occur initially is in one or other liberal democracy. This is because democracies still offer room for free speech and criticism, allowing 'the people' to change their government periodically. But any political shift would require the rapid construction of a new 'common sense' – what Marxists call 'hegemony', based on showing complex national populations what their *common interests* might be. The conditions for this are ripe in the US and elsewhere, as Senator Bernie Sanders shows with his ongoing efforts to make 'democratic socialism' a force for good. But to build a new hegemony requires well-funded organisations to anchor it, and access to current affairs and social media to get the messages out to people from all backgrounds. Without this, people who are feeling 'left out' – as with many members of Trump's MAGA voter base and many Brits who voted to leave the EU – can be persuaded to vilify fellow citizens who are 'different' to them.

To appreciate the importance of building a new progressive hegemony within and far beyond the West, it's useful to briefly consider a much-discussed

Table 0.1 Four possible global futures.

	Planetary sovereignty	*Anti-planetary sovereignty*
Capitalist	Climate Leviathan	Climate behemoth
Non-capitalist	Climate Mao	Climate X

Source: Adapted from Mann and Wainwright (2018).

thought experiment conducted by the geographers Joel Wainwright and Jeff Mann. In 2018, they published a fascinating book about the political effects of anthropogenic climate change. Titled *Climate Leviathan*, in it the authors map-out four possible global futures as the atmosphere gets uncomfortably hot (see Table 0.1). Their key axes are 'capitalism/non-capitalism' and 'planetary sovereignty/anti-planetary sovereignty'. Planetary sovereignty involves one or more very powerful countries taking it upon themselves to 'manage' the Earth's environment for certain sections of humanity (rhetorically for 'all'). Its opposite is either a state of weak or non-existent intergovernmental cooperation (Climate Behemoth) or else a cosmopolitan solidarity among the world's peoples in support of a just and liveable world for all (Climate X). The capitalist version of planetary sovereignty Wainwright and Mann call Climate Leviathan, most likely to be led by the US, China or the two working together (despite ideological differences). The non-capitalist version is Climate Mao, led out of Asia by China as a post-capitalist order in which centralised political power is used to protect 'the people'. Climate X doesn't require a thick sense of shared humanity to be built among the world's peoples. But it does require mutuality and sympathy in order to construct a world that allows many worlds to flourish – a world of 'commonality-in-difference' marked by a capacious *modus vivendi*. This itself would require new local and regional cultural hegemonies to be constructed at the grass-roots level, drawing people away from neoliberalism, right wing populism, deep ethnocentrism, religious fundamentalism, nationalist particularism and other divisive forces.

Climate Leviathan seems the scenario most likely to transpire, so argue Wainwright and Mann. We agree. In it, one or more hegemons will openly 'up-scale' sovereign power, using advanced technology and military might to ensure they call the shots on planetary management, while maintaining opportunities for capital accumulation. Climate Leviathan will be a new elite-led rules-based international order, pursued through the UN and the like, but with the hegemons overseeing radical environmental interventions like SRM measures, probably in the face of opposition, and managing the human consequences to the advantage their domestic populations. Leviathan will thus be an evolution of patterned inequality in causes, impacts and responses as planetary change unfolds. Many people will suffer the lottery of birth and location. Leviathan will operate within the parameters of capitalism, of asymmetric sovereign power and new biophysical realities. At its edges, some territories may suffer civil strife and environmental disasters, becoming what critics call 'sacrifice zones' (see Chapter 6).

Dispiriting as this sounds, it's better than the next plausible alternative, which is Climate Behemoth: a world of largely non-cooperating blocs of countries, trading in various key things, but who may well fight with one another. As indicated above, we don't see anti-capitalist revolution as currently possible in more than a few countries, none of them China, Russia or EU member states. That rules out Climate Mao, and while we can see elements of Climate X emerging in the kinds of networked spaces we explored in Chapter 10, for the time being the Westphalian model of organising human affairs seems too entrenched to enable a fully global Climate X to emerge. So, this leaves us with a 'progressive' Climate Leviathan as the best version of two feasible scenarios. This would be one where America, many EU states and possibly even China move in the direction of a G-DISC programme.

As the social movement slogan has it, 'there's a world to win'. Not *any* world, but a world that's contingently possible given where people and planet currently are in the very long arc of planetary history. Imagine us publishing a 5th edition of this book in 40 years (one of us will be nearly 100 by then). We'd hope to report that a relatively peaceful, tolerant, prosperous, still-habitable world had been built, spurred by the massive Earth System changes that early twenty-first-century humans initially failed to do much about.

If you're among our young readers we urge you to recognise that you possess significant political agency. The world doesn't change unless people demand it. And those demands only influence government policies if they're expressed by large numbers of people, at the ballot box and elsewhere. What sort of 'welcome to the Anthropocene' your older selves and your children will receive – in, say, 2055 – depends on actions taken today. Those actions will reflect the narratives that capture the hearts and minds of billions of people and the small groups who seek to govern them. There is a huge difference between world in which elites use narratives to deceive and indoctrinate the masses, and a world in which narratives are embedded in peoples' hearts and minds like 'cultural DNA'.

References

Gellner, E. 1988. "Introduction." In *Europe and the Rise of Capitalism*, edited by Jean Baechler et al., 1–5. Oxford: Blackwell.

Hajer, M. A., and J. Oomen. 2025. *Captured Futures: Rethinking the Drama of Environmental Politics*. Oxford: Oxford University Press.

Mann, G., and J. Wainwright. 2018. *Climate Leviathan*. London: Verso.

Toal, G. 2024. *Oceans Rise, Empires Fall: Why Geopolitics Hastens Climate Catastrophe*. Oxford: Oxford University Press.

Glossary

Aesthetics Historically, the branch of Philosophy concerned with the study of beauty and taste, and with the feelings, emotions, and attitudes invoked by encounters with works of art. Conventionally, philosophers have debated questions such as whether things (such as works of art or a human face) have inherent, objective beauty, or whether beauty is a purely subjective experience. More broadly, the term 'aesthetics' is often used to describe a particular tradition of representing or visualising certain phenomena, such as landscapes or processes and outcomes of environmental change.

Capitalism One of several systems producing goods and services for large numbers of people. This so-called 'mode of production' has existed since the late eighteenth century (if not earlier) and is now globally dominant. Its distinguishing characteristics are as follows: private producers employ workers to make and sell commodities; competition between producers spurs ceaseless innovation; and profit-making is the system's goal. Capitalism is expansionary, both relying on and penetrating realms outside itself (such as the atmosphere) in search of profits. National governments are the key institutions with responsibility for regulating capitalism (for instance, to prevent business monopolies from forming).

Ecological justice A set of goals and associated practices designed to deliver well-being for the non-human world, in both its biotic and abiotic components. It is consistent with an 'ecocentric' ethic and less anthropocentric than 'environmental justice'. So-called 'inter-species' justice is a sub-domain of ecological justice.

Environmental justice A situation where no one community suffers from polluted or otherwise damaged local environmental conditions. 'Justice' is both a set of principles and goals that, in different ways, are found in all societies. The idea of *environmental* justice presupposes that it's unfair that any community in any country should have to live in 'toxic' or otherwise harmful environments. Justice activists argue for *procedural* justice that ensures the claims of negatively affected communities are properly heard in the domains of government regulation and law. The hoped-for outcome is *substantive* justice, in which remediation and compensation for unjust environmental conditions are achieved. Fundamentally, environmental justice activists made the quality of peoples' immediate, lived environments a matter of justice.

Geoengineering A set of proposed technologies and techniques to alter the course of anthropogenic climate change by intervening in the functioning of the global carbon cycle or atmosphere. Geoengineering proposals are usually divided into two families: (1) 'carbon dioxide removal' technologies, which aim to reduce the amount of carbon dioxide either entering or remaining in the atmosphere, e.g. by scrubbing CO_2 out of power plant emissions and burying it underground, or with machines that pull CO_2 out of the ambient air. Tree-planting done in the name of absorbing atmospheric CO_2 can also be put into this category; (2) 'solar radiation management' technologies which aim to alter the amount of heat in the atmosphere by, for example, injecting particles into the upper atmosphere which can deflect sunlight, or artificially stimulating the regrowth of polar sea ice in order to reflect sunlight back into space.

Great Acceleration The period since the Second World War, which saw a rapid intensification in various human drivers of environmental change – such as energy use and urbanisation – and indicators of resulting environmental change – such as biodiversity loss, global mean temperature, and ocean acidification. For many Anthropocene thinkers, the years 1945 or 1950 represent the start of the Anthropocene. The Acceleration has arguably entered a new and more intense phase since 1990, as China, in particular, has modernised its economy.

Industrialisation A process of socio-economic change defined by the mechanisation, regularisation, and standardisation of processes of production. It also features the geographical separation of spaces of production and spaces of consumption. The onset of industrialisation is commonly traced to the late eighteenth century in northern England. While commonly associated with the onset of the Anthropocene, industrialisation pre-dated the widespread combustion of fossil fuels.

Interspecies justice A set of goals and associated practices intended to ensure that the needs, rights, and entitlements of all non-human living organisms are properly considered. Closely related to ideas of multispecies justice, it is consistent with an 'ecocentric' ethic and 'green politics'.

Mercantilism An economic theory and system that maintained that the wealth of the world was finite, with the task of nations to capture as much of that wealth as possible. This motivated policies aimed at minimising imports and maximising exports in order to grow national monetary reserves and stocks of gold and silver. Considerable amounts of early modern international warfare and colonial expansion have been attributed to the influence of mercantilist ideas. While mercantile activity was typically profit-seeking (merchants would sell commodities for more than the cost of their acquisition), it was not capitalist (see **capitalism**).

Morals Otherwise known as ethics, morals are those concepts, principles, and arguments that are used to justify 'right' or 'proper' action in the world. While individuals are the 'bearers' of morality, morals are deeply social and apply to large groups of people through a mixture of custom, government policy, and law. Morality divides into two areas: the area of *justice*, and the area of *goodness*. It's often said that achieving justice provides a foundation for a society to build towards the 'good life' for all its members. Morals are a matter of judgement and justification, not of science and evidence, even though some believe in a 'naturalistic ethics' (where facts supposedly compel moral beliefs, e.g. 'if an animal can be shown to experience pain it is wrong to cause pain to that animal'). Morals are a key component of **values**.

Narratives An account of a series of events or experiences threaded together to explain, entertain, instruct, and/or moralise. In the discipline of Human Geography, the study of narratives has been closely related to the Foucauldian study of *discourses*, which are broader power-knowledge arrangements that structure how societies collectively make sense of certain phenomena. In this book, we lean on the work of historical geographer William Cronon, who argued that environmental problems are always subject to alternative narratives and therefore explanations, with significant implications for how societies then seek to act on those problems.

Nature A complex word with four meanings in English, namely: the non-human world ('the environment), the intrinsic properties of things ('essential nature'), the animating force that governs the behaviour of things ('super-ordinate nature'), and the totality of phenomena ('universal nature'). Some claim the Anthropocene marks 'the end of nature', but it's important to be clear about which meaning of 'nature' is being referred to. A complicating factor is that calling something 'natural' is not dictated by the things so called: it's a decision made by people. Several non-Western cultures lack a precise equivalent of the word nature.

Politics The various means by which different groups of people vie to have their needs, wants, and desires acknowledged, articulated, and supported in a particular society. Politics arises because of the need to make collective decisions in societies where there are different values and aspirations in play. 'Formal politics', or 'big-P Politics', refers to the work of governments and, in democratic societies, elected representatives are tasked with meeting the needs, wants, and desires of citizens. Small-p politics occurs outside government – in community groups, non-governmental organisations, think tanks, pressure groups, and so on. In democracies, small-p politics shapes big P politics. In more autocratic societies, this is not the case. Societies in states of civil war will normally see politics suspended for the duration of the conflict. Big-p politics necessarily involves the exercise of social **power**, but there are other forms of power that are 'vectors' for politics and which operate outside the domain of government.

Power The capacity of some actors to legitimately or illegitimately impose certain things on many others or to offer (make available) certain things. The things can be resources, values, preferences, opinions, actions, material constraints, material opportunities, information, punishments, violence, and so on. Power is *social*: its use involves many people (e.g. civil servants in democratic societies) and its effects are normally felt by many more people (e.g. all citizens subject to the rule of a new law). Social power is principally exercised in the domains of government (big-P politics), in business (e.g. witness the enormous worldwide impacts of AI since 2022), and in society (as when people vote a new government into power). Social power operates through many different mechanisms, and most societies today are 'force fields' of overlapping forms of power. Social power is widely seen as illegitimate when those wielding it, and those affected by it, are not in mutual agreement about the validity of this power. A key role of all national governments is to exercise social power so as to regulate other forms of social power lying outside government, such as that exercised by large businesses, criminal gangs, or trade unions.

Prefigurative politics A political strategy that aims to put into practice ideas about how the world could be organised differently. It embodies the adage 'be the change you want to see'. For example, people aiming to rebuild the world along eco-socialist lines might decide against pursuing a national or global eco-socialist revolution, and instead work locally to build eco-socialist systems (e.g. at a city level) in a bid to 'prefigure' what could be achieved at wider spatial scales. As with almost all politics, there is a set of moral-ethical beliefs driving prefigurative acts.

Resilience A term taken from systems ecology which emphasises the capacity of systems to withstand and recover from external shocks. Since the 1970s, the term has migrated into fields like urban planning and disaster management, with decision-makers aiming to increase the resilience of social groups or material infrastructures so that they can withstand shocks like extreme weather events. The notion has become controversial among researchers and activists as it can seemingly be used to justify the withdrawal of state support and welfare functions, with emphasis placed instead on the character and capacities of individuals and communities to withstand apparently inevitable shocks. Critics have also argued that the notion gives license to powerful interests to continue inflecting harm on communities (e.g. through fossil fuel pollution) while making the communities themselves responsible for their own survival. However, some geographers have argued that the 'experimental' ethos of resilience planning can have some benefits, as it involves abandoning modernist pretensions of being able to predict and control everything, and potentially opens up space for radically different forms of social life to be developed (see Chapters 8 and 10).

Sacrifice zone A term first used as part of the US federal government's nuclear weapons programme, to refer to places rendered uninhabitable by things like nuclear weapons testing. It was then appropriated and developed by environmental sociologists and environmental justice scholars to describe spaces whose pollution, despoliation, or destruction is seemingly considered an acceptable price to pay or economic or national 'progress'. Decades of research in the US have shown how the proliferation of industrial sacrifice zones has seen minority and impoverished communities bear the brunt of toxic pollution and its health impacts.

Slow violence A concept developed by literary scholar Rob Nixon to describe the violence enacted by things like toxic pollution and climate change. While 'violence' is often imagined as immediate, quick and spectacular, Nixon draws our attention to how things like the slow seepage of toxic chemicals into drinking water, or the gradual diminishment of farm yields by shifting weather patterns, are also forms of violence, but a violence that is attritional, unfolding gradually over time, and which is therefore more difficult to see and witness (and perhaps therefore seek justice for) than more obvious and immediate forms of violence.

Thin/thick geographies The world's geographies are complex and changing. A 'thin' (or anaemic) understanding of them favours generalising and looking for broad trends and patterns among different places and regions. By contrast, a 'thick' (or substantive) understanding attends to the differences and connections, at different scales, that make the world an intricate blend of 'patchwork' and 'meshwork' among the several continents and across the oceans and ice-caps.

Values Those things that individuals and groups consider to be of importance for one or other reason. Values are more substantial than mere 'preferences', which can be fickle. Values animate both secular and religious societies. There are different forms of value – for instance, something is 'instrumentally' valuable if it allows a certain goal to be realised (e.g. a washing machine that cleans clothes), of 'aesthetic value' if it's notably beautiful, or of 'intrinsic value' if it matters for itself (such as a blue whale whose slaughter we might consider unnecessary and cruel). *Social* values contrast with purely *personal* ones and are shared among large groups of people, being 'congealed' in custom, public policies, and laws. A major component of value is moral (or ethical) value: namely, concepts, principles, arguments, and practices relating to (i) what's considered just, and (ii) what's considered good or right. Non-moral values are, however, important – for instance, one might value an SUV even though you recognise the immorality of adding a large per capita amount of carbon dioxide to the atmosphere. We live in a world of plural, competing, and contradictory values. All values involve judgements and are irreducible to questions of evidence and truth. Values help define who people are and are constitutive of identity. Values are also the engine of politics as different people within and outside the domain of government consider what values should guide the collective actions that affect our world. Widely shared values may come under strain due to circumstances, such as a natural disaster or an economic depression. These circumstances 'bend' and narrow people's values for a period of time until normality is restored.

World-ecology A term used by Jason W. Moore to describe the embedding of human (and specifically capitalist) relations within the more-than-human 'web of life'. For earlier historians like Fernand Braudel and Immanuel Wallerstein, the advent of capitalism summoned into being a world-economy defined by global divisions of labour. Moore developed this argument to claim that capitalism also functions as a world-ecology, re-organising flows of matter and life on a planetary scale through the continual pursuit of exploitable 'cheap natures' and profit. See Box 3.2 and **capitalism**.

Index

Note: Page references in *italics* denote figures, in **bold** tables and with "n" endnotes.

For Product Safety Concerns and Information please contact our EU representative GPSR@taylorandfrancis.com Taylor & Francis Verlag GmbH, Kaufingerstraße 24, 80331 München, Germany

Batch number: 10620897

Printed by Printforce, the Netherlands